Developments in Aquaculture and Fisheries Science, 22

FRONTIERS OF SHRIMP RESEARCH

DEVELOPMENTS IN AQUACULTURE AND FISHERIES SCIENCE

1. **FARMING MARINE ORGANISMS LOW IN THE FOOD CHAIN**
   A Multidisciplinary Approach to Edible Seaweed, Mussel and Clam Production
   by P. KORRINGA 1976 xvi + 264 pages

2 **FARMING CUPPED OYSTERS OF THE GENUS *CRASSOSTREA***
   A. Multidisciplinary Treatise
   by P. KORRINGA 1976 x + 224 pages

3 **FARMING THE FLAT OYSTERS OF THE GENUS *OSTREA***
   A multidisciplinary Treatise
   by P. KORRINGA 1976 xiv + 238 pages

4. **FARMING MARINE FISHES AND SHRIMPS**
   A Multidisciplinary Treatise
   by P. KORRINGA 1976 xii + 209 pages

5. **SOUND RECEPTION IN FISH**
   edited by A. SCHUIJF *and* A.D. HAWKINS 1976 viii + 288 pages

6. **DISEASE DIAGNOSIS AND CONTROL IN NORTH AMERICAN MARINE AQUACULTURE**
   edited by C.J. SINDERMANN 1977 xii + 330 pages

7. **MUSSEL CULTURE AND HARVEST: A NORTH AMERICAN PERSPECTIVE**
   edited by R.A. LUTZ 1980 xiii + 350 pages

8. **CHEMORECEPTION IN FISHES**
   edited by T.J. HARA 1982 x + 434 pages

9. **WATER QUALITY MANAGEMENT FOR POND FISH CULTURE**
   by C.E. BOYD 1982 xii + 318 pages

10 **GIANT PRAWN FARMING**
   edited by M.B. NEW 1982 xii + 532 pages

11 **MODERN METHODS OF AQUACULTURE IN JAPAN**
   edited by T. KAFUKU *and* H. IKENOUE 1983 xii + 216 pages

12 **GENETICS IN AQUACULTURE**
   edited by N.P. WILKINS *and* E.M. GOSLING 1983 x + 426 pages

13. **BIOECONOMICS OF AQUACULTURE**
   by P.G. ALLEN, L.W. BOTSFORD, A.M. SCHUUR *and* W.E. JOHNSTON 1984 xvi + 351 pages

14. **RECENT INNOVATIONS IN CULTIVATION OF PACIFIC MOLLUSCS**
   edited by D.E. MORSE, K.K. CHEW *and* R. MANN 1984 xi + 404 pages

15. **CHANNEL CATFISH CULTURE**
   edited by C.,S. TUCKER 1985 xvi + 657 pages

16. **SEAWEED CULTIVATION FOR RENEWABLE RESOURCES**
   edited by K.T. BIRD *and* P.H. BENSON 1987 xiv + 382 pages

17. **DISEASE DIAGNOSIS AND CONTROL IN NORTH AMERICAN MARINE AQUACULTURE**
   edited by C.J. SINDERMANN *and* D.V. LIGHTNER 1988 xv + 412 pages

18. **BASIC FISHERY SCIENCE PROGRAMS: A COMPENDIUM OF MICROCOMPUTER PROGRAMS AND MANUAL OF OPERATIONS**
   by S.B. SAILA, C.W. RECKSIEK *and* M.H. PRAGER 1988 iv + 230 pages

19. **CLAM MARICULTURE IN NORTH AMERICA**
   edited by J.J. MANZI *and* M. CASTAGNA 1989 x + 462 pages

20. **DESIGN AND OPERATING GUIDE FOR AQUACULTURE SEAWATER SYSTEMS**
   by JOHN E. HUGUENIN *and* JOHN COLT 1989 iv + 264 pages

21. **SCALLOPS: BIOLOGY, ECOLOGY AND AQUACULTURE**
   edited by S.E. SHUMWAY 1991 xx + 1095 pages

22. **FRONTIERS OF SHRIMP RESEARCH**
   edited by P.F. DeLOACH, W. J. DOUGHERTY *and* M.A. DAVIDSON 1991 viii + 294 pages

Developments in Aquaculture and Fisheries Science, 22

FRONTIERS OF SHRIMP RESEARCH

Edited by

P.F. DeLOACH

College of Charleston, Charleston, SC 29424, U.S.A.

W. J. DOUGHERTY

Medical University of South Carolina, Charleston, SC 29425, U.S.A.

M.A. DAVIDSON

South Carolina Sea Grant Consortium, Charleston, SC 29401, U.S.A.

ELSEVIER Amsterdam — Oxford — New York — Tokyo 1991

ELSEVIER SCIENCE PUBLISHERS B.V.
Sara Burgerhartstraat 25
P.O. Box 211, 1000 AE Amsterdam, The Netherlands

Distributors for the United States and Canada:

ELSEVIER SCIENCE PUBLISHING COMPANY INC.
655, Avenue of the Americas
New York, NY 10010, U.S.A.

ISBN 0-444-88346-0

This book is printed on acid-free paper.

Printed in The Netherlands

Contents

Part I - Fisheries Science

Part II - Reproductive Biology

Part III - Endocrinology

Part IV - Nutrition

Part V - Pathology

Part VI - Culture Science

Part VII - Information Exchange

Part VIII - Shrimp Research Tomorrow

Acknowledgement

The Sea Grant Program of the National Oceanic and Atmospheric Administration has been a reliable source of funding for shrimp research in the United States. The meeting from which this manuscript was made possible was supported by the Sea Grant College programs of Texas, Louisiana, California, Connecticut, Florida, Hawaii, South Carolina Sea Grant Consortium and the National Sea Grant College Program. The editors wish to acknowledge the assistance of Monica Mulvey in preparing the manuscripts for publication.

The Role of Estuarine Habitats in Regulating Growth and Survival of Juvenile Penaeid Shrimp

Thomas J. Minello and Roger J. Zimmerman
NOAA, National Marine Fisheries Service
Southeast Fisheries Center, Galveston Laboratory
4700 Avenue U, Galveston TX 77551

Abstract

Modifications of estuarine habitats are inevitable, but information on habitat functions can be used to protect those habitats most valuable to fishery species. Density patterns of young brown shrimp in estuaries reflect the importance of macrophytic vegetation. Brown shrimp appear to be obligate carnivores, feeding on epifaunal and infaunal organisms which are frequently abundant in vegetated habitats. The structure of the vegetation also provides protection from fish predators such as the southern flounder and pinfish. A large number of environmental factors have the potential to interact with functions of vegetation, however, and the value of these habitats should be expected to vary within and among estuaries.

Distributions of juvenile white shrimp in estuaries are more variable, and information on habitat interactions is limited. This species appears to be omnivorous, having a better capacity to directly use plant foods than brown shrimp. Obvious protective adaptations to avoid predation are also less pronounced in white shrimp. Rapid growth, however, may be a predator avoidance characteristic of this species.

Introduction

Life cycles of the brown shrimp, *Penaeus aztecus*, and the white shrimp, *Penaeus setiferus*, are integrally connected with estuaries. The young of these species utilize shallow estuarine habitats, and a large part of shrimp production is supported by the productivity of these nursery areas. Unfortunately, estuarine habitats are rapidly being modified through natural phenomena and man's exploitation of coastal regions. Rising sea level, land subsidence, the alteration of fresh-water inflow, dredge and fill activities, and increased channelization and salt-water intrusion all affect estuarine habitats (Boesch *et al.*, 1983; Baumann *et al.*, 1984; Davis, 1986; Titus, 1986). Efforts to minimize the impact of habitat alterations on shrimp stocks require an understanding of the relative value of habitats for shrimp and the mechanisms through which habitats influence growth and survival. Estuaries in the northern Gulf of Mexico support the largest populations of brown shrimp and white shrimp in the U.S. (Klima, 1981), and coastal habitats in this region are especially threatened (Bauman *et al.*, 1984; Titus, 1987). In response, recent investigations in ecology at the Galveston Laboratory of the National Marine Fisheries Service have been directed towards understanding the functional importance of estuarine habitats for penaeid shrimp. This paper is a review and synthesis of research on habitats, growth, and mortality of brown shrimp and white shrimp in estuarine nurseries.

Shrimp Distributions in Estuaries

Density patterns of juvenile shrimp in estuaries appear to be useful as indicators of habitat value. In Mobile Bay, Loesch (1965) reported that small brown shrimp were associated with the submerged vegetation *Ruppia* and *Vallisneria* and that white shrimp were found on nonvegetated bottom with large amounts of organic detritus. Williams (1955) had previously noted an association between white shrimp distributions and detritus-rich sediments in North Carolina estuaries. Stokes (1974) reported that white shrimp were more frequently found on nonvegetated bottom near Laguna Madre, Texas, and that brown shrimp were on both nonvegetated bottom and in seagrass beds. Extensive comparisons of estuarine habitats, however, have been hampered by inefficient sampling methods (Zimmerman *et al.*, 1986). The development of a drop-sampling technique by Zimmerman *et al.* (1984) has improved our ability to quantitatively sample shallow vegetated habitats, including intertidal marsh. Using this technique we have established that during most of the year, young brown shrimp in Galveston Bay, Texas strongly select for *Spartina alterniflora* habitat over nonvegetated bottom (Figure 1). Similar samples taken in the estuaries of both Texas and Louisiana, have shown that brown shrimp frequently select for a wide variety of vegetated estuarine habitats including intertidal marsh vegetation (*Spartina alterniflora*, *S. patens*, *Juncus roemerianus*, and *Scirpus robustus*) and submerged vegetation (*Halodule wrightii* and *Ruppia maritima*). There is now substantial evidence that brown shrimp frequently select vegetated habitats over nonvegetated bottom in estuaries of the northern Gulf of Mexico. Selection

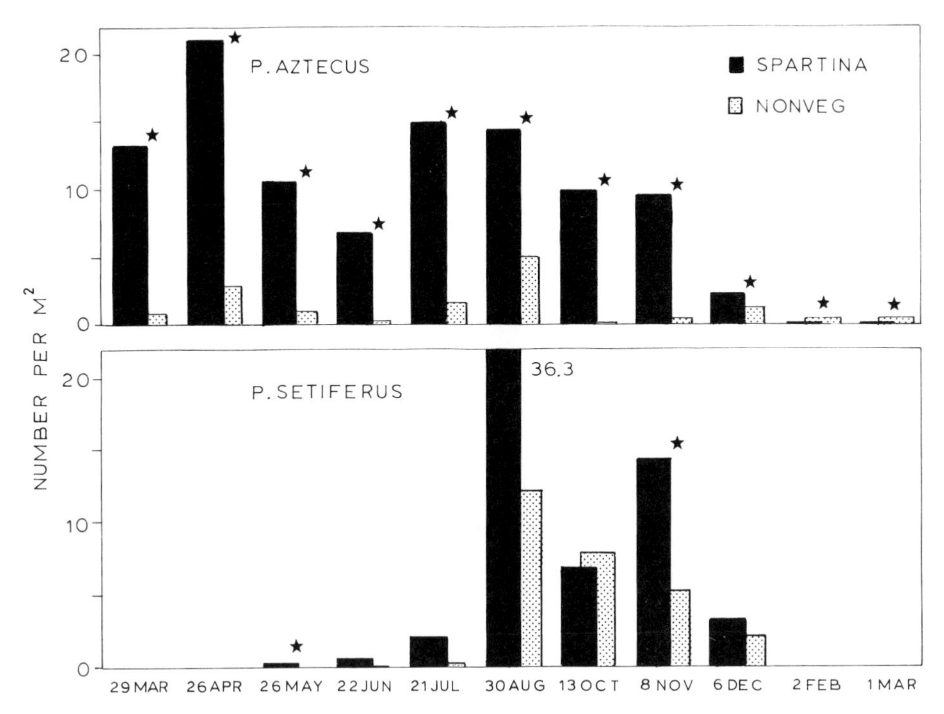

Figure 1. Mean densities of *Penaeus aztecus* and *P. setiferus* collected within intertidal *Spartina alterniflora* habitat and on adjacent nonvegetated bottom. Asterisks indicate significant selection for a habitat based on a comparison of shrimp densities from 12 pairs of drop samples collected on each date (paired t-test, 5% significance level). Data are from Zimmerman and Minello (1984).

patterns for white shrimp are less clear, and in Galveston Bay (Zimmerman and Minello, 1984) white shrimp did not consistently select for *Spartina* or nonvegetated bottom (Figure 1).

Density patterns, however, may not always reflect the value of estuarine habitats. Animals can aggregate in habitats with little food or protective value simply due to current patterns or due to evolutionary selective forces which are no longer in synchrony with habitat value. In contrast, relatively unpopulated habitats may indirectly provide food to organisms or serve as migratory pathways. Determining the importance of various habitats for penaeid shrimp requires information on how habitats function to regulate growth and survival.

Natural Diets and Habitat-related Growth

An understanding of the natural feeding habits and diets of penaeid shrimp in estuaries has been elusive. In part, this can be attributed to variability in diet among species and to ontogenetic changes within species (Stoner and Zimmerman, 1988). Large amounts of detrital material are available in estuaries (de la Cruz, 1965), and detritus has frequently been reported in the gut contents of shrimp (Williams, 1955; Darnell, 1958; Odum and Heald, 1972; George, 1960; Jones, 1973; Chong and Sasekumar, 1981). For this reason, plant detritus and associated bacteria have been believed to be directly important to shrimp nutrition (Moriarty, 1976; 1977). Gleason and Zimmerman (1984), however, have shown that for postlarval brown shrimp, *Spartina* detritus alone does not provide enough nutrition for growth or body maintenance. They also showed that plant material, including epiphytic algae scraped from *Spartina alterniflora* stems, could provide sufficient nutrition for maintenance of small brown shrimp but not enough to sustain normal growth. Considering the apparent requirement for animal protein in shrimp diets (Condrey *et al.*, 1972; Venkataramiah *et al.*, 1975; Zein-Eldin and Corliss, 1976), benthic infauna and epifauna are probably important foods. This conclusion is supported by evidence that meiofauna and small macrofauna (harpacticoid copepods, amphipods and polychaetes) are nearly always present in shrimp guts (Dall, 1968; Moriarty, 1977; Marte, 1980; Hunter and Feller, 1987; Stoner and Zimmerman, 1988), and depletion of these organisms occurs in experimental enclosures with shrimp (Gleason and Zimmerman, 1984; Leber, 1985; Gleason and Wellington, 1988; and Zimmerman *et al.*, in review). These small fauna used as food by shrimp are generally considered to be components of food webs based on aged detritus, algal epiphytes, and macroalgae. Conflicting evidence still exists, however, as to the relative contribution of benthic versus planktonic sources of carbon in shrimp food webs (Fry, 1981; Hughes and Sherr, 1983; Gleason, 1986; Gleason and Wellington, 1988; Stoner and Zimmerman, 1988).

The relationships between food abundance, habitat value, and shrimp growth have been examined in field and laboratory experiments by Zimmerman *et al.* (in review). Brown shrimp enclosed in cages with access to *Spartina alterniflora* marsh surface grew faster than in cages restricting them to nonvegetated bottom (Table 1). By contrast, growth rates of white shrimp in the two habitats were not significantly different. Elevated abundances of small benthic macrofauna have been associated with *Spartina alterniflora* marshes (Rader, 1984), and the availability of these animals was apparently responsible for the increased growth of brown shrimp in the cages with *Spartina*. Cores taken during the experiment showed that populations of peracarid crustaceans and polychaetes were largest within *Spartina* cages, and their populations were depleted in the cages with high densities of brown shrimp (Zimmerman *et al.*, in review). Laboratory experiments conducted during the study confirmed that both brown shrimp and white shrimp fed on infauna and epifauna, but feeding by brown shrimp was more efficient (Table 2).

Such studies demonstrate that there is a nutritional component to habitat selection patterns of juvenile brown shrimp, and that the value of vegetation differs among shrimp species. The data from Zimmerman *et al.* (in review) indicate that brown shrimp are carnivorous, and that amphipods, tanaids, and polychaetes present in *Spartina alterniflora* habitats are a primary source of nutrition for this species. Seasonal differences in the distributions of food organisms may account for the apparent reversal in habitat selection by brown shrimp during the early spring (Figure 1).

White shrimp are less efficient feeders on benthic fauna (Table 2), and appear to incorporate more plant material into their diet. In ponds with mainly phytoplankton present as food, high growth rates reported for white shrimp (Johnson and Fielding, 1956; Wheeler, 1968) suggest that this species has the capacity for direct utilization of plant materials. Laboratory experiments by McTigue and Zimmerman (unpublished) show that the growth response of white shrimp is significantly greater than that of brown shrimp when diatoms are added to animal protein diets. The ability to exploit a wide variety of plant and animal foods might explain the apparent lack of a strong habitat selection pattern by white shrimp. More information on preferred foods and the distribution of these foods is needed to understand interactions between white shrimp and estuarine habitats.

Mortality in Estuarine Nurseries

The role of estuarine habitats in regulating mortality of juvenile shrimp depends upon the magnitude and variability of mortality in estuaries and upon the major causes of mortality. Minello *et al.* (1989) measured brown shrimp mortality using a length-frequency analysis of cohorts from drop sample data in a Galveston Bay salt marsh, and actual mortalities for 2-week periods during the spring of 1982 ranged from 33% to 61% (Figure 2). Other estimates (adjusted to 2-week rates) of 52% for subadult *Penaeus aztecus* (McCoy, 1972) and 65% for juvenile *Penaeus vannamei* (Edwards, 1977) would also indicate that mortality in estuaries can have a large impact on adult populations.

Adverse physical conditions can cause mortality of penaeid shrimp in estuaries, but these catastrophic events appear to be relatively rare. Gunter (1941) and Gunter and Hildebrand (1951) documented cold-related kills of shrimp in Texas estuaries, and similar phenomena have been observed along the southeastern coast of the U.S. (Dahlberg and Smith, 1970). However, cold fronts mainly occur during winter and early spring months, and during most of the time when young shrimp occupy estuarine nurseries, temperatures do not appear to reach levels known to be lethal (Zein-Eldin and Aldrich, 1965; Zein-Eldin and Griffith, 1969; Zein-Eldin and Renaud,

Table 1. Habitat-related summer growth rates of juvenile *Penaeus aztecus* and *P. setiferus* caged together in *Spartina alterniflora* and on non-vegetated bottom. The largest initial size possible (32mm,TL) was used in calculations and growth rates are conservative. (from Zimmerman *et al.*, in review).

	growth in mm/day(1SE)	
	Spartina	Non-vegetated
P. aztecus (high density)	0.98 (0.02)	0.77 (0.03)
P. aztecus (low density)	1.41 (0.05)	1.03 (0.13)
P. setiferus (low density)	1.04 (0.03)	1.05 (0.05)

1986). Persistent hypoxic conditions (dissolved oxygen less than 2 mg/ml), which occur during summer months in estuaries and shallow coastal regions (Gunter, 1942; May, 1973; Turner *et al.*, 1987), can also potentially cause shrimp mortality. Juvenile brown shrimp, however, can survive in waters with dissolved oxygen less than 1 mg/ml (Kramer, 1975), and Renaud (1986) has shown that, at least in the laboratory, both brown shrimp and white shrimp can detect and avoid oxygen depleted water. Juvenile penaeid shrimp have been collected over a wide range of temperatures, salinities, and dissolved oxygen levels, and laboratory studies indicate that these factors are unlikely to be major direct causes of shrimp mortality (see Zein-Eldin and Renaud, 1986 for review). Food supplies in estuaries seem adequate to prevent starvation (Gleason and Zimmerman, 1984; Zimmerman *et al.*, in review, and parasitism and disease do not appear to be a major direct cause of mortality in natural populations (Overstreet, 1973; Couch, 1978). Predation by fishes, however, is common, and many analyses of stomach contents have identified fish predators on shrimp (see Minello and Zimmerman, 1983 for review). The importance of predation as a cause of mortality is supported by results from field caging experiments in which 2-week mortalities of brown shrimp during the spring were less than 3% when predators were excluded (Minello *et al.*, 1989).

Table 2. Depletion of peracarid crustaceans (Amphipoda and Tanaidacea) and annelid worms (Polychaeta and Oligochaeta) from shrimp feeding in 78.5 cm^2 benthic cores taken from a Galveston Bay salt marsh. Individual 28-mm shrimp were placed in 12 cores per treatment, and the values shown are mean numbers of prey organisms remaining after 5 days. In ANOVA's comparing the three treatments, all means were significantly different (5% level) except those indicated by a connecting line.(from Zimmerman *et al.*, in review).

	Treatments		
Prey	Control	*P. setiferus*	*P. aztecus*
Peracarida	13.5	7.1	1.6
Annelida	37.0	10.2	3.9
Total	50.5	17.3	5.5

The dominant predator on penaeid shrimp during the spring in a Galveston Bay marsh was the southern flounder, *Paralichthys lethostigma* (Table 3). Pinfish, *Lagodon rhomboides*, spot, *Leiostomus xanthurus*, and gulf killifish, *Fundulus grandis*, also fed upon penaeids. Young spotted seatrout, *Cynoscion nebulosus*, and red drum, *Sciaenops ocellatus*, were important predators on penaeids during the late summer and fall (Table 3).

Habitat characteristics that modify predation rates of fishes can regulate shrimp mortality, and predation rates on brown shrimp are affected by vegetation (Minello and Zimmerman, 1983; 1985; Minello *et al.*, 1989), substrate type, and water clarity (Minello and Zimmerman, 1984, Minello *et al.,* 1987). In addition, other environmental conditions such as temperature, salinity, and food availability, that may only have a limited direct effect on shrimp mortality, can regulate mortality by mediating the impact of predation.

Estuarine vegetation has been shown to reduce predation by fishes on a variety of prey (Van Dolah, 1978; Coen *et al.*, 1981; Heck and Thoman, 1981; Lascara, 1981; Wilson *et al.*, 1987). Because the dominant estuarine vegetation in the northwestern Gulf of Mexico is intertidal salt marsh, water levels and hydroperiod become important in regulating predation and mortality. Laboratory predation rates on brown shrimp increased with shrimp density and decreased in the presence of *Spartina alterniflora* (Minello *et al.*, 1989). Water levels in the marsh control access to intertidal vegetation, and to some extent control prey densities. On flood tides, brown shrimp are concentrated within protective vegetated habitats, but low tide conditions result in high densities of both shrimp and fish predators on nonvegetated bottom. Extended periods of abnormally low water probably increase brown shrimp mortality, and standing high water which normally occurs during the spring in the northern Gulf of Mexico (Hicks *et al.*, 1983; Bauman, 1987) probably decreases mortality. Thus, seasonal and geographic differences in tidal dynamics and hydroperiod may play an important role in shrimp survival.

Physical conditions and the type and quantity of food available can regulate shrimp growth (Zein-Eldin and Aldrich, 1965; Zein-Eldin and Griffith, 1969; Gleason and Zimmerman, 1984; Zimmerman *et al.*, in review), and growth rates may interact with predation. Size-selection experiments with spotted seatrout

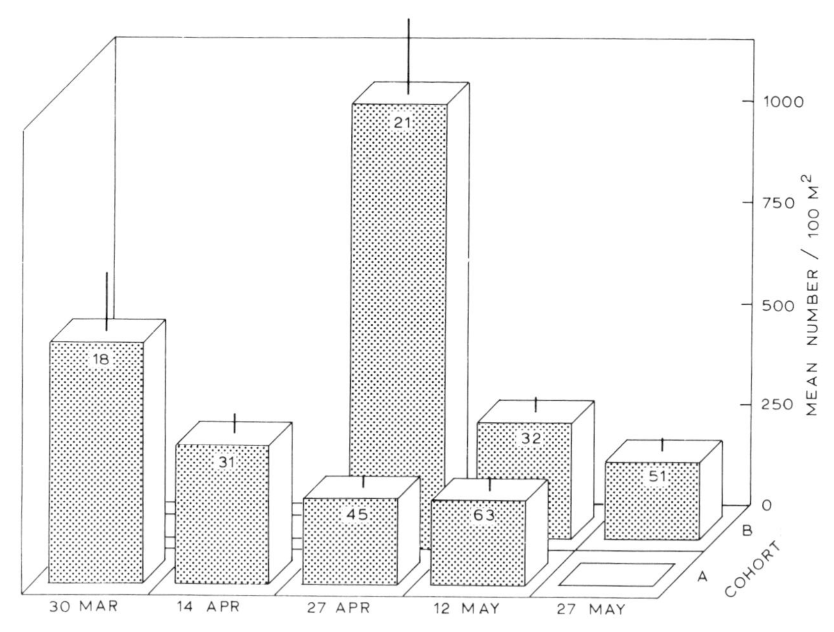

Figure 2. Mean densities of *Penaeus aztecus* in two cohorts identified through a length-frequency analysis. Decreases in density over time represent mortality. Samples were collected in a Galveston Bay salt marsh during the spring of 1982. The mean size (mm, total length) of shrimp in each cohort is shown on the histograms. Error bars are 1 SE of the mean. (from Minello *et al.*, 1989.)

(Minello and Zimmerman, 1983; 1984) and with southern flounder (Minello *et al.*, 1989) indicate that these predators select small shrimp, and the optimal prey size is less than 1/3 to 1/2 the total length of the predator. Rapid growth by shrimp should reduce the time spent at this optimal prey size, and a refuge from predation by juvenile estuarine fishes may be found with increased size. This hypothesis is supported by an apparent reduction in mortality of the larger size classes of young brown shrimp (Figure 2). Environmental conditions and food abundances which affect growth rates may regulate mortality through this mechanism.

The environment can also interact with mortality of shrimp through its effect on protective behaviors. Burrowing in the substratum has been shown to protect brown shrimp from predation by southern flounder and pinfish (Minello *et al.*, 1987). Light intensity at the substrate surface has a large effect on burrowing (Wickham and Minkler, 1975; Lakshmi *et al.*, 1976; Minello *et al.*, 1987), and other environmental conditions including temperature (Fuss and Ogren, 1966; Aldrich *et al.*, 1968), salinity (Venkataramiah *et al.*, 1974; Lakshmi *et al.*, 1976), dissolved oxygen (Egusa and Yamamoto, 1961), and the type of substratum (Rulifson, 1981; Aziz and Greenwood, 1982) can affect burrowing rates. All of these factors have the potential to modify mortality rates of shrimp.

Predator avoidance behaviors appear to be better developed in brown shrimp than in white shrimp. Selection for protective vegetation (Minello and Zimmerman, 1985) and the strong tendency to burrow into the substratum (Wickham and Minkler, 1975) are traits of brown shrimp. White shrimp also burrow, but the burrows are shallow and the frequency is reduced (Williams, 1958). Increased vulnerability to predators from reduced burrowing, may be offset by increased foraging time for white shrimp which should result in increased growth. White shrimp postlarvae generally arrive in estuaries during the late spring and summer (Baxter and Renfro, 1967), when increased water temperatures should also stimulate growth. Reported growth rates for white shrimp are generally higher than those of brown shrimp (see Knudsen *et al.*, 1977; Christmas and Etzold, 1977 for reviews), and this may reduce the time white shrimp are available to predators.

Conclusions

Modifications of estuarine habitats are inevitable, but information on habitat functions can be used to protect those habitats most valuable to fishery species. The importance of estuarine vegetation is strongly indicated for brown shrimp. Brown shrimp appear to be obligate carnivores, feeding on epifaunal and infaunal organisms which are frequently abundant in vegetated habitats. The structure of the vegetation also provides protection from fish predators such as the southern flounder and pinfish. A large number of environmental factors have the potential to interact with functions of vegetation, and the value of these habitats should be expected to vary within and among estuaries.

Information on habitat interactions for white shrimp is limited. This species

Table 3. Dominant fish predators on penaeid shrimp. Stomach contents of fish were analyzed to determine the frequency of feeding on *Penaeus*. (taken from Minello *et al*., 1989).

March - May 1982

Species	Number of fish examined	Size range (mm,TL)	Number with food	Percent[a] with *Penaeus*	Percent[b] of all(11) *Penaeus* eaten
Southern flounder	21	34-143	19	33.3	72.7
Gulf killifish	15	24-85	13	6.7	9.1
Pinfish	254	12-64	252	0.4	9.1
Spot	180	16-75	112	0.6	9.1

March 1982 - September 1984

Species	Number of fish examined	Size range (mm,TL)	Number with food	Percent with *Penaeus*	Percent of all (56) *Penaeus* eaten
Southern flounder	38	34-184	31	31.6	28.6
Spotted seatrout	116	11-135	95	15.5	46.4
Red drum	59	8-131	34	3.4	14.3
Gulf killifish	102	21-88	77	2.0	3.6
Pinfish	483	15-84	459	0.6	5.4
Spot	267	18-110	183	0.4	1.8

[a] Percentage of fish examined having eaten at least one *Penaeus*.
[b] Percentage of the total number of *Penaeus* found in all fish examined (shown in parentheses).

appears to be omnivorous, having a better capacity to directly use plant foods than brown shrimp. Obvious protective adaptations to avoid predation are also less pronounced in white shrimp. Rapid growth, however, may be a predator avoidance

Literature Cited

Aldrich, D.V., C. E. Wood, and K. N. Baxter. 1968. An ecological interpretation of low temperature responses in *Penaeus aztecus* and *P. setiferus* postlarvae. Bulletin of Marine Science 18: 61-71.
Aziz, K. A. and J. G. Greenwood. 1982. Response of juvenile *Metapenaeus bennettae* Racek & Dall, 1965 (Decapoda, Penaeidae) to sediments of differing particle size. Crustaceana 43: 121-126.
Baumann, R. H. 1987. Physical variables. In. Conner, W. and J. Day, Jr. (editors), The ecology of Barataria Basin, Louisiana: An estuarine profile. Biological Rept. 85., USFWS, pp. 8-17.
Baumann, R. H., J. W. Day, Jr, and C. A. Miller. 1984. Mississippi deltaic wetland survival: Sedimentation versus coastal submergence. Science 224: 1093-95.
Baxter, K. N. and W. C. Renfro. 1967. Seasonal occurrence and size distribution of postlarval brown and white shrimp near Galveston, Texas, with notes on species identification. Fishery Bulletin, U.S. 66: 149-158.
Boesch, D. F., D. Levin, D. Nummedal, and K. Bowles. 1983. Subsidence in coastal Louisiana: causes, rates, and effects on wetlands. U.S. Fish and Wildlife Service, Division of Biological Services. Washington, D.C., FWS/OBS-83/26, 30 p.
Chong, V. C. and A. Sasekumar. 1981. Food and feeding habits of the white prawn *Penaeus merguiensis*. Marine Ecology Progress Series 5: 185-91.
Christmas, J. Y. and D. J. Etzold. 1977. The shrimp fishery of the Gulf of Mexico United States: A regional management plan. Gulf Coast Research Laboratory. Technical Rept. Ser., No. 2, 125 p.
Coen, L. D., K. L. Heck, Jr., and L. G. Abele. 1981. Experiments on competition and predation among shrimps of seagrass meadows. Ecology 62: 1484-93.
Condrey, R. E., J. G. Gosselink, and H. J. Bennett. 1972. Comparison of the assimilation of different diets by *Penaeus setiferus* and *Penaeus aztecus*. Fishery Bulletin, U.S. 70: 1281-92.
Couch, J. A. 1978. Diseases, parasites, and toxic responses of commercial penaeid shrimps of the Gulf of Mexico and south Atlantic coasts of North America. Fishery Bulletin, U.S. 76: 1-44.
Dahlberg, M. D. and F. G. Smith. 1970. Mortality of estuarine animals due to cold in the Georgian coast. Ecology 51: 931-933.
Dall, W. 1968. Food and feeding of some Australian penaeid shrimp. FAO Fishery Report 2: 251-58.
Darnell, R. M. 1958. Food habits of fishes and larger invertebrates of Lake Ponchartrain, Louisiana, an estuarine community. Publications of the Institute of Marine Sciences, University of Texas 5: 353-416.
Davis, D. W. 1986. The retreating coast. The Journal of Soil and Water Conservation 41: 146-51.
de la Cruz, A. A. 1965. A study of particulate organic detritus in a Georgia salt marsh-estuarine system. Ph.D.Thesis, University of Georgia, 141 p.
Edwards, R. R. 1977. Field experiments on growth and mortality of *Penaeus vannamei* in a Mexican coastal lagoon. Estuarine, Coastal and Shelf Science 5: 107-121.

Egusa, S. and T. Yamamoto. 1961. Studies on the respiration of the "Kuruma" prawn *Penaeus japonicus* Bate I. Burrowing behavior, with special reference to its relation to environmental oxygen concentration. Bulletin of the Japanese Society of Scientific Fisheries 27: 22-26.

Fry, B. 1981. Natural stable carbon isotope tag traces Texas shrimp migrations. Fishery Bulletin, U.S. 79: 337-346.

Fuss, C. M., Jr and L. H. Ogren. 1966. Factors affecting activity and burrowing habits of the pink shrimp, *Penaeus duorarum* Burkenroad. Biological Bulletin 130: 170-191.

George, M. J. 1960. The food of the shrimp *Metapenaeus monoceros* (Fabricus) caught from the backwaters. Indian Journal of Fisheries 7: 495-500.

Gleason, D. F. 1986. Utilization of salt marsh plants by postlarval brown shrimp: carbon assimilation rates and food preferences. Marine Ecology Progress Series 31: 151-58.

Gleason, D. F. and G. M. Wellington. 1988. Food resources of postlarval brown shrimp (*Penaeus aztecus*) in a Texas salt marsh. Marine Biology 97: 329-337.

Gleason, D. F. and R. J. Zimmerman. 1984. Herbivory potential of postlarval brown shrimp associated with salt marshes. Journal of Experimental Marine Biology and Ecology 84: 235-246.

Gunter, G. 1941. Death of fishes due to cold on the Texas coast, January, 1940. Ecology 22: 203-208.

Gunter, G. 1942. Offatts Bayou, a locality with recurrent summer mortality of marine organisms. American Midland Naturalist 28:631-633.

Gunter, G. and H. H. Hildebrand. 1951. Destruction of fishes and other organisms on the south Texas coast by the cold wave of January 28-February 3, 1951. Ecology 32: 731-36.

Heck, K. L., Jr and T. A. Thoman . 1981. Experiments on predator-prey interactions in vegetated aquatic habitats. Journal of Experimental Marine Biology and Ecology 53: 125-134.

Hicks, S. D., H. A. Debaugh, Jr, and L. E. Hickman. 1983. Sea level variations for the United States 1855-1980. NOAA/NOS Report, National Ocean Survey, Tides and Water Levels Branch., Rockville, 170 p.

Hughes, E. H. and E. B. Sherr. 1983. Subtidal food webs in a Georgia estuary: del carbon analysis. Journal of Experimental Marine Biology and Ecology 67: 227-42.

Hunter, J. and R. J. Feller. 1987. Immunological dietary analysis of two penaeid shrimp species from a South Carolina tidal creek. Journal of Experimental Marine Biology and Ecology 107: 61-70.

Johnson, M. C. and J. R. Fielding. 1956. Propagation of the white shrimp, *Penaeus setiferus* (Linn.) in captivity. Tulane Studies in Zoology 4: 175-190.

Jones, R. R. 1973. Utilization of Louisiana estuarine sediments as a source of nutrition for the brown shrimp *Penaeus aztecus* Ives. Ph.D. Thesis, Louisiana State University, 131 pp.

Klima, E. F. 1981. The National Marine Fisheries Service shrimp research program in the Gulf of Mexico. Kuwait Bulletin of Marine Science 2: 185-207.

Knudsen, E. E., W. H. Herke, and J. M. Mackler. 1977. The growth rate of marked juvenile brown shrimp, *Penaeus aztecus*, in a semi-impounded Louisiana coastal marsh. Proceedings of the Gulf and Caribbean Fisheries Institute. 29: 144-159.

Kramer, G. L. 1975. Studies on the lethal dissolved oxygen levels for young brown shrimp, *Penaeus aztecus* Ives. Proceedings of the World Mariculture Society 6: 157-67.

Lakshmi, G. J., A. Venkataramiah, and G. Gunter. 1976. Effects of salinity and photoperiod on the burying behavior of brown shrimp *Penaeus aztecus* Ives. Aquaculture 8: 327-336.

Lascara, J. 1981. Fish predator-prey interactions in areas of eelgrass (*Zostera marina*). M. A. Thesis, College of William and Mary., Virginia, 81 p.

Leber, K. M. 1985. The influence of predatory decapods, refuge, and microhabitat selection on seagrass communities. Ecology 66: 1951-64.

Loesch, H. 1965. Distribution and growth of penaeid shrimp in Mobile Bay, Alabama. Publications of the Institute of Marine Science, The University of Texas 10: 41-58.

Marte, C. L. 1980. The food and feeding habit of *Penaeus monodon* Fabricius collected from Makato River, Aklan, Philippines (Decapoda Natantia). Crustaceana 38: 225-236.

May, E. B. 1973. Extensive oxygen depletion in Mobile Bay, Alabama. Limnology and Oceanography 18: 353-66.

McCoy, E. G. 1972. Dynamics of North Carolina commercial shrimp populations. North Carolina Department of Natural and Economic Resources., Special Scientific Report No. 21, 53 p.

Minello, T. J. and R. J. Zimmerman. 1983. Fish predation on juvenile brown shrimp, *Penaeus aztecus* Ives: the effect of simulated *Spartina* structure on predation rates. Journal of Experimental Marine Biology and Ecology 72: 211-231.

Minello, T. J. and R. J. Zimmerman. 1984. Selection for brown shrimp, *Penaeus aztecus*, as prey by the spotted seatrout, *Cynoscion nebulosus*. Contributions in Marine Science 27: 159-167.

Minello, T. J. and R. J. Zimmerman. 1985. Differential selection for vegetative structure between juvenile brown shrimp (*Penaeus aztecus*) and white shrimp (*P. setiferus*), and implications in predator-prey relationships. Estuarine, Coastal and Shelf Science 20: 707-716.

Minello, T. J., R. J. Zimmerman, and E. X. Martinez. 1989. Mortality of young brown shrimp *Penaeus aztecus* in estuarine nurseries. Transactions of the American Fisheries Society, 118: 693-708.

Minello, T. J., R. J. Zimmerman, and E.X. Martinez. 1987. Fish predation on juvenile brown shrimp, *Penaeus aztecus*: Effects of turbidity and substratum on predation rates. Fishery Bulletin, U.S. 85: 59-70.

Moriarty, D. J. 1976. Quantitative studies on bacteria and algae in the food of the mullet *Mugil cephalus* L. and the prawn *Metapenaeus bennettae* (Racek and Dall). Journal of Experimental Marine Biology and Ecology 22: 131-43.

Moriarty, D. J. 1977. Quantification of carbon, nitrogen and bacterial biomass in the food of some penaeid prawns. Australian Journal of Marine and Freshwater Research 28: 113-18.

Odum, W. E. and E. J. Heald. 1972. Trophic analyses of an estuarine mangrove community. Bulletin of Marine Science 22: 671-738.

Overstreet, R. M. 1973. Parasites of some penaeid shrimp with emphasis on reared hosts. Aquaculture 2: 105-140.

Rader, D. N. 1984. Salt-marsh benthic invertebrates: small scale patterns of distribution and abundance. Estuaries 7: 413-420.

Renaud, M. L. 1986. Detecting and avoiding oxygen deficient seawater by brown shrimp, *Penaeus aztecus* (Ives) and white shrimp, *Penaeus setiferus* (Linnaeus). Journal of Experimental Marine Biology and Ecology 98: 283-92.

Rulifson, R. A. 1981. Substrate preferences of juvenile penaeid shrimps in estuarine habitats. Contributions in Marine Science 24: 35-52.

Stokes, G. M. 1974. The distribution and abundance of penaeid shrimp in the lower Laguna Madre of Texas, with a description of the live bait fishery. Coastal Fishery Project Report. Texas Parks and Wildlife Department. Technical Series, No. 15, 32p.

Stoner, A. W. and R. J. Zimmerman. 1988. Food pathways associated with penaeid shrimps in a mangrove-fringed estuary. Fishery Bulletin, U.S. 86: 543-51.

Titus, J. G. 1986. Greenhouse effect, sea level rise, and coastal zone management. Coastal Zone Management Journal 14: 147-71.

Titus, J. G. 1987. Saving Louisiana's coastal wetlands., EPA- 230-02-87-026, 102 p.

Turner, R. E., W. W. Schroeder, and W. J. Wiseman, Jr. 1987. The role of stratification in the deoxygenation of Mobile Bay and adjacent shelf waters. Estuaries 10: 13-19.

Van Dolah, R. F. 1978. Factors regulating the distribution of the amphipod *Gammarus palustris* in an intertidal salt marsh community. Ecological Monographs 48: 191-217.

Venkataramiah, A., G. J . Lakshmi, and G. Gunter. 1974. Studies on the effect of salinity and temperature on the commercial shrimp *Penaeus aztecus* Ives, with special regard to survival limits, growth, oxygen consumption, and ionic regulation. U.S. Army Corps of Engineers Waterways Experiment Station, Vicksburg, MS, Contract Report H-74-2, XII. : 1-134.

Venkataramiah, A., G. J. Lakshmi, and G. Gunter. 1975. Effect of protein level and vegetable matter on growth and food conversion efficiency of brown shrimp. Aquaculture 6: 115-125.

Wheeler, R. S. 1968. Culture of penaeid shrimp in brackish-water ponds, 1966-67. Proceedings of the 22nd Conference of the Southeast Association of Game and Fisheries Commission., 5 p.

Wickham, D. A. and F. C. Minkler, III. 1975. Laboratory observations on daily patterns of burrowing and locomotor activity of pink shrimp, *Penaeus duorarum*, brown shrimp, *Penaeus aztecus*, and white shrimp, *Penaeus setiferus*. Contributions in Marine Science 19: 21-35.

Williams, A. B. 1955. A contribution to the life histories of commercial shrimps (Penaeidae) in North Carolina. Bulletin of Marine Science of the Gulf and Caribbean 5: 116-146.

Williams, A. B. 1958. Substrates as a factor in shrimp distribution. Limnology and Oceanography 3: 2 83-290.

Wilson, K. A., K. L. Heck, Jr, and K. W. Able. 1987. Juvenile blue crab, *Callinectes sapidus*, survival: an evaluation of eelgrass, *Zostera marina*, as refuge. Fishery Bulletin, U.S. 85: 53-58.

Zein-Eldin, Z. and D. V. Aldrich. 1965. Growth and survival of postlarval *Penaeus aztecus* under controlled conditions of temperature and salinity. Biological Bulletin 129: 199-216.

Zein-Eldin, Z. P. and J. Corliss. 1976. The effect of protein levels and sources on growth of *Penaeus aztecus*. FAO Tech. Conf. Aquaculture, (FIR:AQ/CONF./ 76/E. 33), 8 p.

Zein-Eldin, Z. P. and G. W. Griffith. 1969. An appraisal of the effects of salinity and temperature on growth and survival of postlarval penaeids. FAO Fishery Report 57 3: 1015-1026.

Zein-Eldin, Z. P. and M. L. Renaud. 1986. Inshore environmental effects on brown shrimp, *Penaeus aztecus*, and white shrimp, *P. setiferus*, populations in coastal waters, particularly of Texas. Marine Fisheries Review 48:9-19.

Zimmerman, R. J. and T. J. Minello. 1984. Densities of *Penaeus aztecus*, *P. setiferus* and other natant macrofauna in a Texas salt marsh. Estuaries 7:421-433.

Zimmerman, R. J., T. J. Minello, and S. Dent. Habitat-related growth and resource partitioning by penaeid shrimp in a salt marsh. Marine Ecology Progress Series (in review).

Zimmerman, R. J., T. J. Minello, and G. Zamora, Jr. 1984. Selection of vegetated habitat by brown shrimp, *Penaeus aztecus*, in a Galveston Bay salt marsh. Fishery Bulletin, U.S. 82: 325-336.

Zimmerman, R. J., T. J. Minello, G. Zamora, Jr., and E. Martinez. 1986. Measurements of estuarine shrimp densities applied to catch predictions. In. Landry, A.M., Jr. and E.F. Klima (editors), Proceedings of the shrimp yield prediction workshop. Texas A&M Sea Grant, Publ. No. TAMU-SG-86-110, pp. 38-55.

REPRODUCTIVE STUDIES CONCERNING NATURAL SHRIMP POPULATIONS: A DESCRIPTION OF CHANGES IN THE SIZE AND BIOCHEMICAL COMPOSITION OF THE GONADS AND DIGESTIVE GLANDS IN PENAEID SHRIMP

F. L. Castille and A. L. Lawrence
Shrimp Mariculture Project
Texas Agricultural Experiment Station
Texas A&M University
P. O. Drawer Q
Port Aransas, TX 78373

Abstract

During gonadal development of *Penaeus aztecus* and *P. setiferus,* changes in organ size and biochemical composition are characterized in the gonads and digestive glands of both males and females, and in the terminal portion of the male reproductive tract (terminal ampoules and vasa deferentia combined). Biochemical changes that occur during reproduction in wild populations provide baseline information about nutritional requirements of penaeid reproduction in captivity and the production of viable offspring. Size and carbohydrate, protein and lipid contents of ovaries increase during gonadal maturation in both species. In digestive glands of females, patterns of change in size and biochemical composition differ between the species. In *P. aztecus*, size, carbohydrate content, and protein content of digestive glands increase during gonadal maturation, but in *P. setiferus*, size and lipid content of digestive glands decrease during gonadal maturation. Data suggest that mobilization of stored lipids from digestive gland to ovaries is more important in *P. setiferus* than in *P. aztecus*. In males, the testes are relatively small and the combined terminal ampoules and vasa deferentia make up a larger proportion of the male reproductive system than the testes. In mature males, the testes are larger in *P. setiferus* than in *P. aztecus*, but the combined terminal ampoules and vasa deferentia are larger in *P. aztecus* than in *P. setiferus*. In *P. setiferus*, the size and protein content of both the testes and the combined terminal ampoules and vasa deferentia are larger in mature males than in developing males.

Introduction

During gametogenesis, the gonads of penaeid shrimp are sites of intensive growth and biochemical synthesis. This paper describes changes in the size and biochemical composition of the gonads of female *Penaeus aztecus* and *P. setiferus,* and both the gonads and the combined vasa deferentia and terminal ampoules of male *P. aztecus* and *P. setiferus.* Changes in the size and biochemical composition of the digestive glands are characterized to determine if nutrients are mobilized from the digestive gland to the gonads during gonadal development.

Male Reproductive System

The male reproductive system of penaeid shrimp has been described in numerous species (King, 1948; Subrahmanyam, 1965; Tuma, 1967; Tirmizi and Javed, 1976; Motoh, 1978; Motoh and Buri, 1980). Internal organs are comprised of the testes, vasa deferentia, and terminal ampoules. Testes consist of lobes of translucent tissue located above the digestive gland. Spermatozoa are transported from the testes into the paired vasa deferentia. The vasa deferentia terminate in paired bulbous, muscular structures called terminal ampoules, which contain the spermatophores. The spermatophores are complex structures containing the spermatozoa and accessory structures that attach the spermatophores to the female. During mating, paired spermatophores are extruded through genital pores at the base of the fifth pair of pereopods.

The Female Reproductive System

The anatomy of the female reproductive system has also been described in numerous shrimp species (King, 1948; Cummings, 1961; Subrahmanyam, 1965; Tuma, 1967; Tirmizi and Javed, 1976; Motoh, 1978; Motoh and Buri, 1980). The internal organs consist of partially fused, paired ovaries that extend the entire length of mature shrimp. Each ovary consists of an anterior lobe that lies along the stomach, a middle lobe, consisting of five to eight projections lying above the digestive gland, and posterior lobe that runs the length of the abdomen and lies dorso-lateral to the intestine. Paired oviducts run from one of the posterior projections of the middle lobe of each ovary to a genital pore located at the base of the third pereopods.

During mating, paired spermatophores are either inserted into or deposited onto an external structure on the female called the thelycum. The thelycum is located between the fifth pair of pereopods of females. The structure of the thelycum has been used to distinguish shrimp of the subgenus *Litopenaeus,* which are referred to as open thelycum shrimp, from shrimp of the other five penaeid subgenera, which are referred to as closed thelycum shrimp. In addition, the structure of the thelycum can be correlated to differences in reproductive behavior between the subgenera (Tuma, 1967; Perez-Farafante, 1969).

In closed thelycum shrimp, the thelycum consists of two flaps or lateral plates separated by a median slit that opens into paired seminal receptacles that lie under the lateral plates. During mating, the spermatophore is squeezed through the anterior V-shaped fissure between the left and right lateral plates and deposited in the seminal receptacles.

In open thelycum shrimp, lateral plates are absent and the thelycum only has ridges and protuberences for attachment of the spermatophore. The thelycum is less specialized than the closed type and the spermatophore is exposed after attachment.

There is also a difference in the sequence of ovarian development, mating, and spawning between open and closed thelycum shrimp. In open thelycum shrimp, mating occurs after ovarian development is complete and spawning occurs within hours of mating. Thus, mating is coordinated with ovarian development. In closed thelycum shrimp, mating is coordinated with the molt cycle of the female and occurs shortly after the female has molted. Closed thelycum shrimp mate prior to ovarian development and spawn after ovarian development is complete. Closed thelycum females may undergo several cycles of ovarian development and spawning after a single mating.

The species of penaeid shrimp native to the Gulf of Mexico and southern Atlantic coast of the United States are *P. aztecus*, *P. duoraum*, and *P. setiferus*. *Penacus aztecus* and *P. duorarum* are members of the subgenus *Farfantepenaeus* and have closed thelyca. *Penaeus setiferus* is a member of the subgenus *Litopenaeus* and has an open thelycum.

Evaluation of Reproduction from Gonadal Development

Determinations of the reproductive state of penaeid shrimp are important for fishery management and culture. One of the most effective methods of evaluating the reproductive state of marine invertebrates is to determine the degree of gonadal maturation, and particularly in shrimp, the degree of ovarian maturation. Because the simultaneous presence of spent animals and mature animals capable of spawning is indicative of reproduction, this method can be used to monitor reproduction without direct observation of mating, spawning, or the presence of eggs or larvae. In addition, the degree and relative abundance of gonadal development in a population provides a quantitative measurement of the reproductive state of the population.

Visual Assessment of Gonadal Development in Female Shrimp

In penaeid shrimp, gonadal maturation can be evaluated from a number of parameters. For females, the most common method has been visual assessment of the ovaries on the basis of ovary size and color (King, 1948; Cummings, 1961; Subrahmanyam, 1965; Tuma, 1967; Rao, 1968; Primavera, 1980). Although the evaluation criteria and number of ovarian development stages are arbitrary and vary in the literature, the most commonly used system divides ovarian maturation into five stages. Using the same stage nomenclature as Rao (1968), the following criteria describe the developmental stages used in this paper for *P. aztecus* and *P. setiferus*.

Immature. Ovaries are thin, translucent, unpigmented, and invisible through the exoskeleton. The diameter of the posterior lobes is smaller than the diameter of the adjacent intestine.

Early maturing. Anterior and middle lobes of the ovaries are either not visible or only faintly visible through the exoskeleton. Upon dissection, ovaries are opaque but with no distinctive coloration. Diameter of posterior lobes is similar to that of intestine. Visibility of ovaries is primarily due to coloration of outer lining of ovaries.

Late maturing. Ovaries are visible through exoskeleton. Development of anterior and middle lobes is complete and diameter of posterior lobes is greater than that of intestine. Upon dissection, ovaries are light yellow in *P. setiferus* and light yellow-green in *P. aztecus*.

Mature. Ovarian lobes are larger than in the preceding stage and clearly visible through exoskeleton. Ovaries are yellow-orange in *P. setiferus* and dark yellow-green in *P. aztecus*.

Spent-recovering. Ovaries are not visible through exoskeleton and on the basis of external appearance, indistinguishable from those in the early maturing stage. Upon dissection, ovaries appear flaccid and are greatly reduced in size. Ovaries remain opaque, but coloration is less distinctive than in late maturing and mature stages. In incomplete spawns, portions of ovaries, particularly posterior lobes, retain the coloration of mature ovaries.

Visual Assessment of Gonadal Development in Male Shrimp

Less emphasis has been placed on the assessment of gonadal development in male shrimp. In terms of the production of spermatozoa, males mature at a smaller size and earlier than females. For this paper, visual evaluation of maturation in males was based on the size of the terminal ampoules and divided into three stages according to the following criteria.

Immature. Terminal ampoules are not visible through exoskeleton.

Developing. Terminal ampoules are small but visible through exoskeleton.

Mature. Terminal ampoules are larger and clearly visible through exoskeleton. Spermataphores can be manually ejected with gentle pressure (King, 1948).

Organ and Biochemical Indices

Another method of assessing gonadal development is to examine changes in the ratio of gonad size to body weight. This ratio, known as the gonad index, has been used to describe maturation in both males and females of the crabs, *Uca annulipes* and *Portunus pelagicus* (Pillay and Nair, 1971), and the shrimps, *Metapenaeus affinis* (Pillay and Nair, 1971), *Parapenaeopsis hardwickii* (Kulkarni and Nagabhushanam, 1979) and *Penaeus setiferus* (Lawrence *et al.*, 1979). In female shrimp, where visual assessments of maturation are related to changes in both size and number of ova, the gonad index can be closely correlated to visual evaluations, and provides a quantitative measurement of gonadal development.

In shrimp, testes are much smaller than ovaries and thus gonad indices are smaller in males than females. In many aquatic species that do not copulate during reproduction but release gametes into the water for external fertilization, the male reproductive cycle is similar to that of females. However, in shrimp, which copulate during reproduction, the male reproductive cycle is less defined and changes in gonad organ index are smaller in males than in females.

In addition to increases in gonadal size that occur during maturation, the gonads are sites of intensive biochemical synthesis. Increases in the organic constituents of the ovaries have been reported in the shrimps, *Metapenaeus affinis* (Pillay and Nair, 1973), *Penaeus duorarum* (Gehring, 1974), and *Penaeus indicus* (Read and Caulton, 1980). In many animals, this biochemical synthesis in gonads is preceded by storage of organic reserves in other tissues. Sizes of storage organs increase prior to gonadal development and then decrease during gonadal development as nutrients are mobilized from storage organs to gonads. Analysis of the biochemical changes that occur in the gonads and storage organs characterize the biochemistry of gonadal maturation and increase the understanding of nutritional requirements for gonadal maturation. Among crustaceans, mobilization of lipids from the digestive glands to the ovaries has been reported in the crabs, *Paratelphusa pelagicus* (Adiyodi and Adiyodi, 1971), *Cancer magister* (Allen, 1972), *Portunus pelagicus* (Pillay and Nair, 1973), and *Carcinus maenas* (Paulus and Laufer, 1982), and the shrimp,

Metapenaeus affins (Pillay and Nair, 1973).

Materials and Methods

For *Penaeus aztecus* and *P. setiferus* collected along the Texas and Louisiana coasts, organ weights and carbohydrate, protein, and lipid contents were determined in gonads and digestive glands of both males and females, and in combined terminal ampoules and vasa deferentia of males. For females, the stage of maturation was visually evaluated on the basis of ovary size and color (King, 1948; Brown and Patlan, 1974). Because there was some degree of gonadal development in all individuals sampled, no organ weights or compositions were determined for immature stage females. For males, maturation was visually evaluated on the basis of the appearance of the spermatophores. No immature *P. setiferus* males and no immature or developing *P. aztecus* males were sampled.

Organ indices were calculated by dividing organ wet weights by whole body weights. Whole body weights were calculated from length measurements by using the regression equations of Fontaine and Neal (1971). Organ wet weights were calculated from the weights of organs after drying and independently determined ratios of dry weight to wet weight for each tissue, species, sex, and maturation stage. Protein, carbohydrate, and lipid contents of organs were expressed as organ biochemical indices, which were calculated as the amounts of carbohydrate, protein, and lipid in the organ divided by the calculated wet weight of the shrimp. Methods used for these biochemical determinations and statistical analysis of indices are described in Castille and Lawrence (1989). Statistical differences were considered significant at the P=0.05 level.

Results and Discussion

Organ indices for *P. aztecus* and *P. setiferus*

During maturation, the size of ovaries increases dramatically in both species (Fig. 1). In *P. aztecus*, the gonad organ index for mature females is almost 5 times that of early maturing females and in *P. setiferus* the gonad organ index of mature females is over 10 times that of early maturing females. In both species, the gonad organ indices of spent-recovering females are intermediate between those of early and late maturing females. These changes in the size of ovaries (from 1.9 to 9.1% in *P. aztecus* and from 0.9 to 9.8% in *P. setiferus*) are comparable to those reported in other crustaceans. In *Parapenaeopsis hardwickii,* the ovarian index increased from 2.80 to 5.85% during maturation (Kulkarni and Nagabhushanam, 1979), and in *Metapenaeus affinis*, the average gonad indices increased from 1.9% in the months of minimal reproductive activity to 9.3% in the months of peak acttivity (Pillay and Nair, 1971).

Although changes in organ size are not as great in the digestive glands of females as in the ovaries, there are significant changes in the digestive gland organ indices (Fig. 1) during maturation. In contrast to ovaries, however, the pattern of change differs between the two species. In *P. setiferus,* digestive glands decrease in size during maturation, but in *P. aztecus*, digestive glands increase in size. In spent-recovering females of both species, digestive gland organ indices are lower than those in late maturing and mature shrimp.

In contrast to females, male gonad indices (Fig. 1) are relatively small and the combined terminal ampoules and vasa deferentia make up a larger portion of the reproductive system than do the testes. In comparing the two species, testes are larger in mature *P. setiferus* than in mature *P. aztecus*; however, the opposite is true for the combined terminal ampoules and vasa deferentia, which are larger in mature *P. aztecus* than in mature *P. setiferus*. In *P. setiferus*, both the testes and the combined terminal ampoules and vasa deferentia are larger in mature than in developing males.

For digestive gland organ indices of males (Fig. 1), there are no significant differences between the two species or between developing and mature *P. setiferus*.

Female Biochemical Indices for *P. aztecus* and *P. setiferus*

As with the gonad organ indices, carbohydrate, protein, and lipid indices of the ovaries (Fig. 2) of *Penaeus aztecus* and *P. setiferus* increase in the ovaries during maturation and decrease after spawning. Similar increases in both lipid and protein

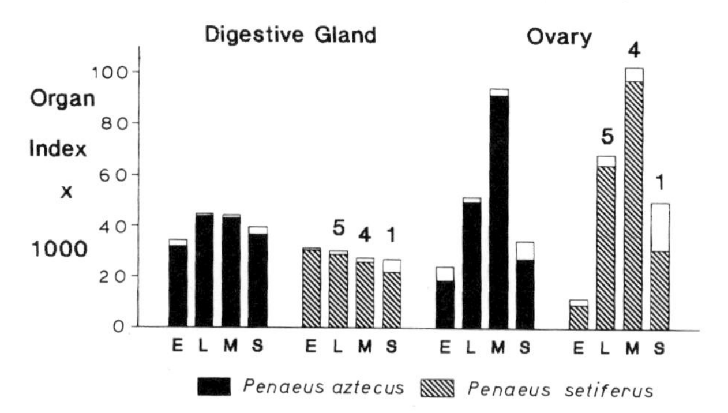

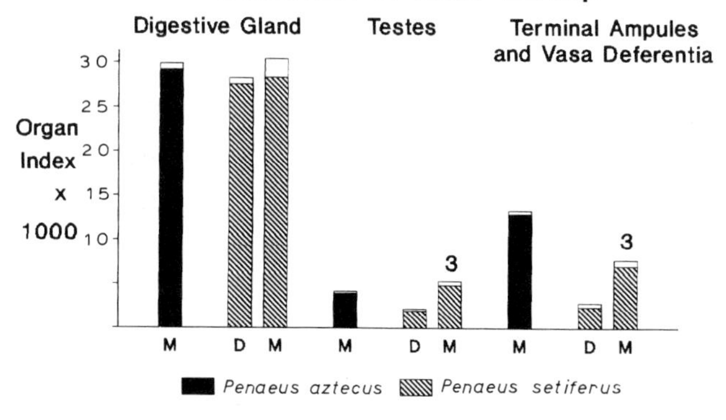

Figure 1. Organ indices for digestive glands, gonads, and combined terminal ampoules and vasa deferentia of *Penaeus aztecus* (solid bars) and *P. setiferus* (shaded bars) collected off the Texas and Louisiana coasts. Solid and shaded bars represent least square means for early maturing (E), late maturing (L), mature (M), and spent-recovering (S) females, and developing (D) and mature (M) males. Open bars represent standard errors for least square means. Organ indices were measured from 4 sampling locations for female *P. aztecus*, 6 sampling locations for female *P. setiferus*, 3 locations for male *P. aztecus,* and 4 locations for male *P. setiferus* except where noted by different numbers of locations above the bars (from Castille and Lawrence, 1989).

during ovarian maturation have been reported in *Metapenaeus affinis* (Pillay and Nair, 1971). In terms of abundance, protein is the largest organic constituent of the ovary and lipid the second largest. Carbohydrate is present at a much lower level than either protein or lipid. In terms of relative change, the increase in lipid is greater than that of protein and carbohydrate.

As with changes in organ indices, changes in biochemical indices (Fig. 2) are not as great in digestive glands as in ovaries. In *P. setiferus,* there is a substantial decrease in the lipid content of digestive glands during maturation. Digestive gland lipid indices of late maturing and mature females are less than half those of early maturing females. Carbohydrate and protein biochemical indices do not significantly differ during gonadal maturation. After spawning, all three digestive gland biochemical indices decrease. Part of this decrease after spawning may be due to the procedure used to obtain spent females. Mature females were held for up to 12 hours after capture until they spawned before removal of ovaries and digestive glands for analysis. Because the shrimp did not feed, this fasting period may have affected the size and biochemical composition of digestive glands. Another explanation is that mobilization of nutrients from the digestive gland to the ovaries may increase shortly before spawning and that mature shrimp in this study were sampled before this mobilization is complete. Thus the digestive gland organ and biochemical indices reported in this study for spent females may more closely represent the actual size and composition of digestive glands shortly before spawning.

In contrast to *P. setiferus*, both carbohydrate and protein indices increase in the digestive glands of *P. aztecus* during maturation. The data suggest that lipid also increases but the increase is not significant at the P=0.05 level. After spawning, digestive gland biochemical indices in *P. aztecus* decline for probably the same reasons as in *P. setiferus*.

A major difference between the composition of digestive glands and ovaries is the higher percentage of lipid in digestive glands. With the exception of late maturing and mature *P. setiferus*, where the lipid levels are decreased in digestive glands, lipid is the largest constituent of digestive glands.

The most interesting difference between the two species is the pattern of change in the lipid contents of digestive glands. The decrease in organ size and amount of

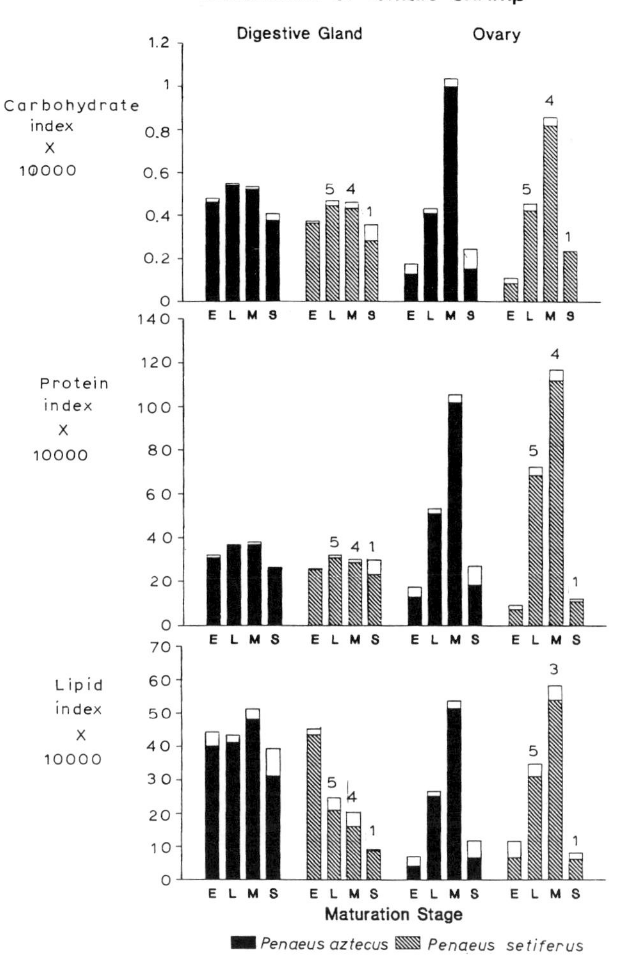

Figure 2. Carbohydrate, protein, and lipid indices for digestive glands and gonads of female Penaeus aztecus (solid bars) and P. setiferus (shaded bars) collected off the Texas and Louisiana coasts. Solid and shaded bars represent least square means for early maturing (E), late maturing (L), mature (M), and spent-recovering (S) females. Open bars represent standard errors for least square means. Organ indices were measured from 4 sampling locations for P. aztecus and 6 sampling locations for P. setiferus except where noted by different numbers of locations above the bars (from Castille and Lawrence, 1989).

lipid concomitant with maturation in *P. setiferus* supports the hypothesis that organic reserves, in this case lipid, are mobilized during maturation from the digestive gland to the ovary. In contrast, there is no evidence that organic reserves are mobilized from the digestive gland during ovarian maturation in *P. aztecus*.

These differences in patterns of mobilization reflect nutritional, habitat, and reproductive differences between the two species. If a species has the ability to store nutrients and then mobilize them for reproduction, reproduction can be timed to take advantage of seasonal factors such as food availability, water temperatures, or optimal conditions for larval survival. Mobilization of stored nutrients is usually seen in species that reproduce seasonally. In omnivorous species that do not experience seasonal food shortages, there is less advantage to seasonal reproduction and the reciprocal relationship between storage organs and the gonads is frequently not observed. In addition, the influence of seasonal factors on reproduction is usually more pronounced in species that reproduce in shallow rather than deep water. Thus, both seasonal reproduction and mobilization of stored nutrients is more frequently seen in species that reproduce at shallow depths. These differences in the ability to mobilize nutrients are consistent with the biology of *P. aztecus* and *P. setiferus*. *Penaeus setiferus*, which can mobilize lipids from the digestive gland to the ovaries, reproduces near shore in shallow water and has a relatively brief spawning season in late spring and early summer. *Penaeus aztecus*, in which mobilization of stored lipids from the digestive gland to the ovaries is either not present or present at a much lower level than in *P. setiferus*, reproduces offshore in deeper water and has continuous reproduction from early spring to early winter. In terms of the nutritional requirements for ovarian development, seasonal availability of food may be more limiting in *P. setiferus* than in *P. aztecus*. In terms of ovarian development under captive conditions, the dependence upon mobilized digestive gland lipids for ovarian maturation means that the nutritional state and history of broodstock prior to captive gonadal maturation are probably more important with *P. setiferus* than with *P. aztecus*.

The potential contribution of nutrients mobilized from the digestive gland to the ovaries in *P. setiferus* can be calculated from the changes in biochemical indices. On the basis of a decrease in digestive gland lipid indices from 43.5 x 10^{-4} for early maturing females to 16.3 x 10^{-4} for mature females and an increase in gonad lipid indices from 6.76 x 10^{-4} for early maturing females to 54.0 x 10^{-4} for mature females, the maximum contribution of mobilized lipid would be 57.5%. The absence of decreases in digestive biochemical indices during gonadal maturation for carbohydrate and protein in *P. setiferus* and for carbohydrate, protein, and lipid in *P. aztecus* indicates that mobilization of these nutrients from the digestive gland is limited and that accumulation of these nutrients in the ovaries is due to either mobilization from other tissues or dietary intake.

Male Biochemical Indices for *P. aztecus* and *P. setiferus*

As in digestive glands of early maturing females, the largest component in digestive glands of males is lipid with protein being the second most abundant component (Fig. 3). The most abundant component in both the testes and combined terminal ampoules and vasa deferentia is protein. Carbohydrate is low in all three tissues.

Differences in biochemical indices due to maturation or species are primarily due to differences in size of the tissues rather than changes in relative percentages of the components. For example, like the organ indices, carbohydrate, protein, and lipid indices are greater in testes of mature *P. setiferus* than in testes of mature *P. aztecus* and less in the combined terminal ampoules and vasa deferentia of mature *P. setiferus* than in the terminal ampoules and vasa deferentia of mature *P. aztecus*. Differences in size and biochemical indices between males of the two species can represent differences in male reproductive organs due to habitat, thelycum structure of females, or subgenus.

Although the protein indices, like the organ indices, for both the testes and the combined terminal ampoules and vasa deferentia are greater in mature than developing *P. setiferus* males, the differences in indices are due to differences in the size, and thus age, of the shrimp. A clearly defined cycle characteristic of gonadal development in females is not observed in mature males.

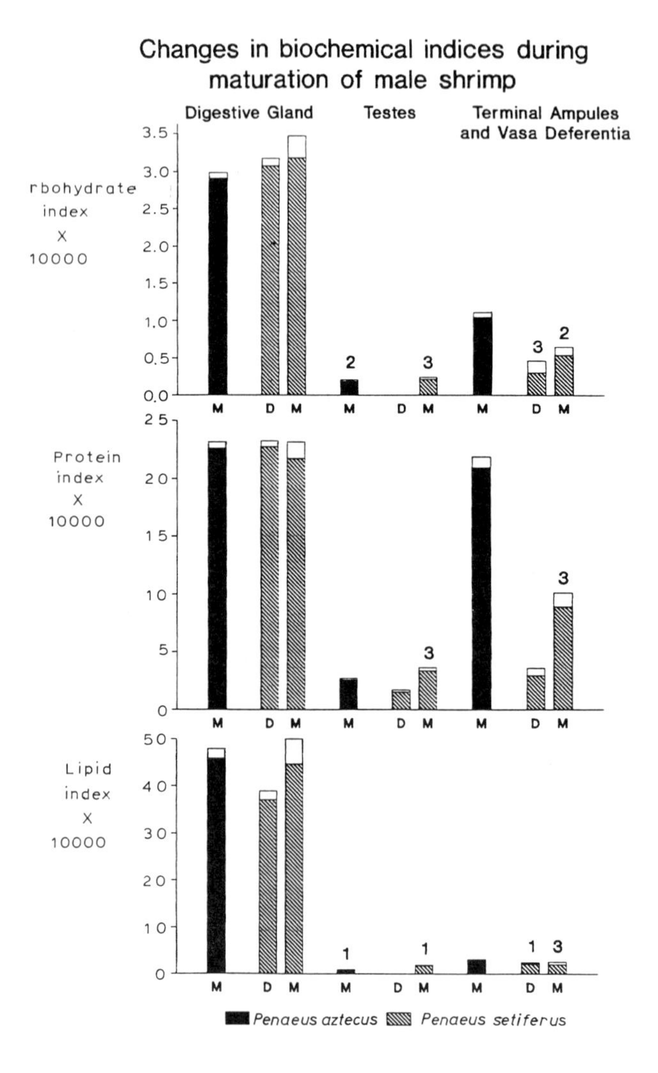

Figure 3. Carbohydrate, protein, and lipid indices for digestive glands, gonads, and combined terminal ampoules and vasa deferentia of male *Penaeus aztecus* (solid bars) and *P. setiferus* (shaded bars) collected off the Texas and Louisiana coasts. Solid and shaded bars represent least square means for developing (D) and mature (M) males. Open bars represent standard error for least square means. Organ indices were measured from 3 locations for *P. aztecus* and 4 locations for *P. setiferus* except where noted by different numbers of locations above the bars (from Castille and Lawrence, 1989).

Future Studies

Further characterization of the biochemical changes that occur during gonadal maturation in wild populations is warranted in order to better understand the importance of nutrition to fecundity and larval viability. Because maturation and reproduction of shrimp in captivity are necessary for the success of commercial shrimp culture in many parts of the world, a thorough understanding of the nutritional requirements of reproduction is vital for the industry. The importance of the composition of dietary lipids to reproduction of penaeid shrimp in captivity suggests that changes in the profiles of ovarian and digestive gland lipids during ovarian maturation in wild populations may be important indicators of dietary lipid requirements necessary to produce viable larvae. In addition, mobilization of stored nutrients from digestive glands to ovaries needs to be examined in other penaeid species to confirm whether differences in mobilization are related to environmental adaptations or phylogenetic relationships.

Literature Cited

Adiyodi, R. G. and K. G. Adiyodi. 1971. Lipid metabolism in relation to reproduction and molting in the crab *Paratelphusa hydrodromous* (Herbst): cholesterol and unsaturated fatty acids. Indian Journal of Experimental Biology 9: 514-515.

Allen, W. V. 1972. Lipid transport in the Dungeness crab, *Cancer magister* Dana. Comparative Biochemistry and Physiology 43(B): 193-207.

Brown, Jr., A. and D. Patlan. 1974. Color changes in the ovaries of penaeid shrimp as a determinant of their maturity. Marine Fisheries Review 36(7): 23-26

Castille, F. L. and A. L. Lawrence. 1989. The relationship between maturation and the biochemical composition of the gonads and digestive glands of the shrimps *Penaeus aztecus* Ives and *Penaeus setiferus* (L.). Journal of Crustacean Biology. 9(2): 202-211.

Cummings, W.C. 1961. Maturation and reproduction in the pink shrimp, *Penaeus duorarum* Burkenroad. Transactions of the American Fisheries Society 90)4): 462-468.

Fontaine, C.T. and R.A. Neal. 1971. Length-weight relations for three commercially important penaeid shrimp of the Gulf of Mexico. Transactions of the American Fisheries Society 100: 584-586.

Gehring, W. B. 1974. Maturational changes in the ovarian lipid spectrum of the pink shrimp, *Penaeus duorarum* Burkenroad. Comparative Biochemistry

and Physiology 49(A): 511-524.
King, J.E. 1948. A study of the reproductive organs of the common marine shrimp *Penaeus setiferus* (Linnaeus). Biological Bulletin 94(3): 244-262.
Kulkarni, G. K. and R. Nagabhushanam. 1979. Mobilization of organic reserves during ovarian development in the marine prawn, *Parapenaeopsis hardwickii* (Miers) (Crustacea, Decapoda, Peanaeidae) Aquaculture 18: 373-377.
Lawrence, A. L., D. Ward, S. Missler, A. Brown, J. McVey, and B.S. Middleditch. 1979. Organ indices and biochemical levels of ova from penaeid shrimp maintained in captivity versus those captured in the wild. Proceedings World Mariculture Socity 10: 453-463.
Motoh, H. 1978. Preliminary histological study of the ovarian development of the giant tiger prawn, *Penaeus monodon.* Quarterly Research Reports 2(4): 4-6.
Motoh, H. and P. Buri. 1980. Development of the external genitalia of the giant prawn, *Penaeus monodon.* Bulletin of the Japanese Society of Scientific Fisheries 46(2): 149-155.
Paulus, J. E. and H. Laufer. 1982. Vitellogenesis in the hepatopancreas and ovaries of *Carcinus maenas.* Biological Bulletin (Woods Hole) 163: 375-376.
Perez-Farfante, I. 1969. Western Atlantic shrimps of the genus Penaeus. Fishery Bulletin 67(3): 461-591.
Pillay, K. K. and N. B. Nair. 1971. The annual reproductive cycles of *Uca annulipes, Portunus pelagicus*, and *Metapenaeus affins* (Decapoda: Crustacea) from the southwest coast of India. Marine Biology 11: 152-166.
Pillay, K. K. and N. B. Nair. 1973. Observations on the biochemical changes in the gonads and other organs of *Uca annulipes, Portunus pelagicus*, and *Metapenaeus affinis* (Decapoda: Crustacea) during reproductive cycle. Marine Biology 18: 167-198.
Primavera, J. H. 1980. Sugpo (*Penaeus monodon*) broodstock. Aqua-Guide Series No. 1, Aquaculture Department, SEAFDEC, Iloila, Philippines. Pp. 10-15.
Rao, P. V. 1968. Maturation and spawning of the penaeid prawns of the southwest coast of India. FAO Fisheries Report 58(2): 285-302.
Read, G. H. L. and M. S. Caulton. 1980. Changes in mass and chemical composition during the moult cycle and ovarian development in immature and mature *Penaeus indicus* Milne Edwards. Comparative Biochemistry and Physiology 66(A): 431-437.
Subrahmanyam, C. B. 1965. On the reproductive cycle of *Penaeus indicus* (M. Edw.). Journal of the Marine Biological Association of India 7(2): 284-290.
Tirmizi, N. M. and W. Javed. 1976. Study of juveniles of *Metapenaeus stebbinal* Nobili (Decapoda, Penaeidae) with particular reference to the structure and development of the genitalia. Crustaceana 30(1): 55-67.
Tuma, D.J. 1967. A description of the development of primary and secondary sexual characters in the banana prawn, *Penaeus merguiensis* deMan (Crustacea: Decapoda: Penaeinae). Australian Journal of Marine and Freshwater Research 18: 73-88.

Shrimp Population Models and Management Strategies: Potentials For Enhancing Yields

Richard E. Condrey
Coastal Fisheries Institute
Center for Wetland Resources
Louisiana State University
Baton Rouge, Louisiana 70803-7503

Abstract

The use of models in management of the U.S. Gulf of Mexico shrimp fishery is reviewed and deficiencies discussed. The historic use has been primarily limited to a continuing reevaluation of two narrowly constructed management measures which were designed to provide moderate increases in yield. Areas of major social or ecological concern and areas in which yield can be dramatically enhanced have received little attention. Specific examples discussed include wetland loss, the use of TEDs, and the excessive growth-overfishing which occurs in some states.

Introduction

The purpose of this paper is to review the historic use of population models in the management of the U.S. Gulf of Mexico shrimp fishery, to address areas in which shrimp yield may be increased, and to address two areas which are receiving recent attention: wetland loss and the impacts of TEDs. I will approach the topic from my role as the primary author of the initial draft of the Gulf of Mexico Fishery Management Council's Shrimp Management Plan (Condrey, 1979). This effort coincided with the beginning of a focused attention within the Gulf scientific community on the population dynamics of shrimp and the development of population models of the fishery.

The review follows the sequential listing of relevant management measures in the shrimp plan. Leary (1985) provides an excellent review of the plan's development, objectives, and management measures.

Management Measure 1: Permanent Closure of The Pink Shrimp Sanctuary

The current state/federal management regime for pink shrimp is described in Management Measure 1 of the shrimp plan. It calls for a permanent closure of shrimping in a portion of the state and federal waters which are generally adjacent to the Florida Everglades. It is intended to increase yield by protecting small pink shrimp from harvest until they have generally reached a size larger than 69 tails to the pound and have migrated into the fishery.

The scientific justification for the inclusion of this management measure in the original draft of the shrimp plan (Condrey, 1979) derives largely from the growth and mortality models of Lindner (1966, M = 0.35 to 0.52 on a monthly basis) and Berry (1970, M = 0.04 to 0.13 on a monthly basis), the migration studies of Costello and Allen (1966), and the size-at-depth equations of Iverson *et al.* (1960). In developing the original draft of the shrimp plan we used these studies to assess the yield implications of the proposed closure. The available data on growth, mortality, and migration of the population supported the concept that the closure would have a positive impact on yield.

After the Council adopted the concept of the closure, a far more elaborate analysis was conducted by Blomo *et al.* (1982) using the bioeconomic model of Grant and Griffin (1979). Blomo used environmentally regulated growth rates, a natural mortality rate of 0.93 on a monthly basis, a demand function which is shrimp-size specific, a heterogeneous fishing fleet, and estimates of the variable costs of operating different vessel classes. He noted that his results were highly dependent upon his choice of a natural mortality rate. He concluded that the federal portion of the closure was too broad and, in comparison to no federal closure, would result in a substantial loss of yield (1.5 million pounds), total revenue (2 million dollars), and quasi-rent (0.9 million dollars). He suggested that the spatial and temporal extent of the federal closure be reassessed if "increased landings, revenue, and quasi-rent are desired."

Grant *et al.* (1981) examined the impact of varying the size of shrimp protected by the closure and of only closing the state and federal waters to shrimping when the greatest concentration of small pink shrimp was expected (November through April). They introduced stochastic variation into their bioeconomic model by allowing the rate of recruitment, natural mortality, growth, movement, and harvest to vary up to 10%, though they noted that the variation in some of these point estimates far exceeded the 10% level. Although they found that these management options did change harvest, revenue, and total rent, they could not strongly recommend any of the three policies examined. At the level of M used (0.87 on a monthly basis) the restriction of harvest to larger shrimp resulted in a significant decrease in harvest, revenue, and rent. While the harvest of smaller shrimp and the seasonal openings did increase total harvest, revenue, and rent, the increases were not significant.

Khilnani and Tse (1980) adapted their Fisheries System Management Model to conditions which described the Florida pink shrimp fishery and examined the benefits of the closure and of two seasonal modifications of the closure. In their sociobioeconomic model, fishing strategy is affected by availability of shrimp and the dockside market price. Prices, in turn, are dependent on the size and (inversely) on the local supply of shrimp. Using an M of 0.48 and finite elasticities for small and large shrimp, Khilnani and Tse found that the closure resulted in a decrease in legal yield and increases in consumer prices, net revenues to the shrimper, and the attractiveness to the individual shrimper of fishing illegally in the closed area. With the two theoretical seasonal openings, the impact of the closure was lessened in direct relationship to the expected abundance of legally harvestable small shrimp.

In discussing their results, Khilnani and Tse made two key points. First, "the closure policy produces (in their model) a mixed bag of benefits and disbenefits," with shrimpers benefiting and consumers losing because of a decreased yield and increased price. Second, the closure resulted in "a tremendous temptation to fish (illegally) in the inshore grounds, thus making enforcement an important issue."

The most recent study by Nichols (1986) suggests that the measure should result in an increase in yield at realistic levels of natural and fishing mortality. At the value of M (0.3 on a monthly basis), which Nichols argues best describes the resource, he concludes that the measure should result in a 7% to 25% increase in yield as F ranges from 0.3 to 0.7.

If M is above 0.3 but still within a realistic range (> 0.4), Nichols notes that the measure still has a positive, but greatly reduced benefit. But if M is below 0.3, further restrictions on the harvest of small shrimp may be justified. Specifically, if M=0.15, a further delay in harvest could result in a 48% to 92% increase in yield and a 146% to 355% increase in ex-vessel value as F ranges from 0.45 to 0.85.

Management Measure 2: The Texas Seasonal Cooperative Closure

Management Measure 2 of the original shrimp plan calls for a seasonal, cooperative closure of the territorial sea of Texas and the adjacent federal waters out to 200 miles. The closure, which normally occurs from June 1 to July 15, is intended to increase yield by protecting the newly emigrating, zero-year-class brown shrimp from harvest until they are 65 tails to the pound. Protection from harvest is only provided once shrimp exit the bays and enter Gulf waters deeper than 4 fathoms.

In developing the yield considerations for the original draft of the plan, we used the historic catch-at-size data, Parrack's (1979) monomolecular growth-rate equation, and a range of instantaneous mortality rates which bracketed expected (but then unpublished) levels. Our analysis indicated that the average change in annual yield would range from plus 4.5 million to minus 1.7 million pounds of tails, as natural mortality ranged from 0.04 to 0.87 on a monthly basis. These yield predictions were based on the assumption that a traditional fishery for white shrimp in Gulf waters less than 4 fathoms would be closed during the time of the brown shrimp closure and that the inshore harvest of brown shrimp would remain at historic levels. These assumptions have not been met: the nearshore white shrimp fishery remains open despite a substantial by-catch of brown shrimp and the inshore harvest is increasing (Leary 1985).

The impacts of the measure on the catch off the coasts of Texas and, to a lesser extent, the other Gulf states have been annually assessed by NMFS. The approach used generally involves a determination of the size distribution of shrimp in the offshore waters during the closure (*e.g.*, Matthews, 1982), an analysis of catch patterns in the fishery (*e.g.*, Klima *et al.*, 1982), yield and economic analyses (*e.g.*, Nichols, 1982; Poffenberger, 1982), and estimates of the impacts on fleet mobility and support facilities (*e.g.*, Jones and Zweifel, 1982; Ward and Poffenberger, 1982). The findings from these individual studies are then synthesized in a summary report (*e.g.*, Jones *et al.*, 1982). The NMFS analyses receive a formal annual review by the Council and its Scientific and Statistical Committee (SSC).

At it's most recent meeting, the SSC concluded with the results of the NMFS studies that the closure does increase yield in years of high recruitment, but has no detectable effect in years of low recruitment. The SSC, therefore, recommended that the Council close the federal waters in years of high recruitment, but leave them open when recruitment is low (GMFMC, 1988).

In addition to the NMFS's analyses, as discussed below Grant, Griffin, and their co-workers have used their bioeconomic model to assess the likely impacts of the Texas Closure on the Texas shrimp fishery. These works are treated separately here since they have not received a formal review by the Council's SSC. This is unfortunate since such presentations would have likely assisted the SSC in reaching their 1988 position at a much earlier date.

George and Grant (1983) developed a stochastic simulation model of the population dynamics of brown shrimp in Galveston Bay. They used this to evaluate the impacts of a variety of management alternatives similar to the Texas closure and of fluctuations in shrimp abundance caused by environmental conditions. At the level of natural mortality used (average M=0.43 on a monthly basis), they found that closure of the inshore waters and delays in the harvest of food shrimp resulted in significant decreases in yield. They also found that yield in their model was far more sensitive to environmental changes and variability in fishing effort than to the management alternatives considered. They noted that "It may be difficult to evaluate the effects of a management alternative in view of the effects of changes in environmental conditions and the effect of unpredictable variation in the fishing effort." While their conclusion that a closure similar to the one off Texas would have an adverse impact on yield was a direct result of the high level of natural mortality used, their early conclusion about the impact of environmental variation was not considered in the original shrimp plan and directly affected the SSC's 1988 position.

Carothers and Grant (1987) used the basic bioeconomic model of Grant and Griffin (1979) to examine how the seasonality of postlarval brown shrimp recruitment impacted yield in the Texas fishery. Their treatment allowed for two types of recruitment patterns: (1) a dominant spring recruitment peak and (2) a seasonal variation in the recruitment peak between spring and fall. They examined the impact of these different recruitment patterns on various temporal and spatial modifications of the Texas closure. They noted a distinct impact at the level of mortality used (M=0.18 on a monthly basis). If the spring recruitment was always dominant, then protection of brown shrimp during June and July resulted in increased ex-vessel value of the harvest and a reduction in fishing effort. If, however, there was a periodic fall recruitment peak, then the June-July closure resulted in net losses of yield in pounds and ex-vessel value.

Krauthamer *et al.*, (1987) developed a sociobioeconomic model as an extension of Grant and Griffin's (1979) bioeconomic model. Sociocultural variables used include age, experience, innovativeness, gratification and work orientation, and vessel ownership status of the captain. The impacts on five types of inshore vessels were simulated in terms of hypothetical management restrictions on harvest. The policies examined were shown to have different, disproportionate "impacts on different groups of fishermen, with revenue and rent of the lowest producers being most sensitive to policy changes."

Management Measure 3: Optimization of State Yields

Despite the considerable attention which has been paid to reevaluation of Management Measures 1 and 2, little attention has been paid to Management Measure 3. This management measure suggests that the states assure that shrimp are being harvested at optimum sizes. The measure points out that the states could "protect critical habitat areas, reduce the waste of shrimp from culling, and increase yield by closing areas where shrimp were smaller than a useful size." The measure suggests that the most sensitive areas are the shallow-water estuaries.

This lack of attention to Management Measure 3 is disheartening since the greatest potentials for increasing yields in the Gulf involve modifications of existing state laws (*e.g.*, Leary, 1985). As specific examples I will discuss the yield impacts of two measures which have been suggested by some prominent Louisiana shrimpers for brown and white shrimp.

With white shrimp there is considerable interest among some Louisiana shrimpers for a seasonal closure of state waters from January to May to protect overwintering white shrimp. Our catch-curve analyses at LSU (Fuller *et al.*, submitted) suggest the real possibility that such a measure might increase yield. In our analyses we used Nichols's (1981) comprehensive growth model for white shrimp. It consists of sex specific, multiple regression equations describing the expected daily incremental growth of white shrimp as a function of size and temperature. The solutions to Nichols's equations predict the fastest daily incremental growth will occur when the smallest shrimp are grown near the maximum temperatures normally encountered in the north central Gulf. As temperature declines from this maximum, the observed maximum growth rates move toward larger shrimp. In addition, Nichols's equations predict that there is a broad range of realistic values of temperature and size over which there will be no growth. In general, this dependence is such that smaller shrimp do not grow at colder temperatures and larger shrimp do not grow at higher temperatures. Nichols's equations are consistent with numerous reports (*e.g.*, Lindner and Anderson, 1956; Klima, 1974) that the growth rate of white shrimp is highly temperature dependent, with maximum growth rates occurring in the summer and minimum rates expected in the winter.

Our catch-curve analyses at LSU indicate that natural mortality during the November to May period should be in the range of 0.15 to 0.28 on a monthly basis. Given the historic mix of sizes of white shrimp caught in 0 to 10 fathoms, if M is in the range of 0.15, a closure from January 1 to June 1 would have a dramatic, positive impact on yield. Conversely, if M is in the range of 0.28, the closure will have a negative impact on yield.

If our mortality estimates for white shrimp are applicable to other areas and seasons, the modelling done by Nichols (1984) suggests drastic changes in white shrimp management are warranted from a yield enhancement point of view. Specifically, Nichols points out that if M is less than 0.3 to 0.6 on a monthly basis, then yield during the major summer-fall period is dramatically increased if harvest is delayed. Our work suggests that M is below 0.3.

Regarding brown shrimp, Louisiana closes its inshore waters to shrimping on December 22, and reopens them at a variable time in May or June. The variable opening date is set in late April and normally coincides with the date at which 50%

of the zero-year-class shrimp are expected to be at least 161 tails to the pound.

Given our currently accepted rates of natural mortality (M=0.2 to 0.3 on a monthly basis) and fishing pressure, there is excessive growth overfishing occurring in Louisiana. We estimate that each weekly delay around the traditional opening date results in a 20% increase in yield. Such an increase should have a major, immediate impact on yield since Louisiana accounts for nearly 50% (by number) of the brown shrimp harvested in the Gulf.

These opportunities to enhance yield with brown and white shrimp are being missed. There appears to be little effort to improve yield-per-recruit in the Gulf shrimp fishery beyond the first two measures of the shrimp plan. Indeed, quite the opposite appears true. Caillouet and his workers noted an increase in the proportion of small shrimp which were landed in the Gulf from 1960 to 1981 (e.g., Caillouet and Koi, 1983). These revealing analyses warrant updating, especially since no state or federal regulations have been passed which would reverse the trend towards a continuing decrease in the yield-per-recruit of shrimp entering the fishery.

Management Measure 4: Habitat Dependence

Management Measure 4 recognized the critical link between habitat protection and shrimp production and established the Council's Habitat Committee.

As discussed by Minello and Zimmerman in the previous chapter, current models being developed by scientists at the NMFS Galveston Lab are addressing the possible impacts of continued wetland loss on shrimp production. At recent meetings of the Council and its SSC, the Galveston team has shown a conceptual link between two well recognized phenomena. The first ties wetland loss and subsidence to the nursery role of flooded marshes for shrimp. The other is a pattern, as of yet unexplained, of increased annual recruitment of juvenile shrimp and menhaden to their respective fisheries from 1960 to the present.

The Galveston workers have elegantly pointed out the obvious possible link between these two trends. That is, if flooded marshes serve as an important habitat for shrimp and the marshes are undergoing subsidence, then what we are experiencing is a temporary stimulation of the carrying capacity of marshes for shrimp. The fear is that when this possible temporary stimulation begins to decline, the fishery and stocks will go into a state of collapse

Management Measure 5: Non-incidential By-catch

Management Measure 5 addresses the incidental capture of finfish and sea turtles. It calls for the development of a TED-like device which will effectively reduce the by-catch of shrimp trawls without unduly impacting shrimp yields.

Griffin *et al.*, (1987) have modeled the potential impact of TEDs on shrimp yields in the offshore fishery. The worst case scenario examined was a 10% loss. They concluded that this would lead to an annual loss of just over $10,000 in net worth to each vessel. Our qualitative work at LSU suggests that some shrimpers will not have losses with some TEDs while other shrimpers will encounter losses of 17% to 27% (Condrey and Day, 1987). Preliminary results of current statistically rigorous studies by NMFS appear to be consistent with our findings. The Georgia TED appeared to function well in tests off the Texas and Georgia coasts. However, "in Louisiana both Georgia and Saunders TEDs showed a high degree of shrimp loss" (Anon, 1988).

Summary

The use of biologic, bioeconomic, and sociobioeconomic models to assess methods of maximizing and/or optimizing yields of U.S. Gulf of Mexico shrimp has been mainly limited to a continuing reassessment of the first two management measures in the Gulf shrimp plan. These two measures were intended to increase yield in the pink shrimp fishery in southwestern Florida and in the brown shrimp fishery off the Texas coast. Analyses to date indicate that the measures increase yields and ex-vessel value, but not dramatically. Minor modifications of both measures have been proposed, but these modifications are expected to have minimal effects on yields.

Little attention has been paid to the use of models to evaluate measures by which the states could increase and/or optimize yield. This is unfortunate since dramatic increases in yields can be expected from prudent management measures in some states and some fisheries. Specific examples discussed are Louisiana's harvest of white and brown shrimp.

Recently modelers have attempted to look at the impacts of factors other than area and seasonal closures. While the earliest workers examined the impact of variation in environmental driving forces on yields, recent attention has been paid to the impacts of wetland loss and TEDs. The work being conducted at the NMFS Galveston Lab suggests that the increasing yields of brown and white shrimp may be a temporary phenomenon associated with wetland loss. Under this hypothesis, a collapse of the fishery is possible. With TEDs, Griffin *et al.* (1987) estimate that a 10% TED-induced shrimp loss will cost the average offshore shrimper $10,000 a year. Initial results of NMFS studies off the Louisiana coast suggest that for some shrimpers these losses may be on the order of 20% to 25%.

The task force which generated the original draft shrimp plan concluded that

"The fishery is not operating at optimum yield" (Condrey, 1979). It appears to this writer that that conclusion is even more correct today. The traditional shrimper is faced with increasing effort, imports, mariculture, wetland loss, and habitat alteration while the average size of the shrimp being caught is steadily declining. In addition, shrimpers have a by-catch which is no longer incidental, a pending mandatory use of TEDs, and the real possibility of a collapsing fishery.

Shrimpers need help. The solutions, if they exist, may involve radical departures (*e.g.*, Blomo, 1981) from the current laissez-faire, open-access, growth-overfished management philosophies.

The apparent reluctance of some state/federal management agencies to examine alternative management regimes may relate to the intense and highly competitive nature of the fisheries which exploit these limited resources. The nature and extent of these conflicts makes it difficult to derive simple answers. Sociobioeconomic models designed along the lines of Khilnani and Tse (1980) or Krauthamer *et al.*, (1987) may be the only practical methods of examining the likely impacts of potential management alternatives on conflicting user groups.

As always, the real world applicability of these models will depend upon the realism of their design and mathematical expression. For example, discussions on the positive or negative yield implications of the Texas and pink shrimp closures still depend on the level of natural mortality selected. Even when these models are made stochastic, the variance used is dwarfed by the available range of natural mortality estimates.

Addendum

The Critical link recognized in management Measure 4 between habitat protection, shrimp production, and fishery collapse has been simulated by Browder *et al.* (1989) for a major portion of Louisiana's brown shrimp harvest. A precipitous decline in brown shrimp production after 1995 is predicted.

Browder *et al.'s* model consists of three major elements. The first is a model of marsh disintegration in which the probability of disintegration is weighted by three factors approximating the natural processes of interior marsh decay, shoreline erosion, and border condition. The model was fit to specific Louisiana marshes by adjusting the weighting factors to simulate verified, land-water distributions olserved in a December 2, 1984 Landsat-5 overflight.

The second part of the Browder *et al.* model consists of extrapolating the simulated pattern of marsh disintegration from the December 2, 1984 date backward to a theoretical state of solid marsh and then forward until all the marsh is theoretically lost. At each interval, the model computed outputs the total length of the marsh-water interface and the percentage of open water. Dimensions of real time are incorporated into the model's hindcasted estimates of open water by linking these with direct measurements made for 1956 and 1978.

The simulations predict that marsh disintegration has resulted in a steady, rapid increase in the total length of the marsh-water interface from 1955 to present. The simulations also predict that this increase will continue until 1995 at which time a precipitous decline is predicted.

In the third part of their model, Browder *et al.* derived a statistical relationship between the historic brown shrimp harvest and their simulated historic levels of marsh-water interface. When they then coupled this dependence to their projections of marsh-water interface, a precipitous decline in brown shrimp harvest was predicted.

While the scope of the model can be improved as by using existing estimates of prerecruits of brown and white shrimp and menhaden, by incorporating existing spawner-recruit relationships, and by linking to existing sociobioenconomic models, it is hoped that the results will be startling enough to focus serious attention on the real-world implications of marsh loss on fisheries production.

Certainly since this paper was delivered, there have been several positive potential changes in the use of models to manage Gulf shrimp which suggest that management is ready to respond to Browder *et al.'s* findings. The state of Texas has adopted a shrimp plan which closely parallels the national plan and which will make full use of the sociobioeconomic model being refined by Griffin and his co-workers. Mississippi is in the process, and Louisiana is gearing up for the process, of preparing shrimp plans which will take full advantage of relevant models to assess management alternatives. The SSC, even without benefit of Browder *et al.'s* paper, has recognized the possibility that shrimp are subject to the possibility of recruitment failure, a position which is at variance with the Councils existing plan. The Council has responded to the SSC's concern by asking NMFS to establish an ad hoc shrimp stock assessment panel. That panel convened in June 1989 at the NMFS Galveston Laboratory.

Management has recognized that the clock on the Gulf shrimp fishery may be ticking. Browder *et al.* suggest that management has little time to respond. The question is, "Will it?"

Acknowledgements

This work was supported by the Louisiana College Sea Grant Program.

Literature Cited

Anon. 1988. Report 2 TED observer program. National Marine Fisheries Services, Southeast Fisheries Center F/SEC6:EFK:mr, Galveston, Texas ,2pp.

Berry, R.J. 1970. Shrimp mortality rates derived from fishery statistics. Proceedings of the Gulf and Caribbean Institute 22:66-78.

Blomo, V.J. 1981. Conditional fishery status as a solution to overcapitalization in the Gulf shrimp fishery. Marine Fisheries Review 43:20-24.

Blomo, V.J., J.P. Nichols, W.L. Griffin, and W.E. Grant. 1982. Dynamic modeling of the eastern Gulf of Mexico shrimp fishery. American Journal of Agricultural Economics 64:475-482.

Browder, J.A., L.N. May, Jr., A. Rosenthal, J.G. Gosselink, and R.H. Baumann. (1989). Modeling future trends in wetland loss and brown shrimp production in Louisiana using thematic mapper imagery. Remote Sensing of Environment 28: 45-59.

Caillouet, C.W., Jr., and D.B. Koi. 1983. Ex-vessel value and size composition of reported May-August catches of brown shrimp and white shrimp from 1960 to 1981 as related to the Texas closure. Gulf Research Reports 7:187-203.

Carothers, P.E., and W.E. Grant. 1987. Fishery management implications of recruitment seasonality: Simulation of the Texas fishery for brown shrimp, *Penaeus aztecus*. Ecological Modelling 36:239-268.

Condrey, R.E. 1979. Draft fishery management plan for the shrimp fishery of the Gulf of Mexico, United States Waters. Center for Wetland Resources, Louisiana State University, Baton Rouge, Louisiana, 269 pp.

Condrey, R.E., and R. Day. 1987. Results of a program for introducing turtle excluder devices (TEDs) into the Louisiana shrimp fishery. Center for Wetland Resources, Louisiana State University, Baton Rouge, Louisiana, 122 pp.

Costello, T.J., and D.M. Allen. 1966. Migrations and geographic distribution of pink shrimp, *Penaeus duorarum*, of the Tortugas and Sanibel grounds, Florida. Fishery Bulletin 65:449-459.

Fuller, D.A., R.E. Condrey, G.H. Itano, and J.P. Geaghan. (submitted). Yield-per-

recruit of overwintering white shrimp. Journal of North American Fisheries Management, 18 pp.
George, L.C., and W.E. Grant. 1983. A stochastic simulation model of brown shrimp (*Penaeus aztecus* Ives) growth, movement, and survival in Galveston Bay, Texas. Ecological Modelling 19:41-70.
Grant, W.E., and W.L. Griffin. 1979. A bioeconomic model of the Gulf of Mexico shrimp fishery. Transactions of the American Fisheries Society 108:1-13.
Grant, W.E., K.G. Isakson, and W.L. Griffin. 1981. A general bioeconomic simulation model for annual crop marine fisheries. Ecological Modelling 13:195-219.
Griffin, W.L., J. Clark, J. Clark, and J. Richardson. 1987. Economic impact of TED on the shrimp industry in the Gulf of Mexico. Agricultrual Economics Department, Texas A & M University, College Station, Texas, 39 pp.
Gulf of Mexico Fishery Management Council. 1988. Minutes of the January 13, 1988 Shrimp Scientific and Statistical Committee Meeting. Gulf of Mexico Fishery Management Council, Tampa, Florida.
Iverson, E.S., A.E. Jones, and C.P. Idyll. 1960. Size distribution of pink shrimp, *Penaeus duorarum*, and fleet concentrations on the Tortugas grounds. United States Fish and Wildlife Service Special Scientific Report 356:1-62.
Jones, A.C., E.F. Klima, and J.R. Poffenberger. 1982. Effects of the 1981 closure on the Texas shrimp fishery. Marine Fisheries Review 44:1-4.
Jones, A.C., and JR. Zweifel. 1982. Shrimp fleet mobility in relation to the 1981 Texas closure. Marine Fisheries Review 44:50-54.
Khilnani, A., and E.T.S. Tse. 1980. Integrated approaches to fisheries policy analysis: A case study of the Tortugas shrimp fishery. Department of Engineering and Economic Systems. Stanford University, Stanford, California.
Klima, E.F. 1974. A white shrimp mark-recapture study. Transactions of the American Fisheries Society 103:107-113.
Klima, E.F., K. N. Baxter, and F. J. Patella, Jr. 1982. A review of the offshore shrimp fishery and the 1981 Texas closure. Marine Fisheries Review 44:16-30.
Krauthamer, J.T., W.E. Grant, and W.L. Griffin. 1987. A sociobio-economic model: The Texas inshore shrimp fishery. Ecological Modeling 35:275-307.
Leary, T.R. 1985. Review of the Gulf of Mexico management plan for shrimp. In. Second Australian National Prawn Seminar, P.C. Rothlisberg, B.J. Hill, and D.J. Staples, Editors. NPS2, Cleveland, Queensland, Australia. p.267-274.
Lindner, M.J. 1966. What we know about shrimp size and the Tortugas fishery. Proceedings of the Gulf and Caribbean Fisheries Institute 18: 18-26.
Lindner, M.J., and W.W. Anderson. 1956. Growth, migration, spawning, and size distributions of shrimp, *Penaeus setiferus*. Fishery Bulletin 56:553-645.
Matthews, G.A. 1982. Relative abundance and size distributions of commercially important shrimp during the 1981 Texas closure. Marine Fisheries Review 44:5-15
Nichols, S. 1981. Growth rates of white shrimp as a function of shrimp size and water temperature. National Marine Fisheries Service, Southeast Fisheries Center, Miami, Florida, 25pp.
Nichols, S. 1982. Impacts on shrimp yields of the 1981 Fishery Conservation Zone

closure off Texas. Marine Fisheries Review 44:31-37.
Nichols, S. 1984. Updated assessments of brown, white, and pink shrimp in the U.S. Gulf of Mexico. National Marine Fisheries Service, Southeast Fisheries Center, Miami, Florida, 37pp.
Nichols, S. 1986. Updated yield-per-recruit information on the Tortugas pink shrimp fishery. North American Journal of Fisheries Management 6:339-343.
Parrack, M.L. 1979. Aspects of brown shrimp, *Penaeus aztecus*, growth in the northern Gulf of Mexico. Fishery Bulletin 76:827-836.
Poffenberger, J.R. 1982. Estimated impacts on ex-vessel brown shrimp prices and values as a result of the Texas closure regulation. Marine Fisheries Review 44:38-43.
Ward, J.M., and J.R. Poffenberger. 1982. Survey of ice plants in Louisiana, Mississippi, and Alabama, 1980-81. Marine Fisheries Review 44:55-57.

Comparative Morphology and Physiology of Egg Activation in Selected Penaeoidea

John W. Lynn[1], Muralidharan C. Pillai[2], Patricia S. Glas[3], and Jeffrey D. Green[3]

[1]Dept. of Zoology and Physiology, Louisiana State University, Baton Rouge, LA 70803

[2]Bodega Marine Laboratory, University California Davis, Bodega Bay, CA 94923

[3]Dept. of Anatomy, Louisiana State University Medical Center, New Orleans, LA 70112

Abstract

Typically, an egg ovulated from a penaeid shrimp, prior to activation, is in meiotic arrest (first meiotic metaphase) and possesses deep extracellular cortical invaginations (cortical crypts) which contain a hetergeneous jelly precursor. Surrounding the cortical crypts and closely apposed to the egg plasma membrane is an extracellular vitelline envelope. Following spawning, the eggs of these species undergo a series of activational changes. These changes include: (1) release of the jelly precursor from the cortical crypts; (2) transformation of the jelly precursor into a homogeneous jelly layer that forms around the egg; (3) resumption and completion of meiotic maturation and; (4) elaboration of an extracellular hatching envelope. The above series of activational events are normally triggered by exposure of the eggs to sea water and do not require fertilization. Sperm-egg interaction, however, is critical for normal embryonic development. Both extracellular enzymes and ions appear to be involved in the regulation of these early activational events. Extracelluar Mg^{+2} and involvement of a serine protease are reported to be necessary for the formation of the jelly layer in the eggs of all penaeid species examined. Similarly, resumption of meiosis and formation of the hatching envelope also require external Mg^{+2}, however, involvement of enzyme(s) in these processes is not yet determined. The above series of early activational events are completed in the absence of external Ca^{+2} if the eggs are fertilized; unfertilized eggs do not activate in the absence of this divalent cation.

Introduction

One of the most critical periods in the life history of an organism is that time span between the ovulation of a relatively quiescent egg and the first few cleavages of the fertilized egg. During this time (though not adhered to in all species) an egg must: 1) achieve a haploid genetic condition, 2) interact and accept a sperm into its cytoplasm for fusion with the female pronucleus (fertilization) restoring the diploid condition, while also preventing the intrusion of an excess number of sperm, 3) organize and regulate metabolic processes essential to the early developmental stages, and 4) survive not only predation but also simple physical threats from the environment.

All of the above processes have intrigued reproductive biologists for many years and extensive knowledge of these processes has been accumulated through diverse studies with a varied group of animals such as the echinoderms (Schuel, 1985), amphibians (Grey *et al.*, 1974), and mammals (Wassarman, 1987; see also Schmell *et al.*, 1983 for a review). Although some of these animal groups have become recognized as model systems for the study of gametes and gamete interaction, it is rather surprising that different animal species have not been exploited to a greater extent to fill the gaps left by the current models and even to allow easier manipulation of identical or similar events. For example, workers during the past decade have collected considerable data on gamete biology among decapod crustaceans (*e.g.*, Clark and Griffin, 1988; Clark and Pillai, 1988). While decapod crustacean gametes do have some distinctive morphological and physiological characteristics (Clark *et al.*, 1984), the ultimate goals in fertilization are similar to those of other species (*e.g.*, echinoderms, amphibians, and mammals). The morphology and physiology of sperm activation (acrosome reaction) in penaeids have been topics of fascination for many years and have been described elsewhere (Griffin, 1987; Clark and Griffin, 1988). In this paper we will compare the morphology and physiology of early stages of egg activation and developmental phenomena in four penaeid species (*Penaeus aztecus*, *Penaeus setiferus*, *Sicyonia ingentis*, and *Trachypenaeus similis*). The results discussed here are taken not only from published works, but also from current investigations in preparation for publication.

Early Morphological Events Associated with Egg Activation

It should be noted that the penaeid shrimp species discussed in this review actually fall into two categories based on: 1) the mode of transfer and/or storage of the sperm after mating, and 2) the arrangement of the female's reproductive specialization (thelycum) for receiving and storing sperm. *Sieyonia ingentis*, *P. aztecus*, and *T. similis* are referred to as closed-thelycum shrimp. After mating, sperm may be stored for extended periods of time as seminal plasm in the paired seminal receptacles of the thelyca in these animals. Mating in closed-thelycum species typically takes place only after a molt. In contrast, *P. setiferus* is an open-thelycum shrimp and the sperm is deposited in the form of a spermatophore directly on the sternum of the female between the fourth and fifth pairs of pereopods. Mating occurs shortly before the female spawns in open-thelycum species. In both cases, at spawning, eggs are mixed with the sperm (released from the seminal receptacles of closed-thelycum species or from the spermatophores of open-thelycum species) and fertilization takes place externally (Clark *et al.*, 1984). The following discussion on the early morphological events associated with egg activation will center on phenomena beginning at the time the egg first contacts the surrounding medium after spawning.

<u>Morphology of the Unactivated Egg</u>

The size of an unactivated egg ranges from 200 to 250 microns in diameter. When viewed with light microscopy, numerous radially arranged translucent areas are apparent in the peripheral region of the egg (Figures 1a-1c). The presence of a "jelly like" substance in the periphery of penaeid eggs was initially described in *P. japonicus* and *P. monoceros* (Hudinaga, 1942). Transmission electron microscopy

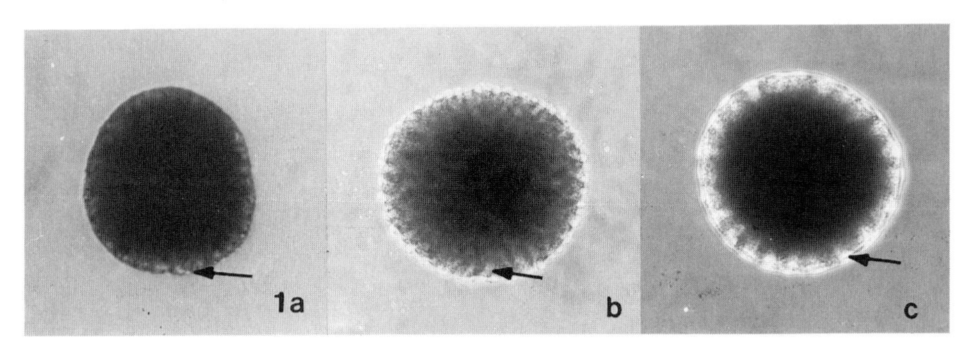

Figure 1. Phase contrast micrographs of unactivated penaeid eggs. a) *P. aztecus*, b) *P. setiferus*, and c) *S. ingentis*. arrows - cortical crypts. x 120.

reveals that these radial striations are actually extracellular crypts formed by the invaginations of the oolemma (Figures 2,3), which contain rod-shaped bodies in the case of *P. aztecus* and *P. setiferus* (Figure 2). These crypts are delimited from the external environment by a thin investment coat, the vitelline envelope. Although a distinct rod-shaped morphology is lacking in *S. ingentis* eggs, the material in the crypts of these three species is composed of feathery elements (Figures 2a,3a). At

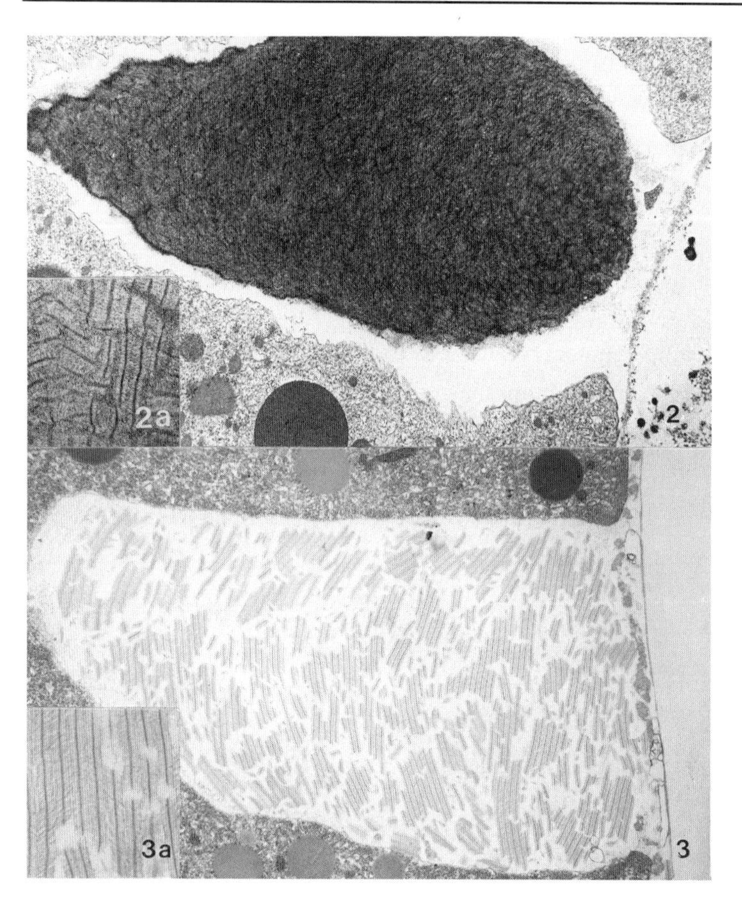

Figure 2. Electron micrograph of a cortical rod in *P. aztecus* egg crypt. x 6000.
Figure 2a. High magnification of feathery elements. x 30,000
Figure 3. Electron micrograph of jelly precursor in crypts of *S. ingentis* egg. x 8,700
Figure 3a. High magnification of feathery elements. x 27,000.

this time, the substructure of the jelly precursor material has yet to be examined in *T. similis*. The feathery elements within the cortical crypts are the precursor material for a homogeneous jelly layer that forms around the egg after spawning (Clark *et al.*, 1980; 1984).

Jelly Layer Formation

Following spawning, the first observable event in the eggs of these species is the rapid release of the jelly precursor material from the extra-oocytic crypts. During the initial phase of the jelly release, the vitelline envelope surrounding the egg is dissipated and lost as the crypt material is externalized and organized into a jelly layer. Initially the precursor material has a heterogeneous appearance and forms a distinct aurora around the egg (Figures 4a-c, 5, 6). Depending upon the temperature of the surrounding medium, this stage of the jelly release may last for 1-2 minutes in the case of *P. setiferus* to several minutes in *S. ingentis*. Once released, the rod-shaped jelly precursors will begin to transform into a more homogeneous jelly layer (Figures 7-10). In *P. aztecus* and *P. setiferus* the jelly precursors begin this transformation with a period of swelling and dissipation at both ends of the released rod-like structures. The total transformation process requires approximately 5-7 min in *P. setiferus* and *P. aztecus* to approximately 10-15 min in *S. ingentis*. Although the actual time period the homogeneous jelly layer remains around the egg is not clear, reports in the literature indicate that the material is present at least until the forma-

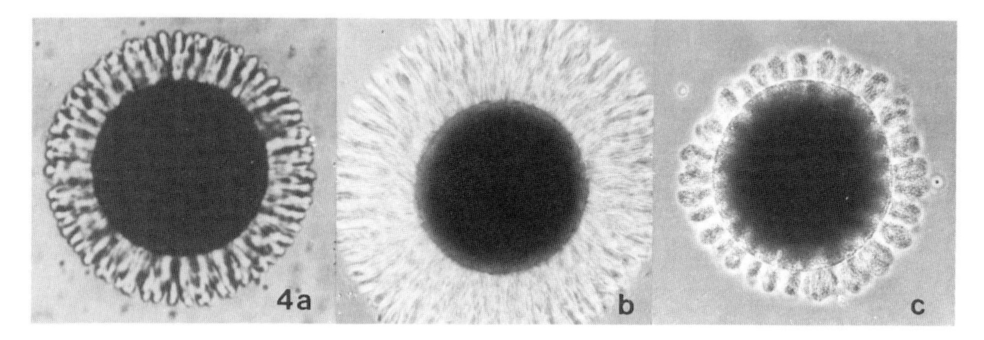

Figure 4. Phase contrast micrographs of penaeid egg releasing jelly precursor material approximately 1 min post spawning. a) *P. aztecus,* b) *P. setiferus*, c) *S. ingentis*. x120. (Figure 4a courtesy Clark *et al.*, 1980, Biol. Bull. 158:175).

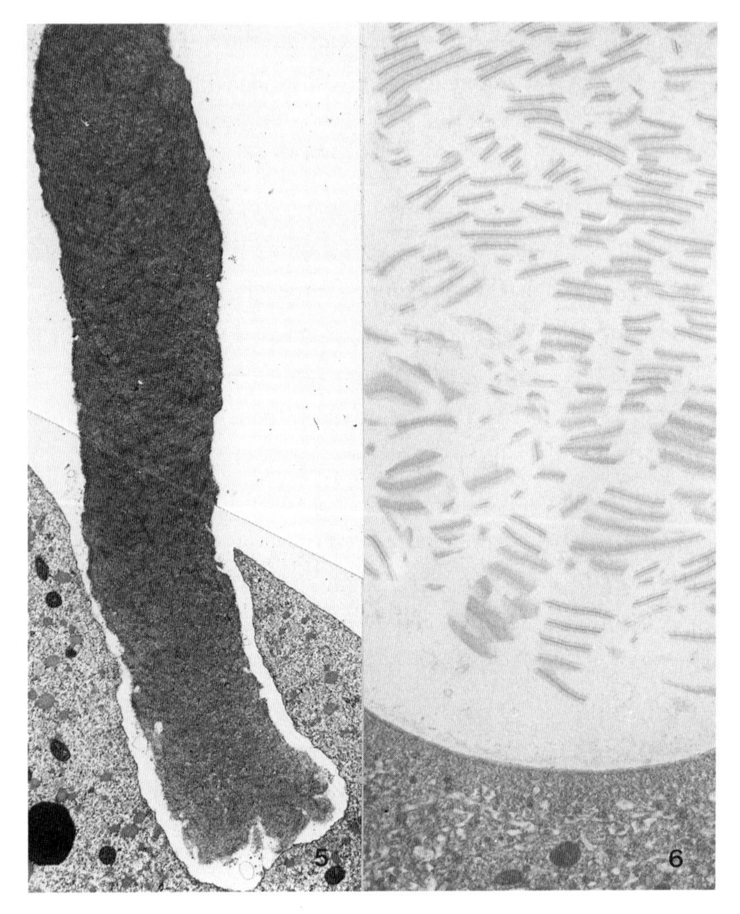

Figure 5. Electron micrograph of an intact jelly precursor rod released from a *P. aztecus* egg. x 4000.

Figure 6. Electron micrograph of intact jelly precursor material released from a *S. ingentis* egg. x 15,500. (Figure 6 courtesy Clark *et al.*, 1980. Biol.Bull.158:175).

tion of the hatching envelope (see description below) in *P. aztecus* and *P. setiferus* and up to the early cleavage stage in *S. ingentis* (Lynn, 1976; Pillai and Clark, 1987).

Meiotic Maturation

Early activation events in animal eggs are characterized by the release of the egg from an arrested meiotic stage. This release follows a stimulus provided by either

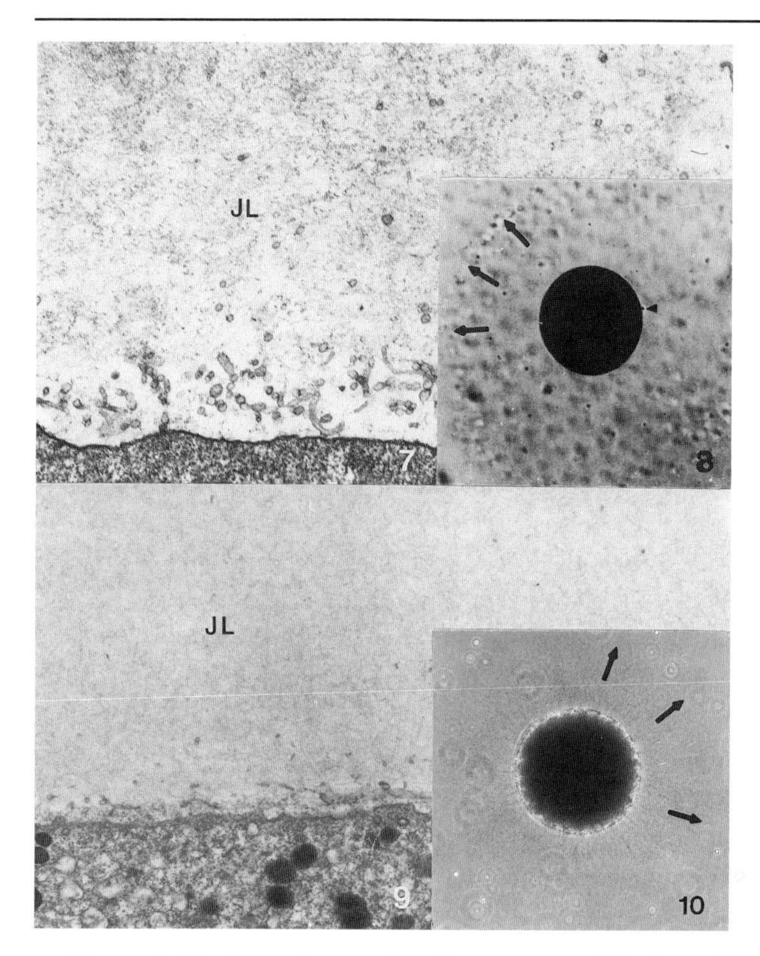

Figure 7. Electron micrograph of homogenous jelly layer (JL) around *P. aztecus* egg. x 26,000.

Figure 8. Phase contrast micrograph of *P. aztecus* egg with formed jelly layer. arrows - margin of jelly layer; arrowhead - first polar body. x 100.

Figure 9. Electron micrograph of homogenous jelly layer (JL) around *S. ingentis* egg. x 24,000.

Figure 10. Phase contrast micrograph of *S. ingentis* egg with formed jelly layer. arrows - margin of jelly layer. x 120.

the ionic conditions of the ambient environment, hormones, sperm-egg interaction, or a combination of all these stimuli. Typically, animal eggs are released from the ovary in one of four meiotic stages: 1) with an intact germinal vesicle indicating the egg is arrested in the diplotene phase of meiosis (*e.g.*, surf clams); 2) with a broken down germinal vesicle with the chromosomes aligned in the first metaphase plate of meiosis (*e.g.*, mussels); 3) with completed first meiotic division but arrested at the second meiotic metaphase (*e.g.*, most mammals); and 4) with both meiotic divisions completed, therefore attaining haploidy (*e.g.*, sea urchins). For *S. ingentis*, germinal-vesicle break down has been documented before ovulation and the eggs are arrested in the first meiotic metaphase until spawning occurs (Pillai and Clark, 1987). Although germinal vesicle breakdown has already occurred in *P. setiferus*, *P. aztecus*, and *T. similis* spawned eggs, the precise timing of this event in relation to ovulation has not been documented, but likely corresponds to the situation in *S. ingentis*. Thus, natantian eggs correspond to the second scenario described above. At spawning, the ovulated eggs are passed down the paired oviducts to the gonopores located at the base of the third pair of pereopods and released into the sur secrounding medium (Pillai *et al.*, 1988).

The appearance of the first polar body follows the formation of the homogeneous jelly layer and indicates the completion of the first meiotic division (Figure 11a). During the elevation of the hatching envelope (see below) the first polar body is lifted off of the egg surface (Figure 11b) and shortly after this the second polar body is formed, thus completing meiotic maturation (Figure 11c). The formation of the

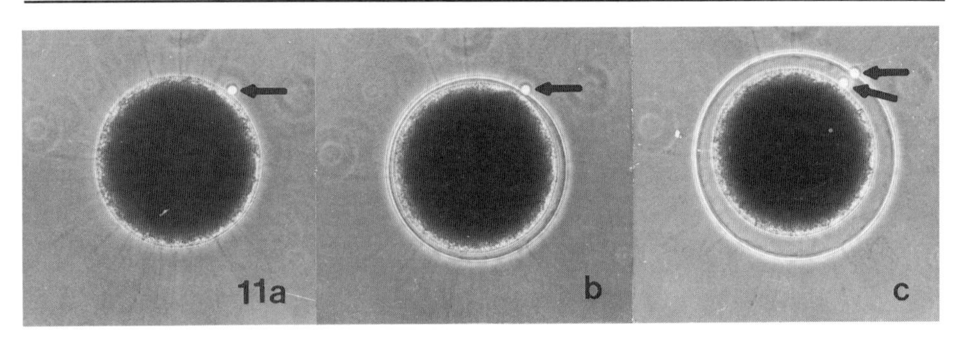

Figure 11. Phase contrast micrograph series of *S. ingentis* egg during 1st and 2nd polar body formation and hatching envelope formation. a) 1st polar body, b) 1st polar body and early hatching envelope, c) 1st and 2nd polar body and hatching envelope. arrow - polar bodies. x 120.

second polar body occurs approximately 5-10 min from the time the first polar body appears in *S. ingentis*. There is no confirmation of the time to second polar body formation in *P. setiferus* and *P. aztecus* eggs. Formation of two polar bodies, following spawning, has also been observed in *P. stylirostris*, *P. vannamei* and *T. similis* (Lynn and Green, unpublished data).

Hatching Envelope Formation

Between the time of first and second polar body formation, an additional egg investment coat is assembled which has been termed the "hatching envelope" (HE) (or "hatching membrane" in older literature; Clark *et al.*, 1980; see also Figures 11c, 12-15). Although formation of the HE has been reported in several species of Penaeoidea, the morphology of its formation has been described in detail only for *S. ingentis* (Pillai and Clark, 1988). In *S. ingentis*, the elevation of the HE occurs at about 40-45 min post-spawning as a result of an extensive exocytosis of two morphologically distinct populations of cortical vesicles, and is thus analogous to the cortical reaction and formation of the "fertilization envelope" described in other species (Schmell *et al.*, 1983). In *S. ingentis*, the cortical vesicles appear sequentially in the ooplasm after the release of the jelly precursor material and jelly layer formation. These vesicles comprise 1) the "dense vesicles" (which are apparently the result of fusion between Golgi derived granules) containing an electron dense material; and 2) the "ring vesicles" that are formed by fusion of cisternal elements containing ring-shaped inclusions. Thirty to 35 min after spawning these vesicles undergo sequential exocytosis, with the exocytosis of the dense vesicles followed by that of the ring vesicles, and the released contents coalesce with a fibrous surface coat to form the HE (Pillai and Clark, 1988; Pillai and Clark, 1990)).

Control of Egg Activation

The control of morphological changes that occur in the egg after spawning has not been exhaustively investigated, but current data implicates regulator mechanisms involving ions and/or extracellular enzymes.

Ionic Requirements

Clark and Lynn (1977) reported that externalization and transformation of the

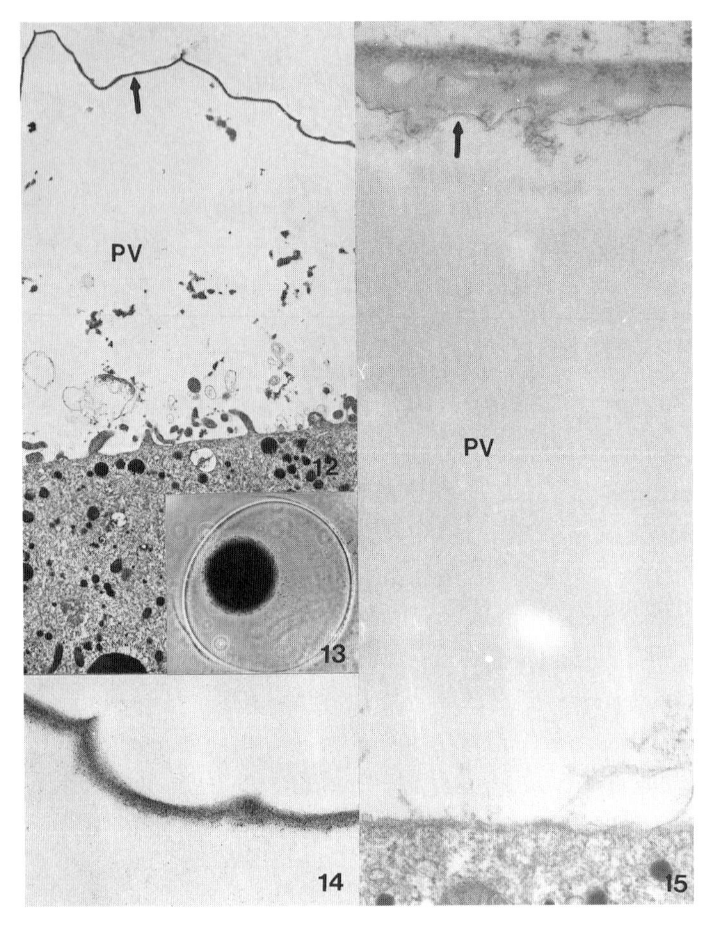

Figure 12. Electron micrograph of the fully formed hatching envelope in *T. similis* showing the envelope and the perivitelline space (PV). arrow - hatching envelope. x 5100.

Figure 13. Phase contrast micrograph of hatching envelope fully formed and elevated in *T. similis*. x 75.

Figure 14. High magnification electron micrograph of hatching envelope of *T. similis*. Note outer dense layer, inner flocculent layer and spikes on envelope. x 33,000.

Figure 15. Electron micrograph of fully formed hatching envelope around *S. ingentis*. Note the electron dense outer layer and the less electron dense inner layer. x 28,000.

rod-like jelly precursors in the eggs of *P. setiferus* and *P. aztecus* were inhibited in the absence of external Mg^{+2}. Externalization of the jelly precursor resumes when the inhibited eggs are returned to normal sea water, but transformation of the precursor into a homogeneous state does not occur. Transformation of the jelly precursor is also inhibited in *S. ingentis* eggs spawned into Mg^{+2} free sea water, but externalization does not appear to be affected (Pillai and Clark, 1987).

Involvement of ions in the regulation of morphological changes is not restricted to the release and transformation of the jelly precursors. Pillai and Clark (1987) have recently demonstrated that the resumption of meiosis in *S. ingentis* is dependent on both Ca^{+2} and Mg^{+2} in the medium. In the absence of either of these divalent cations, resumption of meiosis is inhibited in unfertilized eggs. When returned to normal sea water, unfertilized eggs resumed meiosis. In contrast, the early activational events including resumption of meiosis proceeded in the absence of external Ca^{+2} if the eggs were fertilized but external Mg^{+2} was still required.

Finally, the formation and elevation of the HE also demonstrates ionic requirements. In *S. ingentis*, extracellular Mg^{+2} is required for the formation of the envelope. Extracellular Ca^{+2}, on the other hand, is required for the envelope formation only if the eggs are unfertilized (Pillai and Clark, 1987). In addition, *P. aztecus* and *P. setiferus* eggs spawned into Mg^{+2} free sea water frequently fail to elevate an HE, but this failure is more likely an indirect effect on the formation of the envelope. When eggs released into Mg^{+2} free sea water are gently shaken to dislodge the jelly precursor material from the surface of the egg, an envelope will form and components of the jelly precursor material do not appear to be involved (Lynn, 1976; Pillai and Clark, 1988). Although the work on the regulation of HE elevation is preliminary, experiments with *P. setiferus* and *T. similis* suggest that both Na^{+} and Cl^{-} ions are also required for proper structuralization of the envelope to occur. Under either of these conditions, the envelope will initially elevate, but the morphology differs from that of control eggs held in normal sea water and the envelope rapidly collapses back to the surface of the egg. Collapse is accompanied by the loss of the extracellular material normally contained in the perivitelline space of these eggs.

Enzymatic Control

The mechanism of egg activation involving Mg^{+2} ions is not clear but may involve effects on serine proteases released at the time the egg contacts the sea water. Both release and dissipation of the jelly precursor in *P. setiferus* and *P. aztecus* can be irreversibly inhibited by the serine protease inhibitors N-α-p-tosyl-L-lysine chloromethyl ketone (TLCK) and soybean trypsin inhibitor (SBTI) (Lynn and Clark, 1987). These inhibitors also prevent the transformation of jelly precursor in *S. ingentis* eggs (Griffin and Clark, personal communication). Although a definitive correlation between the enzyme inhibition and dependence on Mg^{+2} ions in the above processes has not been made, the coincidental effects of both the protease inhibitors and lack of Mg^{+2} ions argues a likely relationship between ion requirement and enzyme activity.

The apparent involvement of serine proteases in the release and transformation of the jelly precursor, however, gives an immediate clue to the possible biochemical composition of the precursor material. Lynn and Clark (1987) have demonstrated that the precursor from the eggs of *P. aztecus* and *P. setiferus* contain a high percentage of protein in comparison to carbohydrate (70-75% protein; 25-30% carbohydrate). Little or no sialic acid residues were detectable. Sulfhydryl reducing agents such as dithiothreitol and mercaptoethanol had no noticeable effect on the morphology of the isolated precursor material, and with the exception of proteolytic enzymes (*e.g.*, trypsin, chymotrypsin, and papain), enzymatic treatment had no apparent effect. A detailed analysis of the biochemical composition of the jelly precursor in *S. ingentis* has not been reported, but the protein:carbohydrate ratio is reported to be similar (Clark, unpublished).

Discussion

As described above, upon activation, the eggs of penaeid shrimps undergo discrete sequential morphological and physiological changes (release of the jelly precursor and formation of the jelly layer, resumption of meiotic maturation, and formation of the HE). The extent of the changes in the egg surface are not only dramatic, but also represent a serious problem with regard to the *in vitro* manipulation of the gametes of these animals. Since the eggs are fertilized externally, one of the first obstacles that the male gamete encounters and must deal with is the release of the jelly precursor material. In early attempts at *in vitro* fertilization of *P. aztecus* eggs, Clark *et al.* (1973) reported only a 10% success rate determined by the number of eggs that cleaved normally. In retrospect, the low success rate can be attributed to the release of the jelly precursor and the subsequent sweeping of the non-motile spermatozoa from the surface of the egg before they could bind to the egg. Even given the investigator's ability to control the release of the jelly precursors, the sperm is still faced with the arduous task of interacting with the surface of the egg and penetrating to the cytoplasm before the formation of the HE is completed. Manipulation of either jelly release or HE elevation requires not only the ability to prevent these reactions but also the ability to control the physiological progression of events. Some comparisons of our present state of knowledge and the possibility of applying that knowledge to the Penaeoidea are discussed below in relation to our knowledge of other animal species.

An important aspect of the early activational events discussed above is that they are not dependent on the entry of sperm. For example, early egg activation in *S. ingentis* is interesting since it is unlike the fertilization-dependent egg activation described in other species such as crabs (Goudeau and Becker, 1982), lobsters (Talbot and Goudeau, 1988), sea urchins (Schuel, 1978), and amphibians (Grey *et al.*, 1974). The presence of the sperm is critical, however, in that later stages of development (particularly early cleavage) will not proceed normally without sperm incorporation (Pillai and Clark, 1987). Recently, resumption of meiotic maturation in oocytes of the caridean prawn *Palaeomon serratus* has also been shown to occur in the absence of sperm (Goudeau and Goudeau, 1986). Interestingly, however, a dependency on external Mg^{+2} has been demonstrated in both *P. serratus* (Goudeau and Goudeau, 1986) and *S. ingentis* (Pillai and Clark, 1987). Information on the alteration of the egg surface and other activational events (*e.g.*, HE formation) were not reported for *P. serratus*.

While eggs of other animal species may also undergo a massive release of material from the egg in response to activation (see below), the natantian shrimp appear to have developed a unique format for jelly release and subsequent formation of the jelly layer, as well as a unique mode of regulation. This initial egg activation event is not analogous to the exocytotic events in other animals. An event in which a distinct jelly-like layer is formed around the egg has been reported in *N. limbata*

(Wilson, 1892; Lillie, 1911; Novikoff, 1939; Costello, 1949; Fallon and Austin, 1964). Release of the jelly-like material in nereid eggs requires a membrane fusion event and is initiated by the penetration of the sperm into the egg cytoplasm (Fallon and Austin, 1964). In contrast, the release of the jelly precursors in penaeids occurs from extraoocytic crypts not requiring membrane fusion and occurs independently of sperm at the egg surface. Therefore, release of jelly precursors in shrimp eggs is not analogous to the cortical reaction reported in *N. limbata*. The composition of the jelly layer around shrimp eggs also suggests an analogy with egg jellies in other animal eggs. The greatest differences are the protein to carbohydrate moieties ratio and the types of subunits of each.

The function of the jelly release is still unclear but there are several possibilities based on purported functions of jelly layers in eggs of other animal species. These include: 1) protection of the early zygote from the environment; and 2) induction of the acrosome reaction of the sperm. Early zygote protection may be afforded in two ways: either mechanically, biochemically or both. Mechanically, the jelly layer would act to cushion the egg until and even after the formation of the HE. Preliminary biochemical evidence from the jelly precursors of *P. aztecus* suggest that the jelly could act as a bacteriostatic agent. Application of isolated jelly precursors to agar plates of shrimp exoskeleton bacteria resulted in inhibited areas of bacterial growth (Lynn, 1976).

A role in the induction of the sperm acrosomal reaction is suggested by the experiments of Griffin *et al.* (1985). In these experiments, a component of egg jelly induced the exocytotic phase of the acrosome reaction. More recently, Griffin (1987) has demonstrated that while acrosomal exocytosis may be dependent on one or more components of the jelly, it is also related to the release of a proteolytic enzyme from the egg. Induction of the sperm acrosome reaction by components derived from egg investments has also been reported in sea urchins (SeGall and Lennarz, 1979), sturgeon (Cherr and Clark, 1985), and in numerous other species (see Lopo, 1983 for review). Interestingly, the component of sea urchin egg jelly responsible for the acrosome reaction is a fucose sulfate polymer. The jelly precursor of *P. aztecus* contains a similar sugar component, 6-deoxyhexose, although the sulfated, polymerized form does not appear to be present. Recent evidence also suggests that the jelly layer in *S. ingentis* eggs contains L-fucose as established by specific lectin (UEA-1) binding patterns (Pillai and Clark, unpublished).

Generally, massive cortical reactions in other animal eggs involve exocytosis of membrane-bound cortical vesicles (cortical granules) which is typically Ca^{+2} sensitive and in many species may be sperm initiated (Schuel, 1985 for a review). The result of these exocytotic processes is the formation of a fertilization envelope by the combination of cortical granule exudate and the surface coat (*e.g.*, vitelline envelope, chorion, zona pellucida) or a portion of the jelly layer (Anderson, 1968; Grey *et al.*, 1974; Hart, 1980; Brummet and Dumont, 1981; Wassarman, 1987). The formation of the HE in the Penaeoidea is at least partially analogous to the reactions seen in other animal eggs in that membrane-bound vesicles exocytose and the exudates combine with a surface coat (although the jelly layer of the shrimp ovum does not appear to play a significant role in this process; Pillai and Clark, 1988). Two major differences in the Penaeoidea, however, are: 1) the formation of the hatching envelope independent of sperm attachment to the surface of the egg, and 2) the considerable period of time which elapses between the time the eggs are released from the female and the time the hatching envelope begins to form (40-45 min in most species). Although the function of the fertilization envelopes in other species has been described as a polyspermy preventing mechanism (Jaffe and Gould, 1985) a similar role for the hatching envelope in penaeid eggs has yet to be determined.

Future of Penaeoidea as a Research Animal - Concluding Remarks

The eggs of penaeid shrimp provide us with a unique opportunity to study early egg activational events which are both similar to other animal species and in some cases quite different. The penaeid shrimp have the desirable characteristics of an excellent laboratory animal since the abundance of gametes is considerable. We are continually learning ways to manipulate them in the lab and from the economical viewpoint, individual species appear to have the same or at least a very similar physiological basis for those early activational events which have been examined. This last fact presents a tremendous advantage since it is then possible to study the details of these events in whichever particular penaeid species is available and then transfer that fundamental knowledge to other penaeid species which may be of economic or commercial importance. While it is not possible to rule out that some differences may exist, the great similarity between species thus far studied is promising.

We are left, nevertheless, with a tremendous void of information on many other activational events in the Penaeoidea egg. For example, currently there are no data on the changes in protein synthesis in relation to those activational events discussed in this paper or in relation to the entry of the sperm into the egg cytoplasm. Neither do we have any information on changes in oxygen consumption, ion fluxes across the plasmalemma, or changes in the membrane potential accompanying ion transport. How, we might ask, do any of these events remotely concern the commercial use of these animals? Current investigators are manipulating the reproductive cycles of commercially important species with hormones and techniques such as eyestalk ablation. Yet the ultimate effects on the viability of the gametes produced are poorly understood at best. How will we be able to tell quickly and efficiently whether eggs produced by these techniques are consistently viable? Furthermore, how will the researcher or the commercial producer be able to cross-fertilize between species (especially crosses between closed-thelycum and open-thelycum species) if we have no idea of how the egg reacts to fertilization, and how will we be able to manipulate those events designed by nature to prevent the entry of excess sperm into the egg's cytoplasm? We have no way at present to predict the time involved in understanding the processes of sperm-egg interaction and egg activation in the penaeid shrimp. Because of the lack of data we can expect that several years of intensive research will be necessary to obtain an adequate grasp of the physiology of egg activation.

Acknowledgements

The authors wish to express their gratitude to the Louisiana State Department of Wildlife and Fisheries and Dr. A.L. Lawrence at the Texas A&M University at Port Aransas for their help in supplying some of the animals used in studies discussed in this paper. In addition, we would like to acknowledge the years of effort and research carried out in Dr. Wallis H. Clark's laboratory part of which has formed the basis for this review. The above research was supported in part by California Sea Grant NA85AA-D-SG140 R/A-61 and Louisiana Sea Grant NA85AA-D-SG141 R/SA-1.

Literature Cited

Anderson, E. 1968. Oocyte differentiation in the sea urchin, *Arbacia punctulata* with special reference to the cortical granules and their participation in the cortical reaction. J. Cell Biol. 37:514-539.

Brummet, A. R. and J. N. Dumont. 1981. Cortical vesicle breakdown in fertilized eggs of *Fundulus heteroclitus*. J. Exp. Zool. 216:63-79.

Cherr, G. N. and W. H. Clark, Jr. 1985. An egg envelope component induces the acrosome reaction in sturgeon sperm. J. Exp. Zool. 234: 75-85.

Clark, W. H. Jr. and F. J. Griffin. 1988. Morphology and physiology of the acrosome reaction in the sperm of *Sicyonia ingentis*. Dev. Growth Differ. (in press).

Clark, W. H. Jr. and J. W. Lynn. 1977. A $Mg+^2$ dependent cortical reaction in the eggs of the penaeid shrimp. J. Exp. Zool. 200:177-183.

Clark, W. H. Jr., J. W. Lynn, A. I. Yudin, and H. O. Persyn. 1980. Morphology of the cortical reaction in the eggs of *Penaeus aztecus*. Biol. Bull. 158: 175-186.

Clark, W. H. Jr. and M. C. Pillai. 1990. Egg production, release and activation in the shrimp *Sicyonia ingentis*. In. Crustacean Issues. A.M. Wenner and A. Kuris (eds.). Balkema Press, Rotterdam (in press).

Clark, W. H., P. Talbot, R. A. Neal, C. R. Mock, and B. R. Salser. 1973. *In vitro* fertilization with nonmotile spermatozoa of the brown shrimp *Penaeus aztecus*. Mar. Biol. 22:353-354.

Clark, W. H. Jr., A I. Yudin, F. J. Griffin and K. Shigekawa. 1984. The control of gamete activation and fertilization in the marine Penaeidae, *Sicyonia ingentis*. Advances in Invertebrate Reproduction Vol. 3, (W. Engels *et al.*, eds), Elsevier Science Publishers, B.V. pp. 459-472

Costello, D. P. 1949. The relations of the plasma membrane, vitelline membrane and jelly in the egg of *Nereis limbata*. J. Gen. Phys. 32:351-366.

Fallon, J. F. and C. R. Austin. 1964. Fine structure of gametes of *Nereis limbata* before and after interaction. J. Exp. Zool. 166:225-242.

Goudeau, M. and H. Goudeau. 1986. The resumption of meiotic maturation of the oocytes of the prawn *Palaeomon serratus* is regulated by an increase in extracellular Mg^{+2} during spawning. Dev. Biol. 118:361-376.

Goudeau, M. and M. Becker. 1982. Fertilization in a crab. II. Cytological aspects of the cortical reaction and fertilization envelope formation. Tissue and Cell 14:273-282.

Grey, R. D., D. P. Wolf and J. L. Hedrick. 1974. Formation and structure of fertilization envelope in *Xenopus laevis*. Dev. Biol. 36:44-61.

Griffin , F. J. 1987. Induction and control of the acrosome reaction in the sperm of *Sicyonia ingentis*. Ph.D. Dissertation, University of California, Davis. pp. 122.

Griffin, F. J., A. I. Yudin and W. H. Clark, Jr. 1985. The effects of a jelly associated protease on the acrosome reaction. J. Cell Biol. 101:227a.

Hart, N. H. and S. Yu. 1980. Cortical granule exocytosis and cell surface reorganization in eggs of *Brachydanio*. J. Exp. Zool. 213:137-159.

Hudinaga, M. 1942. Reproduction, development and rearing of *Penaeus japonicus* Bate. Jap. J. Zool. 10:304-393.

Jaffe, L. A. and M. Gould. 1985. Polyspermy preventing mechanisms. In. Biology of Fertilization, Vol. 3, C.B. Metz and A. Monroy, eds. Academic Press, New York, pp. 223-250.

Lillie, F. R. 1911. Studies of fertilization in *Nereis*. I. The cortical changes in the egg. II. Partial fertilization. J. Morph. 22:361-393.

Lopo, A. C. 1983. Sperm-egg interaction in invertebrates. In. Mechanism and control of Fertilization. (J.F. Hartman, ed.) Academic Press, New York, pp. 269-324.

Lynn, J. W. 1976. The physiology of the cortical reaction in the eggs of penaeid shrimp. Master's Thesis. University of Houston, Houston, Texas. pp. 84.

Lynn, J. W. and W.H. Clark, Jr. 1987. Physiological and biochemical investigations of the egg jelly release in *Penaeus aztecus*. Biol. Bull. 173:451-460.

Novikoff, A. B. 1939. Changes at the surface of *Nereis limbata* eggs after insemination. J. Exp. Biol. 16:403-408.

Pillai, M. C. and W.H. Clark, Jr. 1987. Oocyte activation in the marine shrimp *Sicyonia ingentis*. J. Exp. Zool. 244:325-329.

Pillai, M. C., F. J. Griffin and W. H. Clark, Jr. 1988. Induced spawning in the decapod crustacean *Sicyonia ingentis*. Biol. Bull. 174:181-185.

Pillai, M. C. and W. H. Clark, Jr. 1988. Hatching envelope formation in *Sicyonia ingentis* ova: Origin and selective exocytosis of cortical vesicles. Tissue and Cell 20: 941-952.

Pillai, M.C. and W. H. Clark, Jr. 1990. Development of Cortical Vesicles in *Sicyonia ingentis* ova: Their heterogeneity and role in elaboration of the hatching envelope. Mol. Reprod. Dev. 26: 78-89.

Schuel, H. 1985. Functions of egg cortical granules. In. Biology of Fertilization, Vol. 3, The fertilization response of the egg. C.B. Metz and A. Monroy, eds. Academic Press, New York, pp. 1-43.

Schuel, H. 1978. Secretory functions of egg cortical granules: A critical review. Gamete Res. 1:299-332.

Schmell, E. D., B.J . Gulyas and J. L. Hedrick. 1983. Egg surface changes during fertilization and the molecular mechanism of the block to polyspermy. In. Mechanism and Control of Animal Fertilization. J.F. Hartman, ed. Academic Press, New York, pp. 365-413.

SeGall, G. K. and W. J. Lennarz. 1979. Chemical composition of the component of the jelly coat from sea urchin eggs responsible for the induction of the acrosome reaction. Dev. Biol. 71:22-48.

Talbot, P. and M. Goudeau. 1988. A complex cortical reaction leads to formation of the fertilization envelope in the lobster *Homarus*. Gamete Res. 19:1-18.

Wassarman, P. M. 1987. Early events in mammalian fertilization. Ann. Rev. Cell Biol. 3:109-142.

Wilson, E. B. 1892. Lineage of *Nereis*. J. Morph. 6:361.

Endocrine Control of Reproduction In Shrimp and Other Crustacea

Hans Laufer
Department of Molecular and Cell Biology
University of Connecticut
Storrs, CT 06269-3125
and
The Marine Biological Laboratory
Woods Hole, MA 02543

Matthew Landau
Department of Marine Science
Stockton State College
Pomona, NJ 08240

Abstract

A comparison of insects with crustaceans suggests that Crustacea may have juvenile hormone-like substances. Crustacea were found to synthesize methyl farnesoate (MF). The mandibular organ (MO) is the source of MF. Every species examined, including several species of shrimp, was found to produce MF when the MO was tested *in vitro*. MF fluctuated during the vitellogenic cycle of *Libinia emarginata;* the highest levels coincided with maximum vitellogenin synthesis. The effect of the MF on reproduction may be causative since MO implants stimulate ovarian development in juvenile, non-reproductive female *L. emarginata.*

Neuropeptides from the eyestalk and other parts of the CNS inhibit or stimulate MF production by the MO. In the crayfish MO *in vitro* these are PDH and RPCH, respectively. The effects of the neuropeptides are part of a modulator-activator-inhibitor system. In addition, the MO is influenced by biogenic amines. These factors also influence reproductive behavior by modulating postural as well as copulatory-aggressive behavior in lobsters and possibly shrimp.

We conclude that there seems to be a cascade of regulatory activators and inhibitors which are elaborated by the nervous and endocrine systems. It is these two systems which mediate between external environmental and internal signals to translate them into a cohesive reproductive response in crustaceans and which are now helping us to understand these vital processes in shrimp.

Introduction: A Comparative Approach

The approach of comparative endocrinology has a great deal to offer when it comes to understanding the hormonal regulation of shrimp reproduction and reproductive behavior. For example, both insects and crustaceans, including shrimp, utilize identical molecules, 20-hydroxyecdysone, as the primary molting hormone. In addition, juvenile hormone (JH) plays a major role in the regulation of insect development, controlling ovarian growth and vitellogenesis in adult females as well as accessory reproductive structures in male insects. The most commonly active form of JH is JH III, an expoxidated sesqui-terpinoid (Downer and Laufer, 1983).

The level of JH present in insects is regulated by neuropeptides from the CNS; allatotropins increase JH production, and allatostatins and allatohibins decrease hormone levels (Girardie, 1983). In addition to the regulation of JH synthesis by neuropeptides in the insects, esterases and other enzymes metabolize and alter JH structure and function. Here we review the evidence for the existence of a JH-like compound in crustaceans in general, and indicate where the information may be useful for the control of reproduction in shrimp.

The Mandibular Organ and JH-like Compounds

There has been much interest in JHs of crustaceans ever since JH and its analogs, methoprene and hydroprene, were first shown to affect barnacle metamorphosis (Gomez *et al.*, 1973). These results were confirmed by Tighe-Ford (1977). Landau and Finney (1977) also used several analogs, but showed that mevalonic acid, a precursor of JH, had no activity. Landau and Rao (1980) showed that precocene II, which prevents JH production in insects by destroying the corpora allata (CA), the site of hormone synthesis, could strongly inhibit the hatching of barnacle embryos early in their development (stage 1) but had no effect on hatching if development had reached stage II. Development of larval decapods, too, was shown to be the target of JH analogs (Bookhout and Costlow, 1974).

JH and its analogs can also affect crustacean reproduction. For example, treatment of the water flea, *Daphnia magna*, with JH analogs has produced complete sterility (Templeton and Laufer, 1983). Some of these analogs have been shown to inhibit spermatogenesis and oogenesis in the mud crab, *Rhithropanopeus harrisi* (Payen and Costlow, 1977), and the spider crab (Hinsch, 1981). When Paulus and Laufer (unpublished) and Paulus (1984) injected methoprene into *Carcinus maenas* females, either with or without eyestalks, the ovaries were appreciably enlarged. Whether these effects reflect a true endocrine response or a non-specific toxic response is not clear. The low levels of the compound required, however, are consistent with the former explanation.

Studies on the role of JH-like compounds in crustacean reproduction and development are complicated by the role(s) of the eystalk in these processes. Peptides secreted from the eyestalk may control the release of the JH-like compounds. Costlow (1968) removed the eyestalks of larval blue crabs; those that survived continued to molt but developed into intermediate larval forms. His experimental results suggest that there is a mechanism in the eyestalk that controls the frequency of molting, and a "second endocrine site or storage organ" which is "developed within the zoeal eyestalk." This mechanism "regulated stages" and the "normal transition to the final megalops stage." Hinsch (1972) also found that eyestalk ablation of immature spider crabs, *Libinia*, resulted in crabs that molt but do not mature, forming giant juveniles. Charmantier *et al.* (1988) showed that, if treated early, larval lobsters would molt into an intermediate stage when eyestalk ablated or injected with JH I.

The results of eyestalk extirpation experiments support the hypothesis that the secretion of a JH-like compound is under the inhibitory control of an eyestalk factor in Crustacea. If this were the case, sublethal concentrations of JH or its analogs should not slow down or block the progress of metamorphosis, but result in the formation of intermediate developmental stages, effects similar to those seen with eyestalk removal (Costlow, 1968; Charmantier *et al.*, 1988).

Despite the evidence indicating that JH-like compounds are active in Crustacea, it has not been possible until recently to determine both the nature and the source of an endogenous JH. Extracts of crustacean eyestalks have detectable JH activity when used in bioassays on insects (Schneiderman and Gilbert, 1958), but the chemical nature of the activity was not identified. More recently, implants of crustacean thoracic ganglia or mandibular organs (MOs) into immature crabs have been shown to enhance vitellogenesis (Hinsch and Bennet, 1979; Hinsch, 1980). MOs were suggested, based on their morphology, as the site of synthesis for JH-like compounds in lobsters (Byard *et al.*, 1975).

Such observations suggested that an endocrine compound functioning in a manner similar to JH may exist in crustaceans. However, prior to the research described below, there was no definitive identification of crustacean JH. The lack of a defined compound impeded work on many aspects of the physiology of Crustaceans.

Using techniques of gas chromatography/mass spectrometry (GC/MS) and selected ion monitoring (SIM), we have found that methyl farnesoate (MF) is the major JH-like product in the hemolymph of the spider crab, *Libinia emarginata* (Laufer *et al.*; 1987a). Using *in vitro* methods originally developed to study the insect CA, the MO was identified as the source of this factor in the spider crab, lobster, crayfish, and several other decapod crustaceans including penaeid shrimp (*Penaeus semisulcatus, P. vannamei* and *P. duorarum*); MF was also synthesized by some tissue in a barnacle, *Balanus nubilis*, although in this case the MOs have not been identified as discrete glands. Wherever the gland was found, its activity always appeared to be significantly related to the reproductive state of the organism (Laufer *et al.*, 1986; 1987a; Borst *et al.,* 1987), further suggesting a homology to the CA of adult insects. The relationship between MF and reproduction appears to be that of cause and effect since transplantation of the MO from adult male to juvenile female *L. emarginata* initiates vitellogenin (VG) synthesis and ovarian development (Hinsch, 1980). Thus, the MO seems to be producing a "gonad-stimulating hormone," probably MF. The MO of penaeid shrimp is found at the base of the mandibular tendon, synthesizing in non-reproductive specimens approximately 0.01 pM MF/hr in *P. duorarum* and about 0.001 pM MF/hr in *P. vannamei* (the identification of MF was confirmed after normal and reverse phase HPLC). The tissue that we used in culture resembles that reported by others for shrimp MOs, which are small and difficult to locate in non-reproductive animals (Taketomi and Kawano, 1985; Bell and Lightner, 1988) but appear to be larger and more active in reproductive specimens.

<u>Inhibitory Factors</u>

A "gonad-inhibiting hormone" (GIH), sometimes called the "vitellogenesis-inhibiting hormone" (VIH), seems to be produced by the sinus gland - X organ complex in alternation with a "molt-inhibiting hormone" (MIH). Molting and reproduction are generally held to be antagonistic processes in malacostracan crustaceans (Anilkumar and Adiyodi, 1980; Quackenbush and Herrnkind, 1981; Chang, 1984). In fact, it was suggested by Fyhn *et al.* (1977) that in barnacles 20-hydroxyecdysone itself may function as GIH. MIH and GIH can be separated chromatographically (Quackenbush and Herrnkind, 1983; Meusy and Charniaux-Cotton, 1984).

The observation has been made that eyestalk ablation can increase ovarian growth, presumably because of the removal of a gonad inhibitory factor (Panouse, 1943; Brown and Jones, 1949), but eggs often do not develop properly following ablation (Anilkumar and Adiyodi, 1985). In the shrimp, *P. canaliculatus*, ablated females spawn more frequently than intact females, but the number of eggs produced and the hatching success is better when the eyestalks are not removed (Choy, 1987). The production of GIH has been shown to be seasonal and is responsible for the period of ovarian rest during the non-breeding season (Bomirski and Klek, 1974; Klek-Kawinska and Bomirski, 1975). GIH is produced by juveniles as well as by resting adults (de Leersnyder and Dhaihaut, 1978). GIH is thought to exert its effects directly on the ovary and hepatopancreas (HP) *in vivo* since eyestalk extracts will inhibit protein synthesis by cultured ovaries. Cyclic AMP can mimic this inhibition suggesting its function as an intermediate (Eastman-Reks and Fingerman, 1984). Recently, Quackenbush and Keeley (1986) reported that a partially purfied GIH inhibits *in vitro* vitellin synthesis by the ovary and the HP, the site of extra-ovarian vitellogenin synthesis (Paulus, 1984; Paulus and Laufer, 1987).

We have previously found an eyestalk factor(s) which inhibits MF synthesis (Laufer *et al.*, 1986; 1987a,c) and, because of the role that the MO seems to play in reproduction, that factor(s) may be considered a GIH, for it may affect the HP, the ovary, as well as the MO. Alternatively, there may be different factors for different target tissues. When purified ovarian and/or HP-affecting GIHs become available it will be interesting to see if they will affect other targets including MOs in culture. Much progress has been made recently on the isolation of GIHs (Quackenbush, personal communication; Soyez *et al.*, 1987; Meusy *et al.*, 1987). The factor which inhibits MF production by the MO is tentatively called the "mandibular organ-inhibiting hormone" (MO-IH); it is water soluble and heat stable (Laufer *et al.*, 1986), and inhibits MF production (Table 1). When the optic ganglion is removed from the eyestalk, placed in culture for two hours, and then removed from the culture medium, and that medium is added to an *in vitro* MO preparation, the MF synthetic activity of the gland is strongly reduced. This suggests that the eyestalk factor can diffuse from the eyestalk and remain active. Since the MO is homologous to the insect CA, the MO-IH may be a homolog of allatostatin, and may resemble allatostatin in structure. A preliminary report in the literature suggests that the eyestalk factor may work through a cGMP intermediate (Tsukimura *et al.*, 1986).

Table 1. The effect of eyestalk extract (3/4 eyestalk equivalent per gland) on MF production by MOs *in vitro* compared to MF production by contralateral MOs from juvenile female *Libinia.*

Treatment of MO	N	pM MF per gland per hr.	% Change
a. Eyestalks ablated 8 days; control gland	7	124.0 ± 40.6**	
b. Eyestalks ablated 8 days; Eyestalk extract added to gland	7	48.7 ± 18.4**	-57.8 + 18.3

** a is significantly different from b [Student's t-test (P<.01)].

Stimulation of Reproduction

In addition to MF, which is produced by the MO, there seem to be other hormones, particulary neuropeptides, that stimulate the reproductive process. The thoracic ganglion is thought to produce a factor(s), the "gonad-stimulating hormone" (GSH), which has received little attention and is therefore poorly understood. The GSH acts to stimulate ovarian development. The concept of the GSH-GIH "bihormonal system" was proposed by Otsu (1960; 1963). Otsu showed that the removal of the eyestalks would cause precocious ovarian growth in adult crabs but not in juveniles, so he reasoned that the removal of the suspected inhibitory factor (GIH) was not enough to begin the reproductive cycle, and that there also had to be a stimulatory factor which was assumed to be missing in the juveniles. Indeed, when Otsu implanted thoracic ganglia of adult crabs into juveniles the ovaries began to grow. Gomez (1965) confirmed that there was an active factor in the thoracic ganglion but found a brain factor as well that could stimulate reproduction in both sexes. Hinsch and Bennet (1979) were able to repeat these experiments using juvenile spider crabs, *Libinia emarginata*. Takayanagi *et al.* (1986a) found that in the shrimp, *Paratya compressa*, the brain was more active than the thoracic ganglion in stimulating ovarian growth *in vivo* and *in vitro*, although the reproductive state of the donor animals was not described. Extracts of the thoracic ganglia of reproductive *Uca pugilator* stimulate ovarian development in adult crabs (intact and ablated), while extracts from nonreproductive *U. pugilator* had no effect on normal crabs and actually inhibited ovarian growth in ablated crabs (Eastman-Reks and Fingerman, 1984).

In an effort to identify peptides that regulate reproduction in crustaceans, we surveyed the literature concerned with insect neuropeptides affecting reproduction. Applebaum and Moshitzky (1986) reported that a peptide, which is extracted from the corpora cardiaca (CC) and reacts with an antibody to adipokinetic hormone (AKH), acts to inhibit yolk production in the migratory locust, *Locusta migratorias.* We hypothesized that this factor may have inhibited the production of JH. Since the amino acid sequence of insect AKH and crustacean red pigment-concentrating hormone (RPCH) are almost identical (Table 2), it was thought that RPCH might inhibit the MO's synthetic abilities in culture.

There are additional reasons for comparing CC secretions to those of the crustacean eyestalk. The CC is an endocrine structure that, in addition to containing neurosecretory cells, receives axons from other neurosecretory cells in the brain. Most importantly, the CC has connections to the CA and also is a neuralhemal organ

Table 2. Comparison of the structures of crustacean RPCH and AKH from the migratory locust, *Locusta*.

Crustacean RPCH
GLU-LEU-ASN-PHE-SER-PRO-GLY-TRP-NH_2
Lucusta AKH (form I)
GLU-LEU-ASN-PHE-THR-PRO-ASN-TRP-GLY-THR-NH_2
Lucusta AKH (Form II)
GLU-LEU-ASN-PHE-SER-ALA-GLY-TRP-NH_2

which has secretory axon termini that release their products into the blood. Thus, the CC is similar in a number of respects to the crustacean eyestalk which contains neurosecretory cells and receives neurosecretory substances. For example, RPCH is localized in the eye-stalk sinus gland - X organ complex (Mangerich *et al* ., 1986). The RPCH is of further interest because, as mentioned above, it may serve other functions in addition to the control of erythrophores. The initial experiments using pure synthetic RPCH in cultures of MOs provided a surprise. At 5 x 10^{-7} M, RPCH stimulated MF synthesis in *Libinia* MOs by about 25% (in 5 out of 6 cases) rather than inhibiting synthesis. At 10^{-6} M, RPCH nearly doubled MF synthesis in *Procambarus clarkii* MOs. Thus, RPCH may be a model compound for "MO-stimulating hormone" (MO-SH).

Because RPCH seems to stimulate an increase in intracellular Ca^{2+} in pigment cells (Lambert and Fingerman, 1979), we cultured MOs with the synthetic calcium ionophore A23187 or with lanthanum, an ionic calcium channel blocker. The ionophore, at 10^{-4} M, significantly stimulated MF snynthesis; similar results have been observed in cultures of insect CA (Dale and Tobe, 1988). If calcium is replaced on a molar basis by lanthanum in the culture medium, MF synthesis is strongly inhibited in a dose dependent manner (Laufer *et al.*, 1987b).

There was an apparent paradox that emerged if RPCH was equivalent to the MO-SH or GSH; RPCH is produced in the eyestalk, yet ablation of the eyestalk enhances ovarian development. We therefore reasoned that there may be alternative sources of RPCH or RPCH-like hormones. The recent finding of Mangerich *et al.* (1986) resolved this problem using AKH antibodies; they showed that in addition to the eyestalk, the AKH antibodies also reacted to an antigen present in the thoracic ganglia. It is not known if this "RPCH activity" cycles with the reproductive cycle. However, the experimental results suggest that RPCH-like peptides appear to have MO stimulating activity in P. *clarkii* and *Libinia emaginata*.

We reasoned that if RPCH stimualtes MF synthesis *in vitro*, then crustacean "pigment-dispersing hormone" (PDH) might inhibit synthesis. When synthetic PDH, obtained from Dr. K. Ranga Rao, University of West Florida, was assayed, it was found to inhibit almost all MF synthesis by MOs of *Procambarus clarkii* at 10^{-7} M. Interestingly, Quackenbush and Herrnkind (1983) reported that partially purified GIH could not be separated from pigment cell dispersing activity. Thus, in some cases the two activities of pigment dispersion and gonad inhibition may be performed by the same or similar molecules (Laufer *et al.*, 1987b). We also know that there is some species specificity to MO stimulation and inhibition since RPCH and PDH have no effect on lobster MOs.

Mechanisms of Hormone Action and Gland Activation

The release of RPCH and black pigment concentrating hormone (BPDH) stored in CNS have been studied in the context of controlling pigment dispersion; some of the results reported may find application in the study of crustacean reproduction. RPCH appears to be released in *Uca* when the eyestalk is exposed to the biogenic amine, dopamine, and to met-enkephalin. RPCH is also released in the presence of FMRFamide (Butler and Fingerman, 1985).

Serotonin (5-HT) acts to stimulate PDH release (Rao and Fingerman, 1970; Fingerman and Rosenberg, 1986), as may norepinephrine (Fingerman *et al.*, 1981). Mattson and Spaziani (1985a) showed that the release of molt-inhibiting hormone (MIH) is stimulated by 5-HT, although dopamine, GABA, acetycholine, octopamine, and norepinephrine had no effect. Homola *et al.* (1989) have shown that 5-HT strongly inhibits MF synthesis by dissociated MOs from *Libinia* at concentrations as low as 10^{-8} M. These results are of considerable interest since 5-HT seems to play a role in the reproductive and postural behavior of lobsters and crayfish, (Livingstone *et al.*, 1980) and may influence similar responses in shrimp. Clark (this volume) indicated that female shrimp assume a typical extended posture during oviposition. When 5-HT is injected into lobsters and crayfish there is a sustained flexion with the legs held high under the body and the claw is held open and straight out, while injection of octopamine has an opposite effect causing the animals to "lie down" and raise the tail; these are the same positions that are taken by male and female lobsters, respectively, as the male approaches before mating during courtship behavior. Thus, serotonin and octopamine not only affect the mating behavior of lobsters (Beltz, 1988), an external manifestation of reproductive activity, but affect MF production, which may be an internal manifestation of gonad maturation. These same substances may play a role in shrimp reproduction.

The cellular events that are associated with the mechanisms of action of crustacean hormones have been studied in a few cases and are worth considering. Cyclic nucleotides and calcium appear to also play a role in regulating crustacean reproduction. In insect cells cAMP levels are raised by a number of primary hormonal messengers including AKH (Gade and Holwerda, 1976), the "puparium tanning factor" (Fraenkel *et al.*, 1977), and PTTH (Vedeckis *et al.*, 1976). Cyclic AMP levels in the CA are raised by several biogenic amines (Gole *et al.*, 1987; see also Lafon-Cazal and Baehr, 1988). The allatostatin(s) from the insect brain may control JH synthesis in the CA via cAMP (Aucoin *et al.*, 1987). Cyclic GMP is less widely found and its role, though not well defined, is thought to be associated with tr-eclosion (Truman *et al.*, 1979). Cyclic nucleotides also act as second messengers in crustaceans. Cyclic GMP and calcium are probably involved in the action of RPCH (Lambert and Fingerman, 1979) and cAMP appears to regulate the effect of MIH in the Y-organ (Mattson and Spaziani, 1985b). Recently, Tsukimura *et a*l. (1986) have suggested that an extract of the sinus gland is able to increase the amount of cGMP, but not cAMP, in the MO of the lobster.

Other Possible Hormonal Factors Regulating Reproduction

Pheromones, substances released into the water by crustaceans to attract a mate, are well documented in the literature. Takayanagi *et al.* (1986b) showed that female shrimp, *Paratya*, would normally delay ovarian development unless males were present, but if extracts of the male testis or vas deferens were added to the water, maturation would take place, suggesting the release of a pheromone by the male to stimulate ovarian development. In the blue crab, *Callinectes sapidus*, males will not respond to the female pheromone if the male has been eyestalk ablated (Gleeson *et al.*, 1987).

Ecdysteroids have a role in the reproduction of insects and it has been speculated that they may also be critical to crustacean reproduction. We have already referred to an apparent antagonism between molting and reproduction. In the amphipod, *Orchestia gamarellus*, destruction of the site of ecdysone synthesis, the Y-organ, at the beginning of the molt cycle results in no ovarian growth, and if the Y-organ is destroyed during ovarian growth the synthesis of vitellogenin stops (Meusy and Charniaux-Cotton, 1984). The Y-organ seems to also be required for oocyte growth in the isopod, *Armadillium* (Suzuki, 1986). Chang (1984) hypothesized that the pleopod glands of lobsters may send a signal to the animal preventing molting when embryos are attached to the pleopods.

Other steroids may be involved in reproduction as well. A number of sex steriods, including testosterone, progesterone, and pregnenolone have been identified in the gonads and serum of lobsters and crayfish (Burns *et al.*, 1984; Ollevier *et al.*, 1986). Of these progesterone may be of particular interest; injections of progesterone and 17-alpha-hydroxyprogesterone have been reported to stimulate vitellogenesis in penaeid shrimp (Yano, 1985; 1987). Couch *et al.* (1987) found estradiol in a number of lobster tissues, but the highest levels were observed in the MO, and progesterone levels in immature animals were low or undetectable in all tissues tested except the MO.

There are reports of other "vertebrate" hormones affecting crustacean reproduction. Human chorionic gonadotrophin (HCG) is a placental glycoprotein that has been reported to stimulate vitellogenin synthesis in the isopod *Idotea balthica* (Souty and Picaud, 1984) and the shrimp *Crangon crangon* (Bomirski and Klek-Kawinska, 1976). *Crangon crangon* ovary development is also stimulated by two pituitary hormones, follicle-stimulating hormone (FSH) and luteinizing hormone (LH) (Zukowska-Arendarczyk, 1981).

The androgenic gland of crustacea is responsible for the masculinization of the animal. These glands seem to produce a number of compounds, including farnesylacetone, a molecule similar in structure to MF (Ferezou *et al.*, 1978).

Farnesylacetone will inhibit ovarian lipovitellin synthesis *in vitro* (Berreur-Bonnenfant and Lawrence, 1984).

Conclusions

The finding of MF synthesis by the MO in Crustacea, including shrimp, opens up the possibility that reproduction is driven by JH-like substances in arthropods other than insects. The endocrinology of shrimp reproduction can be visualized as a cascade , initiated by environmental cues such as food, photoperiod, and temperature. The CNS, mediated through the eyestalk neuroendocrine system, produces stimulatory and inhibitory factors which target their effects either directly to the gonads and/or indirectly to the MO and hepatopancreas. More experiments are needed to determine how many factors are involved in the cascade, how and when they act, and which tissues are responding to the stimuli transducing environmental cues and coordinating the effects into the maturation of the accessory reproductive tissues as well as the production of the gametes.

Acknowledgements:

The research reported in this paper was supported in part by Grants from the Sea Grant College Program, NOAA, NA-85AA-D-SG101; The U.S. Department of Agriculture, Grant 85-CRCR-1-1839; The Lady Davis Trust; A National Research Service Award from the National Institutes of Health, 1-F-33-NS-08334-01.

Literature Cited

Anilkumar, G. and K.G. Adiyodi. 1980. Ovarian growth, induced by eyestalk ablation during the prebreeding season, is not normal in the crab, *Paratelphusa hydrodomous* (Herbst). Int. J. Invert. Reprod. 2: 95-105.

Anilkumar, G. and K.G. Adiyodi. 1985. The role of eyestalk hormones in vitellogenesis during the breeding season in the crab *Paratelphusa hydrodromous* (Herbst). Biol. Bull. 169: 689-695.

Applebaum, S.W. and P. Moshitzky. 1986. The involvement of brain factors in regulating vitellogensis in the African migratory locust (*Locusta migratoriodes*). In. Advances in Invertebrate Reproduction 4, M. Porchet, J.C. Andries, and A.Dhainaut (editors). Elsevier Science Publications.

Amsterdam, pg 489.
Aucoin, R. R., S. M. Rankin, B. Stay, and S. S. Tobe. 1987. Calcium and cyclic AMP involvement in the regulation of juvenile hormone synthesis in *Diploptera punctata*. Insect Biochem. 17: 965-969.
Bell, T. A. and D. V. Lightner. 1988. A handbook of normal penaeid shrimp histology. World Aquaculture Soc. Baton Rouge, LA, 114 pages.
Beltz, B.S. 1988. Crustacean neurohormones. In. Endocrinology of Selected Invertebrate Types. Invertebrate Endocrinology Vol. 2. H. Laufer and R. Downer (editors). Alan R. Liss, New York, pp. 235-258.
Berreur-Bonnenfant, J. and F. Lawrence. 1984. Comparative effect of farnesylacetone on macromolecular synthesis in gonads of crustaceans. Gen. Comp. Endocrin. 54: 462-468.
Bomirski, A. and E. Klek. 1974. Action of the eyestalk on the ovary in *Rhithropanopeus harrisii* and *Crangon crangon* (Crustacea: Decapoda). Mar. Biol. 24: 329-337.
Bomirski, A. and Klek-Kawinska. 1976. Stimulation of oogenesis in the sand shrimp, *Crangon crangon* by human gonadotropin. Gen. Comp. Endocrin. 30: 239-242.
Bookhout, C. G. and J. D. Costlow, Jr. 1974. Crab development and effects of pollutants. Thalassia jogosl. 10: 77-87.
Borst, D. W., H. Laufer, M. Landau, E. S. Chang, W. A. Hertz, F. C. Baker, and D. A. Schooley. 1987. Methyl farnesoate and its role in crustacean reproduction and development. Insect Biochem. 17: 1123-1127.
Brown, F. A., Jr., and G. M. Jones, 1949. Ovarian inhibition by a sinus-gland principle in the fiddler crab. Biol. Bull. 96: 228-232.
Burns, B. G., G. B. Sangalang, H. C. Freeman, and M. McMenemy. 1984. Isolation and identification of testosterone from the serum and testes of the American lobster (*Homarus americanus*). Gen. Comp. Endocrin. 54: 429-432.
Butler, T. A. and M. Fingerman. 1985. Effects of dopamine and neuropeptides on the isolated eyestalk of the fiddler crab. Amer. Zool. 25: 102A.
Byard, E. H., R. R. Shivers, and D. E. Aiken. 1975. The mandibular organ of the lobster, *Homarus americanus*. Cell Tiss. Res. 162: 13-22.
Chang, E. S. 1984. Ecdysteroids in crustacea: role in reproduction, molting, and larval development. In. Advances in Invertebrate Reproduction 3, W. Engles (editor). Elsevier Science Publishers, New York. pp. 223-230.
Charmantier, G., M. Charmantier-Daures, and D.E. Aiken. 1988. Larval development and metamorphosis of the American lobster *Homarus americanus* (Crustacea, Decapoda): effect of eyestalk ablation and juvenile hormone injection. Gen. Comp. Endocrin. 70: 319-333.
Choy, S. C. 1987. Growth and reproduction of eyestalk ablated *Penaeus canaliculatus* (Olivier, 1811) (Crustacea: Penaeidae). J. Exp. Mar. Biol. Ecol. 112: 93-107.
Costlow, J. D., Jr. 1968. Metamorphosis in crustaceans. In. Metamorphosis. W. Etkin and L.I. Gilbert (editors), Appleton. pp. 3-41.
Couch, E. F., N. Hagino, and J.W. West. 1987. Changes in estradiol and progeste-

rone immunoreactivity in tissues of the lobster, *Homarus americanus*, with developing and immature ovaries. Comp. Biochem. Physiol. 87A: 765-770.

Dale, J. F. and S. S. Tobe, 1988. The effect of a calcium ionophore, a calcium channel blocker and calcium-free medium on juvenile hormone release *in vitro* from corpora allata of *Locusta migratoria.* J. Insect Physiol. 34: 451-456.

de Leersnyder, M. and A. Dhainaut. 1978. Evolution du metabolisme ovocytairie apres ablation des pedoncules oculaires chez le crabe *Eriocheir sinensis.* Etude autoradiographique. Arch. Zool. exp. gen. 119: 399-407.

Downer, R. G. H. and H. Laufer. 1983. Endocrinology of Insects. Invertebrate Endocrinology Vol. I. Alan R. Liss, New York. 707 pages.

Eastman-Reks, S. and M. Fingerman. 1984. Effects of neuroendocrine tissue and cyclic AMP on ovarian growth *in vivo* and *in vitro* in the fiddler crab, *Uca pugilator.* Comp. Biochem. Physiol. 79A: 679-684.

Ferezou, J P., M. Barbier, and J. Berreur-Bonnenfant. 1978. Biosynthese de la farnesylacetone-(E,E) par les glandes androgenes du crabe *Carcinus maenas.* Helv. Chim. Acta 61: 669-674.

Fingerman, M., M. M. Hanumante, S. W. Fingerman, and D. C. Reinschmidt. 1981. Effects of norepinephrine and norepinephrine agonist and antagonist on the melanophores of the fiddler crab, *Uca pugilator.* J. Crust. Biol. 1: 16-27.

Fingerman, M. and M. Rosenberg. 1986. Control of the integumentary chromatophores of the crab, *Pachygrapsus marmoratus.* Amer. Zool. 26: 28A.

Fraenkel, G., A. Blechl, J. Blechl, P. Herman, and M. Seligman. 1977. 3',5'-cyclic AMP and hormonal control of puparium formation in the fleshfly *Sarcophaga bullata.* Proc. Natl. Acad. Sci. 74: 2182-2186.

Fyhn, U.E.H., H.J. Fyhn, and J.D. Costlow. 1977. Cirriped vitellogenesis: effect of ecdysone *in vitro.* Gen. Comp. Endocrin. 32: 266-271.

Gade, G. and D. A. Holwerda. 1976. Involvement of adenosine 3',5'-cyclic monophosphate in lipid metabolism in *Locusta migratoria.* Insect Biochem. 6: 535-540.

Girardie, A. 1983. Neurosecretion and Reproduction. In. Endocrinology of Insects. Invertebrate Endocrinology Vol. I. R. Downer and H. Laufer (editors). Alan R. Liss, New York. pp. 305-317.

Gleeson, R. A., M. A. Adams, and A. B. Smith, III. 1987. Hormonal modulation of pheromone-mediated behavior in a crustacean. Biol. Bull. 172: 1-9.

Gole, J. W. D., G. L. Orr, and R. G. H. Downer. 1987. Pharmacology of octopamine-, dopamine-, and 5-hydroxytryptamine-stimulated cyclic AMP accumulation in the corpus cardiacum of the American cockroach, *Periplaneta americana* L. Arch. Insect Biochem. Physiol. 5: 119-128.

Gomez, E. D., D. J. Faulkner, W. A. Newman, and C. Ireland. 1973. Juvenile hormone mimics: effects on cirriped crustacean metamorphosis. Science 179: 813-814.

Gomez, R. 1965. Acceleration of development of gonads by implantation of brain in the crab *Paratelphusa hydrodromous.* Naturwissensheftan. 52: 217.

Hinsch, G. W. 1972. Some factors controlling reproduction in the spider crab, *Libinia emarginata.* Biol. Bull. 143: 358-366.

Hinsch, G. W. 1980. Effects of mandibular organ implants upon the spider crab ovary. Trans. Amer. Micros. Soc. 99: 317-322.

Hinsch, G.W . 1981. Effects of juvenile hormone mimics on the ovary in the immature spider crab, *Libinia emarginata.* Int. J. Invert. Reprod. 3: 237-244.

Hinsch, G. W. and D. C. Bennet. 1979. Vitellogenesis stimulated by thoracic ganglion implants into destalked immature spider crabs, *Libinia emarginata.* Tiss. Cell 11: 345-351.

Homola, E. M. Landau, and H. Laufer. 1989. An *in vitro* bioassay for the regulation of methyl farnesoate synthesis in disaggregated mandibular organ cells from *Libinia emarginata* and the effect of serotonin. Biol. Bull. 176: 69.

Klek-Kawinska, E. and A. Bomirski, 1975. Ovary-inhibiting hormone activity in the shrimp (*Crangon crangon*) eyestalks during the annual reproductive cycle. Gen. Comp. Endocrin. 25: 9-13.

Lafon-Cazal, M. and J. C. Baehr. 1988. Octopaminergic control of corpora allata activity in an insect. Experientia 44: 895-986.

Lambert, D. T. and M. Fingerman. 1979. Evidence implicating calcium as a second messenger for red pigment-concentrating hormone in the prawn, *Palaemonetes pugio.* Physiol. Zool. 52: 497-508.

Landau, M. and C. Finney. 1977. Insect juvenile hormones and their mimics: applications in crustaceana research. FL Sci. 40 (Suppl.), 14.

Landau, M. and K. R. Rao. 1980. Toxic and sublethal effects of precocene II on the early developmental stages of the brine shrimp *Artemia salina* (L.) and the barnacle *Balanus eburneus* Gould. Crustaceana 39: 218-221.

Laufer, H., M. Landau, D. Borst, E. Homola. 1986. The synthesis and regulation of methyl farnesoate, a new juvenile hormone for crustacean reproduction. In. Advances in Invertebrate Reproduction 4, M. Porchet, J.C. Andries, and A. Dhianaut (editors). Elsevier Science Publications, Amsterdam. pp.135-143.

Laufer, H., D. Borst, F. C. Baker, C. Carrasco, M. Sinkus, C. C. Reuter, L. W. Tsai, and D. A. Schooley. 1987a. Identification of a juvenile hormone-like compound in a crustacean. Science 235: 202-205.

Laufer, H., E. Homola, and M. Landau. 1987b. Control of methyl farnesoate synthesis in crustacean mandibular organs. Amer. Zool. 27: 69A.

Laufer, H., M. Landau, E. Homola, and D. W. Borst. 1987c. Methyl farnesoate: its site of synthesis and regulation of secretion in a juvenile crustacean. Insect Biochem. 17: 1129-1131.

Livingstone, M. S., R. M. Harris-Warrick, E. A. Kravits. 1980. Serotonin and octopamine produce opposite postures in lobsters. Science 208: 76-79.

Mangerich, S. R. Keller and H. Dirchsen. 1986. Immunocytological identification of structures containing putative red pigment concentrating hormone in two species of decapod crustaceans. Cell Tiss. Res. 245: 377-386.

Mattson, M. P., and E. Spaziani. 1985a. 5-hydroxytryptamine mediates release of the molt-inhibiting hormone activity from isolated crab eyestalk ganglia. Biol. Bull. 169: 246-255.

Mattson, M. P. and E. Spaziani. 1985b. Cyclic AMP mediates the negative regulation of Y-organ ecdysteroid production. Mol. Cell. Endocrin. 42: 185-189.

Meusy, J. J. and H. Charniaux-Cotton. 1984. Endocrine control of vitellogenesis in malacostraca crustaceans. In. Advances in Invertebrate Reproduction 3. W. Engles (editor). Elsevier Science Publishers, New York, pp. 231-241.

Meusy. J. J., G. Martin, D. Soyez, J. E. van Deijen, and J.M. Gallo. 1987. Immunochemical and immunocytochemical studies of the crustacean vitello-genesis-inhibiting hormone (VIH). Gen. Comp. Endocrin. 67: 333-341.

Ollevier, F., D. De Clerck, H. Diederik, and A. De Loof. 1986. Identification of non-ecdysteroid steroids in the hemolymph of both male and female *Astacus leptodactylus* (Crustacea) by gas chromatography-mass spectrometry. Gen. Comp. Endocrin. 61: 214-228.

Otsu, T. 1960. Precocious development of the ovaries in the crab, *Potamon dehaani,* following implantation of the thoracic ganglion. Annot. Zool. Jap. 33: 90-96.

Otsu, T. 1963. Bihormonal control of sexual cycle in the freshwater crab, *Potamon dehaani*. Embyologia 8: 1-20.

Panouse, J. B. 1943. Influence de l'ablation de pedoncle oculaire sur la croissance de l'ovaire chez la crevette *Leander serratus.* C.R. acad. Sci. 217: 553-55.

Paulus, J. E. 1984. Hormonal control of vitellogenesis in decapod crustacea (Brachyura). Univ. of Connecticut Ph.D.

Paulus, J. E. and H. Laufer. 1987. Vitellogenocytes in the hepatopancreas of *Carcinus maenus* and *Libinia emarginata.* Int. J. Invert. Reprod. Develop. 11: 29-55.

Payen, G. G. and J. D. Costlow. 1977. Effects of a juvenile hormone mimic on the male and female gamnetogenesis of the mud crab, *Rhithropanopeus harrisii* (Gould) (Brachyura; Xanthidae). Biol. Bull. 152: 199-208.

Quackenbush, L. S. and W. F. Herrnkind. 1981. Regulation of molt and gonadal development in spiny lobster *Panulirus argus* (Crustacea: Palinuridae): effects of eyestalk ablation. Comp. Biochem. Physiol. 69A: 523-527.

Quackenbush, L. S. and W. F. Herrnkind. 1983. Partial characterization of eyestalk hormones controlling molt and gonadal development in the spiny lobster, *Panulirus argus*. J. Crust. Biol. 3: 34-44.

Quackenbush, L. S. and L. L. Keeley. 1986. Vitellogenesis in the shrimp, *Penaeus vannamei*. Amer. Zool. 26: 56A.

Rao, K. R. and M. Fingerman. 1970. Action of biogenic amines on crustacean chro-matophores. II. Analysis of the response of erythrophores in the fiddler crab, *Uca pugilator*, to indolealkylamines and an eyestalk hormone. Comp. Gen. Pharmacol. 1: 117-126.

Schneiderman, H. A. and L. I. Gilbert. 1958. Substances with juvenile hormone activity in crustacea and other invertebrates. Biol. Bull. 115: 530-535.

Souty, C. and J. L. Picaud. 1984. Effet de l'injection d'une gonadotropine humaine sur la synthese et la liberation de la vitellogenine pa le tissu adipeux du crus-tace isopode marin *Idotea balthica basteri.* Gen. Comp. Endocrin. 54: 418-421.

Soyez, D., J. E. Van Deijnen, and M. Martin. 1987. Isolation and characterization of a vitellogenesis-inhibiting factor from sinus glands of the lobster,

Homarus americanus. J. Exp. Zool. 244: 479-484.

Suzuki, S. 1986. Effect of Y-organ ablation on oocyte growth in the terrestrial isopod, *Armadillium vulgare*. Biol. Bull. 170: 350-355.

Takayanagi, H., Y. Yamamoto, and N. Takeda. 1986a. An ovary stimulating factor in the shrimp, *Paratya compressa*. J. Exp. Zool. 240: 203-209.

Takayanagi, H., Y. Yamamoto, and N. Takeda. 1986b. Ovary-stimulating pheromone in the freshwater shrimp, *Paratya compressa*. J. Exp. Zool. 240: 397-400.

Taketomi, Y. and Y. Kawano. 1985. Ultrastructure of the mandibular organ of the shrimp, *Panaeus japonic*us, in untreated and experimentally manipulated individuals. Cell. Biol. Int. Rep. 9: 1069-1074.

Templeton, N. S. and H. Laufer. 1983. The effects of a juvenile hormone analog (Altosid ZR-515) on the reproduction and development of *Daphnia magna* (Crustacea: Cladocera). Int. J. Invert. Reprod. 6: 99-110.

Tighe-Ford, D. J. 1977. Effects of juvenile hormone analogues on larval metamorphosis in the barnacle *Elminius modestus* Darwin (Crustacea: Cirripedia). J. Exp. Mar. Biol. Ecol. 26: 163-176.

Truman, J. W., S. M. Mumby, and S. K. Welch. 1979. Involvement of cyclic GMP in the release of stereotyped behavior patterns in moths by a peptide hormone. J. Exp. Zool. 84: 201-212.

Tsukimura, B., T. M. Tanji, F. I. Kamemoto. 1986. Sinus gland activiation of cyclic GMP in the mandibular organ of *Homarus americanus*. Amer. Zool. 26: 91A.

Vedeckis, W. V., W. E. Bollenbacher, and L.I . Gilbert. 1976. Insect prothoracic glands: a role for cyclic AMP in the stimulation of alpha-ecdysone secretion. Mol. Cell. Endocrin. 57: 360-370.

Yano, I. 1985. Induced ovarian maturation and spawning in greasyback shrimp, *Metapenaeus ensis*, by progesterone. Aquaculture 47: 223-229.

Yano, I. 1987. Effect of 17-alpha-hydroxy-progesterone on vitellogenin secretion in Kuruma prawn, *Penaeus japonicu*s. Aquaculture 61: 49-57.

Zukowska-Arendarczyk, M. 1981. Effect of hypophyseal gonadotropins (FSH and LH) on the ovaries of the sand shrimp *Crangon crangon*. Mar. Biol. 63: 241-247.

Crustacean Molting Hormones: Cellular Effects, Role in Reproduction, and Regulation By Molt-inhibiting Hormone

Ernest S. Chang
Bodega Marine Laboratory, University of California,
P.O. Box 247, Bodega Bay, CA 94923

Introduction

Rapid, controlled growth is a primary goal of all shrimp culturists. In order for growth to occur, shrimp (and indeed all arthropods) must first periodically loosen the connectives between their epidermis and the extracellular cuticle, rapidly escape from the confines of this rigid cuticle, take up water to expand the new, flexible exoskeleton, and then quickly harden it with minerals and proteins for purposes of defense and locomotion (Figure 1). If a crustacean has initial difficulties in shedding its exoskeleton, subsequent progress in ecdysis (the molt cycle stage involving escape from the old exoskeleton) is made even more difficult due to the initiation of expansion of the tissues covered by the soft new exoskeleton.

In addition to the potential physical problems associated with ecdysis, crustaceans are also at great risk during this period of their molt cycle because they release a molting fluid composed of amino acids, enzymes, and other organic components of the partially digested old exoskeleton. This molting fluid is a very strong feeding stimulant and attractant. The newly molted animal is often unable to move or defend itself. Considering these problems, it is remarkable that shrimp and other crustaceans have been so evolutionarily successful.

Ecdysis itself is only the completion of a complex process which was initiated days or even weeks prior. Essentially every tissue is involved in preparation of the impending molt. Lipid reserves are mobilized in the midgut gland (combination digestive and storage organ). There are increases in cell division. New mRNA is transcribed followed by the appearance of new proteins. Behavior is modified. Not surprisingly, these complex processes, which require close coordination over a prolonged time course, are hormonally mediated.

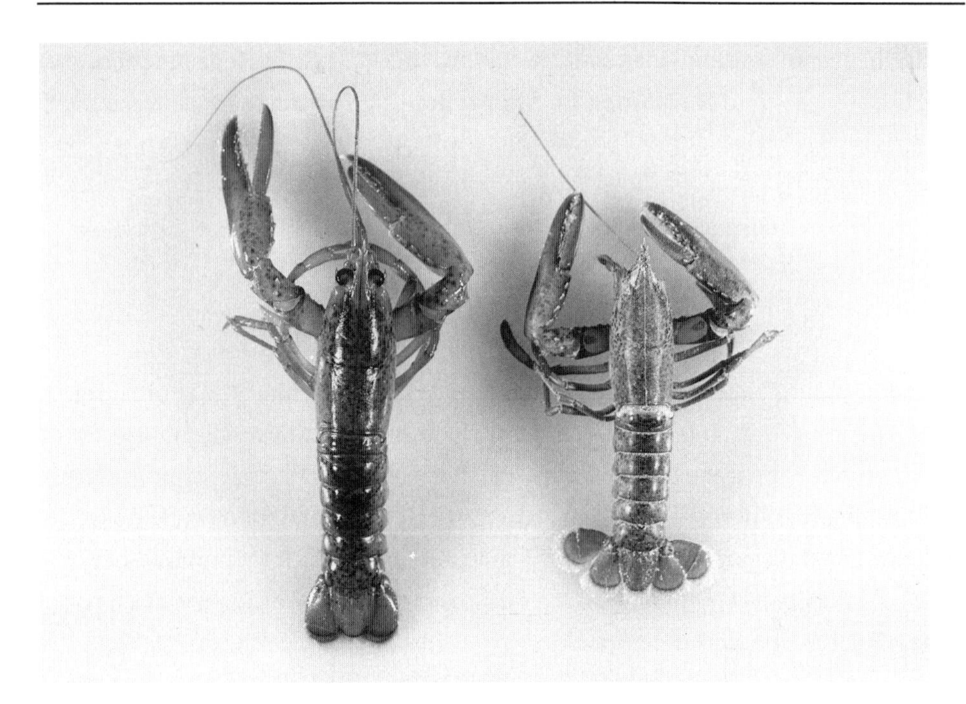

Figure 1. Old exoskeleton (right) of a newly molted lobster (*Homarus americanus*) (left).

The molt cycle is often divided into several stages, some with substages. Different species often have their own set of stage definitions (see Skinner, 1985, for a more thorough discussion). For the purposes of simplicity in this paper, I will mention only four general stages. Postmolt is the stage just following exuviation (shedding of the old exoskeleton). It is the period in which the exoskeleton expands due to an increased hemolymph volume as a result of water influx. Water influx occurs across the epidermis, gills, and gut. After several days, the new exoskeleton begins to harden to an extent that it retains its rigidity. Intermolt is the next stage. During this stage, the exoskeleton obtains the majority of its hardness through the deposition of minerals and proteins. Premolt is the stage just prior to exuviation. It is defined by the separation of the old exoskeleton from the epidermal layer underneath it. The old exoskeleton begins resorption and energy reserves are mobilized from the midgut gland. Entry into premolt is mediated by a dramatic increase in the circulating concentration of the molting hormone. Ecdysis, as a stage, only lasts a few minutes. It begins with the opening of the old skeleton (at the dorsal junction of the thorax and abdomen in decapod crustaceans) and is completed when the animal has escaped from its confines. There is a noticeable lack of information concerning the endocrine regulation of molting in shrimp. This is in part likely due to the observation that conspecific variation in crustaceans is unusually large. This is confounded by the molt and reproductive cycles. This situation will likely change with the intense interest in shrimp aquaculture and the availability of cultured stocks.

Although I have included the endocrinological work conducted on marine shrimp whenever possible, by necessity this review will include data obtained from a variety of decapod Crustacea. There is no reason to suspect that the conclusions made from the data are not generally applicable to shrimp. I will first briefly summarize a few historical aspects of the identification of the molting hormone and its metabolism. Greater emphasis will then be placed upon recent data from my laboratory concerning the *in vitro* action of ecdysteroids and the regulation of the circulating concentration of the hormone by a neuropeptide. I will also describe some preliminary studies from my laboratory that implicate the involvement of ecdysteroids in reproduction and embryonic development. Several other reviews have appeared recently and should be consulted for other aspects of this general topic (Kleinholz and Keller, 1979; Spindler *et al.*, 1980; Kleinholz, 1985; Skinner, 1985; Mattson, 1986; Chang and O'Connor, 1988; Chang, 1989).

Ecdysteroids: Steroid Molting Hormones

Identification and Source of Ecdysteroids

Much of the initial biological and chemical work on the characterization of the arthropod molting hormones (ecdysteroids) was conducted on insects. Using similar techniques, the circulating and active form of the molting hormone was isolated from the spiny lobster, *Jasus lalandii,* and identified as 20-hydroxy ecdysone (2ß, 3ß, 14 α 20R, 22R, 25-hexahydroxy-5ß-cholest-7-en-6-one; Hampshire and Horn, 1966; Horn *et al* , 1966; Figure 2). It appears likely that 20-hydroxyecdysone is the active ecdysteroid from all of the crustacean species thus far examined.

Analogous to the insect situation, the precursor to 20-hydroxyecdysone, ecdy-

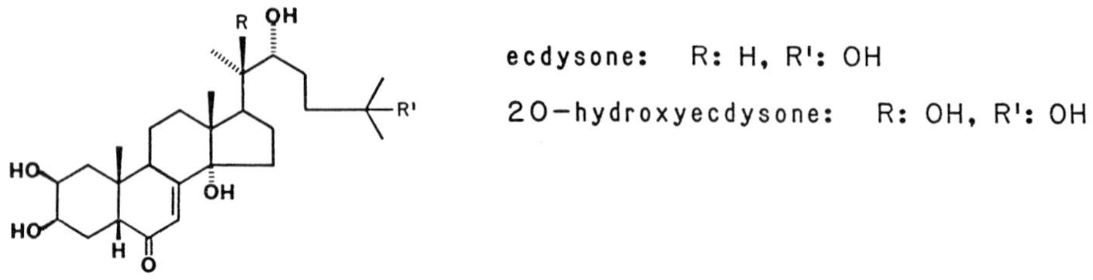

ecdysone: R: H, R': OH

20-hydroxyecdysone: R: OH, R': OH

Figure 2. Chemical structures of ecdysone and 20-hydroxyecdysone.

sone (2ß, 3ß, 14α, 22R, 25-penta hydroxy-5ß-cholest-7-en-6-one, Figure 2),was determined to be the actual molecule secreted by the molting gland (Y-organ) (Chang and O'Connor, 1977; Keller and Schmid, 1979). (However, recent data by Warren *et al.*, 1988, indicate that this paradigm may be more complex in insects.)

Additional ecdysteroids have since been identified in various tissues of different crustacean species (see Chang, 1989 for review). Various synonyms have appeared in previous literature: α-ecdysone for ecdysone; ß-ecdysone, crustecdysone, and ecdysterone for 20-hydroxyecdysone; ecdysones for ecdysteroids. By convention, the preferred terms are ecdysone, 20-hydroxyecdysone, and the generic term ecdysteroids (for compounds with molting hormone activity and similarity to the polyhydroxylated steroid structure of the parent compound ecdysone).

Exogenously applied ecdysteroids have a number of effects on various shrimp tissues. Most of these effects are consistent with the expected action of a molting hormone at premolt. For example, injections of 20-hydroxyecdysone into the grass shrimp, *Palaemonetes pugio*, resulted in separation of the old exoskeleton from the epidermis and generation of a new exoskeleton (Freeman and Bartell, 1976).

Exogenous ecdysteroids also promoted various aspects of premolt development in *Penaeus japonicus* (Kurata, 1968), *Palaemon elegans* (Webster, 1983), and the snapping shrimp, *Alpheus heterochelis* (Mellon and Greer, 1987).

Regulation of Circulating Concentrations of Ecdysteroids

Not unexpectedly, the concentration of the molting hormone varies dramatically during the course of the molt cycle. At postmolt, just after ecdysis, the titer is negligible. In some species, such as the crab *Pachygrapsus crassipes*, there is a small peak at the latter part of postmolt followed by lower basal levels during intermolt (Figure 3). In other species, such as the lobster *Homarus americanus*, there are no small postmolt peaks and the titers remain low throughout postmolt and intermolt (Figure 4). The penaeid shrimp, *Sicyonia ingentis*, has a pattern similar to lobsters (Figure 5) in that there is no significant postmolt peak. This is in contrast to some crabs and the shrimp *Palaemon serratus* (Baldaia, 1984). The significance of this small postmolt peak is unknown. In all cases examined, however, there is a dramatic increase in ecdysteroids

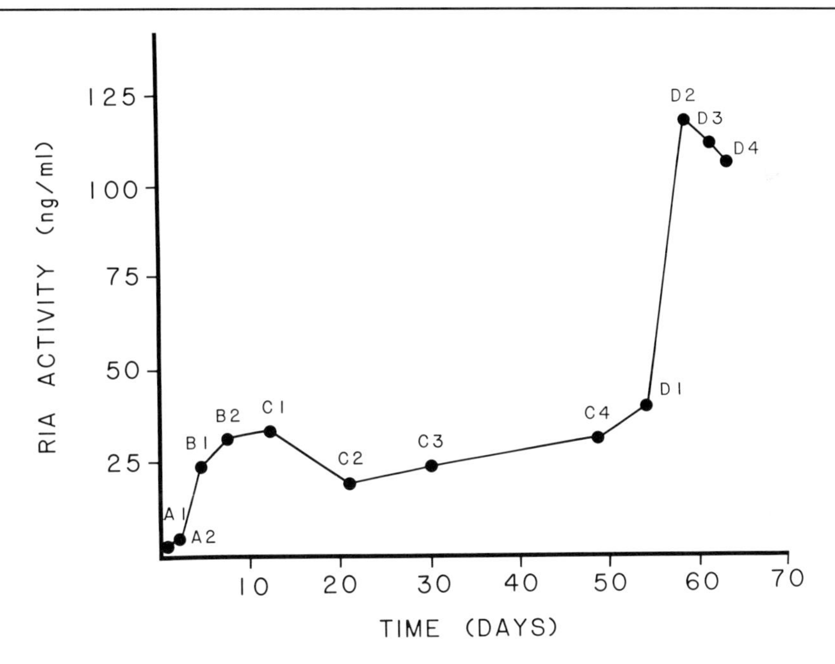

Figure 3. Hemolymph ecdysteroid titers as determined by RIA during the course of the molt cycle of adult *Pachygrapsus crassipes*. The male crabs were staged according to the method of Hiatt (1948) and then 10 ul of hemolymph were removed for radioimmuno-assay (RIA). The abscissa represents the average number of days to the end of each of the molt cycle stages. The stage is indicated above each point. Substages A and B are postmolt, C is intermolt, and D is premolt. The points represent mean values. N ranged from 2 to 24 (modified from Chang and O'Connor, 1978).

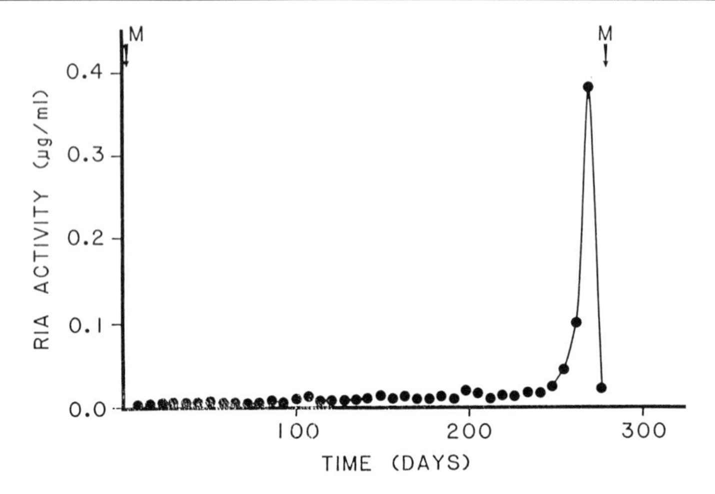

Figure 4. Hemolymph ecdysteroid titers as determined by RIA during the course of the molt cycle of an adult female *Homarus americanus*. There were no reproductive events during this molt cycle. (M) Indicates the time of molt (from Chang, 1984).

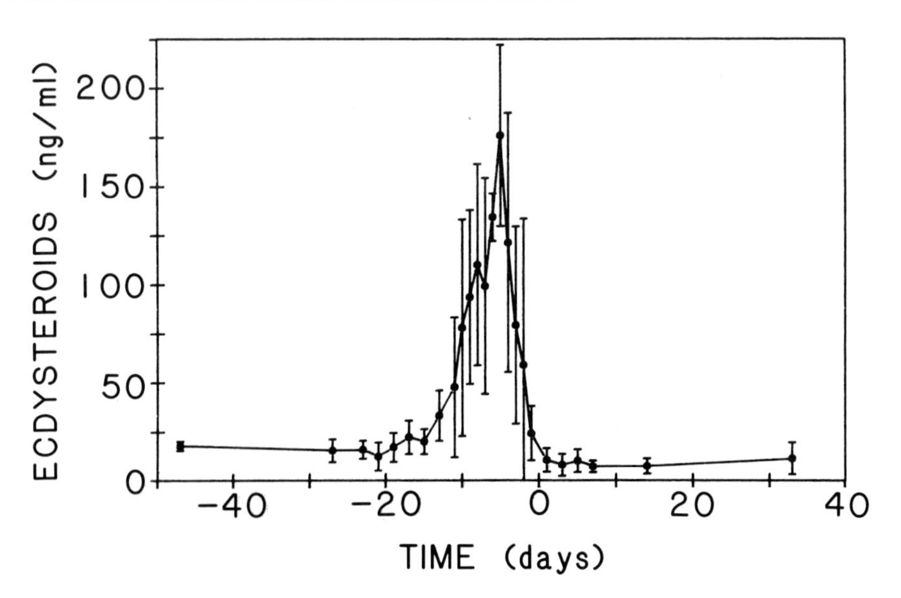

Figure 5. Ecdysteroid titers (mean + S.D.) of the penaeid shrimp *Sicyonia ingentis* at various times before (negative days) and after molt (positive days). Molt occurred on day 0. Hemolymph (25 ul) was removed and extracted with 75% methanol and analyzed by RIA (n = 3 to 16) (Hertz and Chang, unpublished).

in premolt which drops precipitously just prior to ecdysis (Figures 3-5).

It is apparent, however, that crustaceans must possess some sort of mechanism to regulate the circulating, and presumably active form of the molting hormone (20-hydroxyecdysone). These alterations in circulating concentrations could be regulated by either changes in the synthetic and/or secretory rates of ecdysone by the Y-organ, the rate of hydroxylation of ecdysone to 20-hydroxyecdysone, and the rates of degradation and clearance of the hormone from the hemolymph.

From a number of indirect studies, it appears that alterations in the rate of synthesis and/or secretion of ecdysone by the Y-organ may be the predominant means of controlling circulating ecdysteroids (Chang and O'Connor, 1978). The endocrine factor that regulates the Y-organ is discussed more fully below.

There are also data from our laboratory indicating that there are both differential

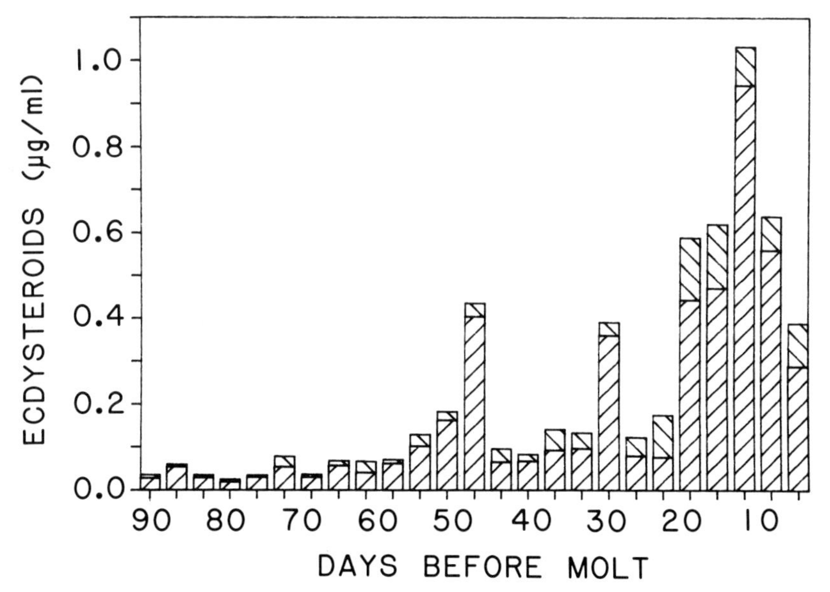

Figure 6. Ecdysteroid concentrations in the urine of an adult female lobster (*Homarus americanus*) during the latter half of a molt cycle. The upper bars represent urinary ecdysteroids. The lower bars represent hemolymph ecdysteroids. Lobsters were carefully removed from the water and suspended in air. This stimulated the lobsters to urinate from the nephridiopore. Urine was collected (range of 0.4 to 8.4 ml) every 3 or 4 days by directing the stream into a test tube. It was reduced to dryness and the residue extracted with methanol. Ecdysteroids were assayed by RIA (Snyder and Chang, unpublished).

rates of ecdysteroid hydroxylation (Chang and O'Connor, 1978) and excretion during the molt cycle. Figure 6 illustrates the alterations that occur in levels of ecdysteroids in the urine from a female adult lobster during the course of a molt cycle. Over 95% of this ecdysteroid radioimmunoassay activity is due to a single peak (when analyzed by high-performance liquid chromatography; HPLC) that is more polar than 20-hydroxyecdysone. We are presently investigating the possible role of urinary ecdysteroids in terms of molt cycle regulation and chemical communication.

In Vitro Study of Ecdysteroid Action

For several years our research group has been attempting to establish continuously dividing cell lines derived from shrimp, lobsters, and crayfish. Our best results have been obtained from lobster and crayfish gonadal and hematopoietic tissue (Brody and Chang, 1989). The animals were first surface sterilized with 20% Wescodyne. They were then rinsed with ethanol and air dried in a laminar flow hood. Tissues were placed in a drop of medium (Table 1) in sterile petri dishes and minced with iris scissors. To dissociate the tissue, fragments were incubated with collagenase. The dispersed cells were pelleted and washed with fresh medium.

As seen in Figure 7A, lobster testicular cells appear healthy and undergo mitoses after several months in culture. The cells in these primary cultures are also responsive to physiological concentrations of ecdysteroids. Some cells become necrotic and disperse following addition of the molting hormone (Figure 7B), whereas other groups proliferate (Figure 7C). These differential results most likely reflect the *in vivo* situation in which different cell types display different responses to ecdy-

Table 1. Culture conditions and additions to Medium 199 buffered to atmospheric CO_2 concentrations for crayfish and lobster primary cell cultures. All cultures had 400 mg/l of $NaHCO_3$ and 10% fetal bovine serum added. They were maintained at 20°C at pH 7.5 (from Brody and Chang, 1989).

Species	Osmolarity (mosmoles)	NaCl (g/l)	Proline (mg/l)	MgCl2 (mg/l)
Crayfish	400-450	2.8	40	190
Lobster	950-1000	22.0	60	200

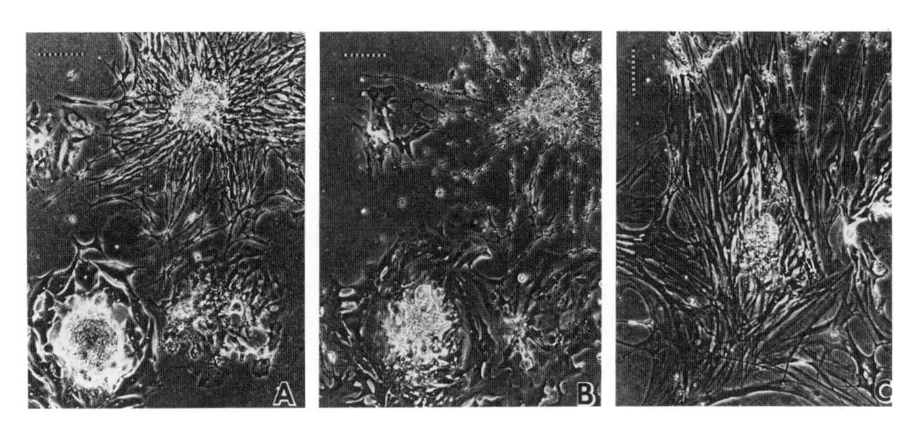

Figure 7. Primary culture of lobster testes after 2 months without any hormone addition (A). 20-Hydroxyecdysone (10^{-8} M) was then added. Six days after hormone addition (B), the cells in the upper colony have started to die, whereas the cells in the lower colony are proliferating. The hormone was then withdrawn. The upper colony has disappeared and the cells of the lower colony have continued their proliferation 21 days after hormone withdrawal (C). Scale bars = 200 μm (Chang and Brody, 1989).

steroids during the course of the molt cycle.

These preliminary data indicate that it is possible to establish long term primary crustacean cell cultures and that these cells are responsive to physiological concentrations of 20-hydroxyecdysone. These cultures will likely be useful for not only endocrinological studies on shrimp and other crustaceans but also for the fields of nutrition, genetics, and pathology.

Ecdysteroids and Reproduction

The molting hormone may mediate several different aspects of crustacean reproduction and development. There are reports that ecdysteroids may be involved in vitellogenesis since they have been isolated from ovaries in relatively high concen trations (Lachaise *et al.*, 1981; Chaix and De Reggi, 1982; Snyder and Chang, unpublished).

There are also recent reports describing changes in the levels of ecdysteroids during embryogenesis in a number of different crustacean species (reviewed in Chang, 1989), including the shrimp, *Palaemon serratus* (Spindler *et al.*, 1987). It was observed that embryonic ecdysteroids were low after fertilization and rose rapidly following the appearance of the embryonic Y-organ. A different pattern has been observed in our laboratory for the crab, *Cancer anthonyi*. Just after fertilization, instead of low levels, we observed that the concentrations of ecdysteroids were high (about 9.5 ng/mg) and steadily decreased throughout embryogenesis to about 2 ng/mg prior to hatching (Figure 8). These data indicate to us that there is initially a maternally-derived investment of ecdysteroids in the maturing eggs which is reflected by the high levels observed at fertilization. The embryo may utilize these maternal ecdysteroids (primarily 20-hydroxy ecdysone as determined by HPLC analyses) for the coordination of the various developmental processes. The levels of the hormone decrease until hatching. Presumably the larval Y-organ in this species

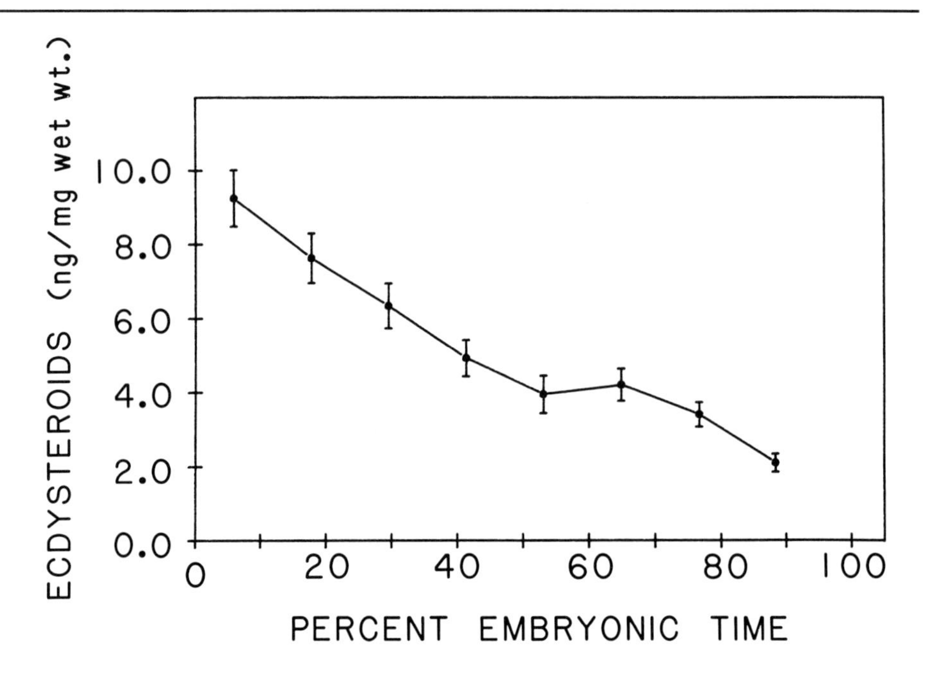

Figure 8. Ecdysteroid concentrations (mean + S.E.) in extracts of crab (*Cancer anthonyi*) embryos. The crabs' embryos (n = 8) were sampled during the course of their development. Embryos were homogenized and the ecdysteroids extracted with methanol. The extracts were analyzed for ecdysteroids by RIA (Okazaki and Chang, 1990).

does not begin to synthesize endogenous ecdysteroids until after hatching.

There are interesting correlations in the female's circulating titers of ecdysteroids during embryogenesis. Just after egg extrusion, the circulating level of ecdysteroids in the female crab (*Cancer anthonyi*) is low (about 0.1 ng/μl). The concentration then steadily increases to a maximum just prior to hatch (Figure 9). Immediately following hatch, the female undergoes a subsequent round of ovarian development without an intervening molt. As the ovaries develop, her circulating level of ecdysteroids decreases to a minimum just prior to the subsequent round of egg extrusion (Figure 10).

We hypothesize that the increasing concentration of maternal circulating ecdysteroids either promotes vitellogenesis and/or will serve as a source of maternally-derived ecdysteroids for investment in the next clutch of eggs.

The decrease in maternal circulating ecdysteroids during the interbrood period

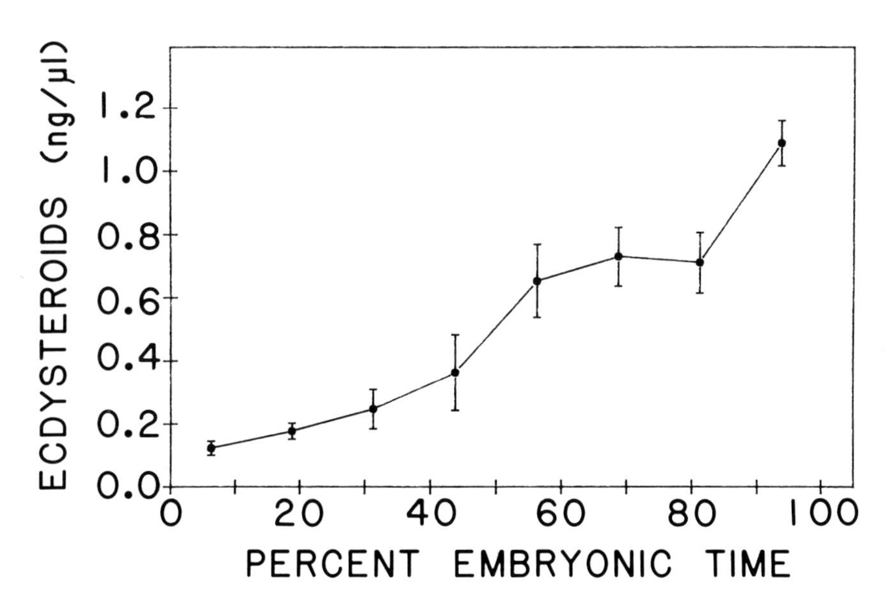

Figure 9. Circulating ecdysteroid concentrations (mean + S.E.) in the hemolymph of female crabs (*Cancer anthonyi*) during the course of a brooding period (n = 8). The hemolymph was analyzed for ecdysteroids by RIA (Okazaki and Chang, 1990).

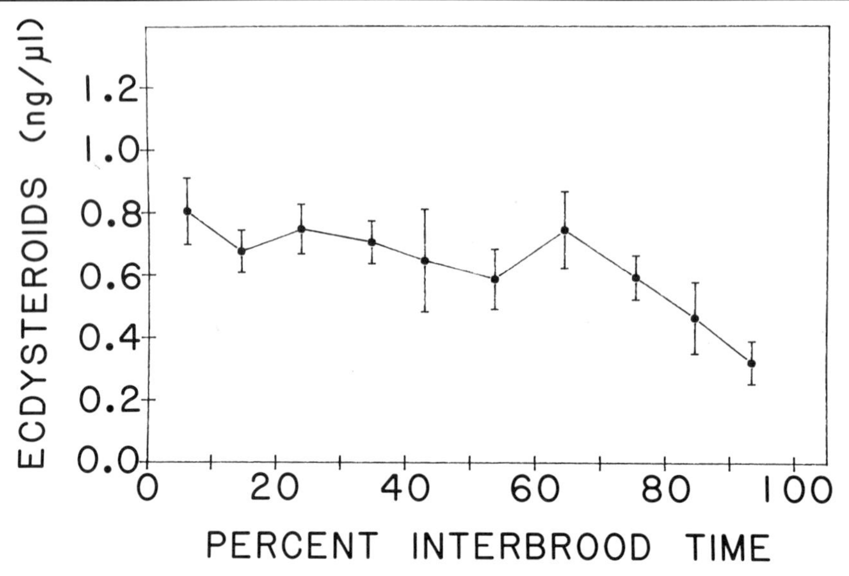

Figure 10. Circulating ecdysteroid concentrations (mean + S.E.) in the hemolymph of female crabs (*Cancer anthony*) during the course of the non-brooding period immediately following the brooding period of the crabs used for Fig. 9 (n = 8). The hemolymph was analyzed for ecdysteroids by RIA (Okazaki and Chang, 1990).

Ablation Studies

Early in this century, a simple experiment was performed. Zeleny (1905) removed both of the stalked eyes of the fiddler crab, *Uca pugilator*, and observed a decrease in the length of the molt cycle interval. This observation plus the replacement experiment of implanting eyestalk contents (Brown and Cunningham, 1939), led to the postulation of an endocrine factor present in the eyestalks that normally inhibits molting - a molt-inhibiting hormone (MIH). This simple procedure of eyestalk ablation has been repeated on a large number of different crustaceans, especially decapods (see Chang, 1989, for review).

Bliss (1951) and Passano (1953) described a neurohemal organ (storage gland for neurosecretory products) in the eyestalks of the decapod crustaceans examined. This organ, called the sinus gland, which serves as a storage site for neurosecretory products (Figure 11), is actually comprised of the enlarged endings of a group of neurosecretory neurons collectively called the X-organ (Bliss and Welsh, 1952; Passano,

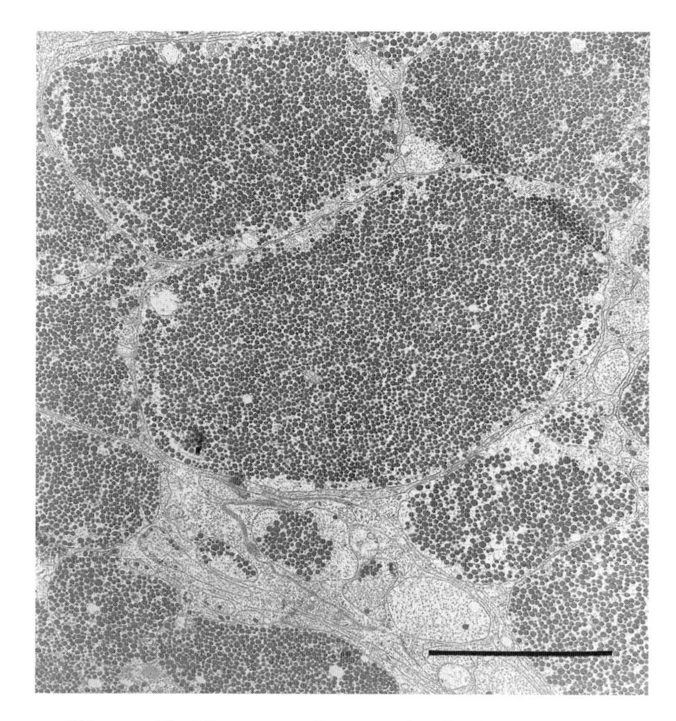

Figure 11. Electron micrograph of a lobster sinus gland. The micrograph shows the distal portions of several neurosecretory neurons. These nerve endings are filled with electron-dense granules which likely contain MIH and other neuropeptide hormones. Scale bar is 5 um. Whole eyestalks were fixed for 15 min in 2.5% glutaraldehyde in phosphate buffer (0.2 M with 2% sucrose, pH 7.4). The gland was then removed and fixed for 2 additional hours. It was postfixed in 1% osmium tetroxide (in 0.2 M phosphate buffer with 2% sucrose), dehydrated in an acetone series, and then embedded in resin prior to sectioning (Aronstein and Chang, unpublished)

may reflect the removal of the molting hormone(s) from the blood and storage in the developing eggs.

Molt-Inhibiting Hormone: Peptide Regulator of Molting

1953). It has been postulated that some of the stored neurosecretory material is MIH and that it is responsible for inhibiting the secretion and/or synthesis of ecdysone by the Y-organ.

In addition to the shortening of the molt cycle interval following eyestalk ablation, other effects have been observed that implicate the sinus gland-X-organ complex as a source for MIH. It has been observed in crayfish (Jegla *et al.*, 1983), lobsters (Chang and Bruce, 1980), and crabs (Chang *et al.*, 1976), that eyestalk removal results in a dramatic and rapid elevation in the concentration of circulating ecdysteroids (Figure 12). Our assay for MIH is based upon these and other observations (Keller and O'Connor, 1982).

Chemical Studies

Sinus glands and non-sinus gland eyestalk neural tissue for controls were obtained from juvenile *Homarus americanus* lobsters that were hatched and raised

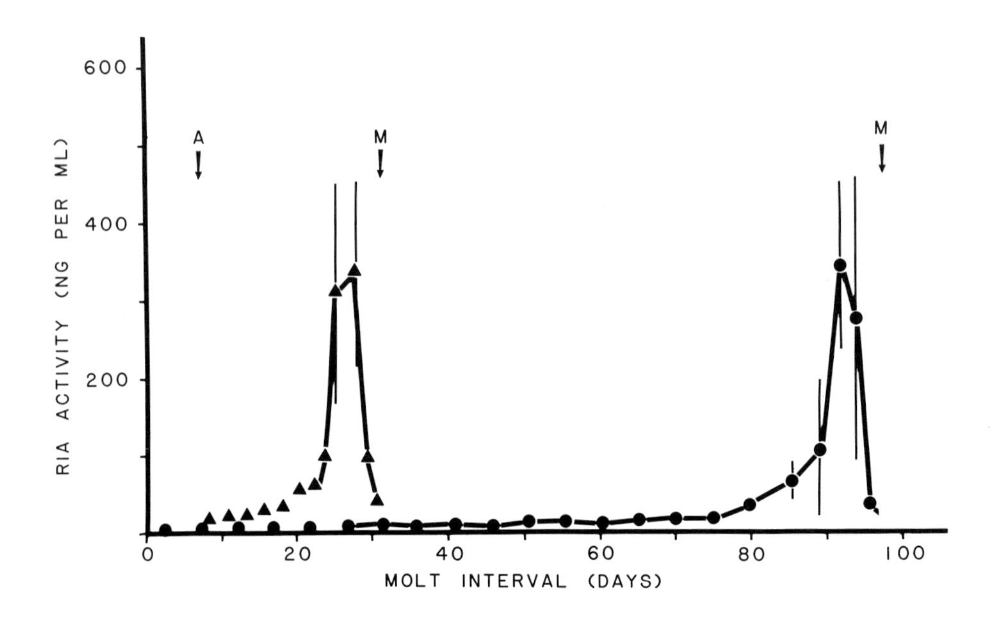

Figure 12. Hemolymph ecdysteroid activity as determined by RIA during the course of the molt cycle of juvenile lobsters (*Homarus americanus*). Lobsters were either eyestalk-ablated 7 d postmolt (triangles) or left intact (circles) and kept at 20°C. Vertical lines represent standard deviations. The arrows indicate the times of ablation (A) or molt (M) (from Chang and Bruce, 1980).

in our laboratory (Chang and Conklin, 1983; Conklin and Chang, 1983; Hedgecock, 1983). The glands were excised and homogenized in 0.1 N HCl for subsequent chemical characterizations.

For the assays, the eyestalk-ablated lobsters were siblings obtained from a narrow hatching period to synchronize the subsequent molt cycles. Seven or 8 days after ablation, 10 ul of hemolymph were withdrawn followed immediately by an injection of a chromatographic fraction (details in Chang *et al.*, 1987). These injections were repeated 18 hr later. Twenty-four hours after the initial sampling, a second hemolymph sample was obtained. Hemolymph samples were processed for ecdysteroid determination as previously described (Chang and O'Connor, 1979; Bruce and Chang, 1984) and assayed by radioimmunoassay (RIA). Changes in ecdysteroid titers in the lobsters following the injections were compared to the levels before the injections and were statistically analyzed (Figure 13).

Following HPLC separation of the sinus gland peptides, eight predominant peaks were identified (Figure 14). These peaks were tested for MIH activity and only Peak VIII was found to contain any (Chang *et al.*, 1987). Several nanomoles of Peak VIII

Table 2. Amino acid composition (in moles) of Peak VIII (putative lobster molt-inhibiting hormone).

Amino Acid	Moles	Amino Acid	Moles
Ala	3	Lys	4
Arg	5	Met	1
Asx	12	Phe	4
Cys	+	Pro	1-2
Glx	9	Ser	4-5
Gly	3	Thr	3
His	0	Trp	++
Ile	1	Tyr	3
Leu	8	Val	9

The data are given in moles and are the averages of 10 determinations. Standard errors were similar to those observed previously (Newcomb, 1983). (+) Indicates presence as determined by sequence analyses. (++) Indicates presence as determined by absorption at 280 nm and by its intrinsic fluorescence (data from Chang *et al.*, 1987, and Chang and Bruce, unpublished).

were purified for subsequent analyses.

Amino acid analyses were conducted using the precolumn o-pthaldialdehyde derivatization method as previously described (Newcomb, 1983). The data (Table 2) were derived from 3 separate preparations of the purified peptide. A molecular

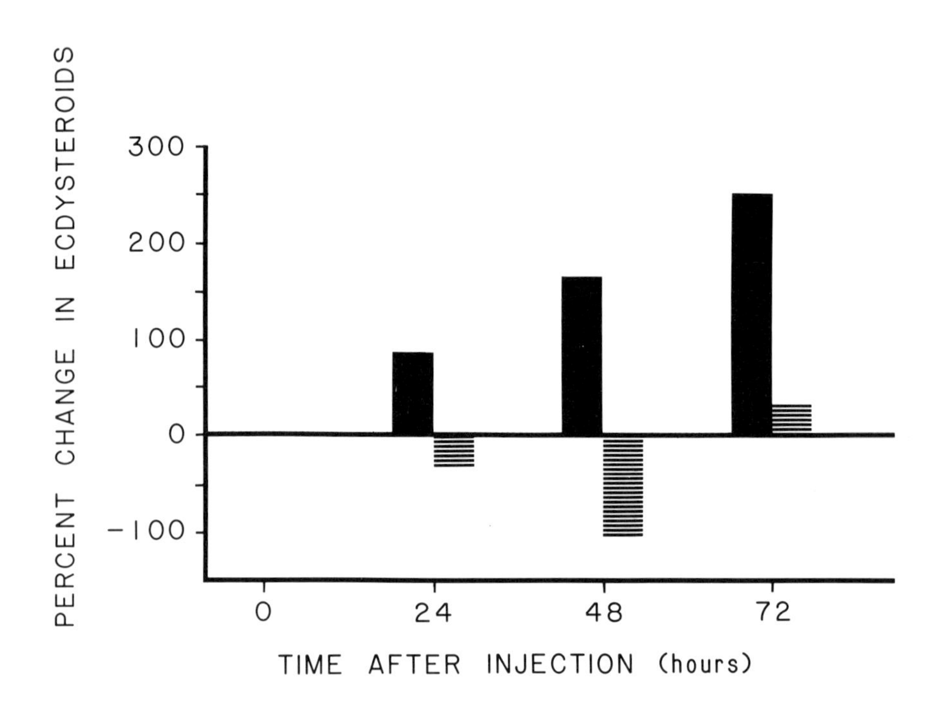

Figure 13. Effects of injections of sinus gland extracts on circulating ecdysteroids in lobsters. The animals were 7 months old and were eyestalk-ablated 48 hr following molt. Six days later, 10 μl of hemolymph were taken for RIA of ecdysteroids. Immediately afterwards, an acid extract of 2.0 sinus glands was injected (hatched bars). Control lobsters were eyestalk-ablated siblings that were injected with an equivalent mass of non-sinus gland eyestalk neural tissue (solid bars). They were then assayed by RIA for circulating ecdysteroids at 24, 48, and 72 hr post-injection. Separate groups of animals were used for each time point. The data are presented as the percent change from the level at the time of injection (0 hr). The data are significant at $P < 0.001$, $P < 0.05$, and $P < 0.01$ for the 24, 48, and 72 hr time points, respectively. Number of lobsters used ranged from 5 to 15 (from Chang, 1989).

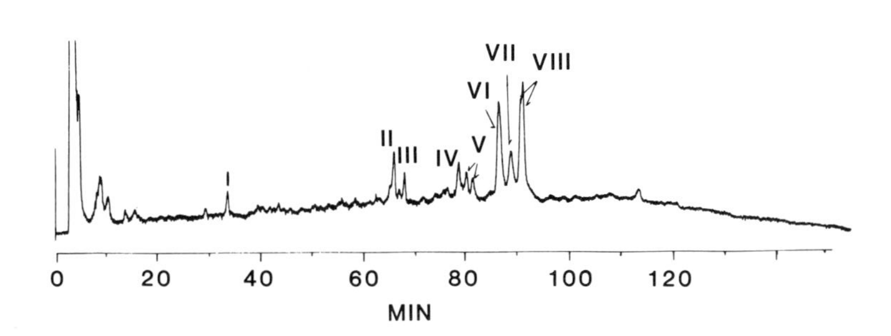

Figure 14. Chromatogram of 2 adult lobster sinus glands that were extracted in hot (80°C) HCl (0.1 N) for 5 min. The high-performance liquid chromatography (HPLC) column (4.6 x 250 mm) was packed with Vydac TP C_{18} and eluted with a linear gradient of 0-50% acetonitrile in 0.1% trifluoroacetic acid for 2 hr at a flow of 1.0 ml/min. Detection was by post-column fluorescence (modified from Chang *et al.*, 1987)

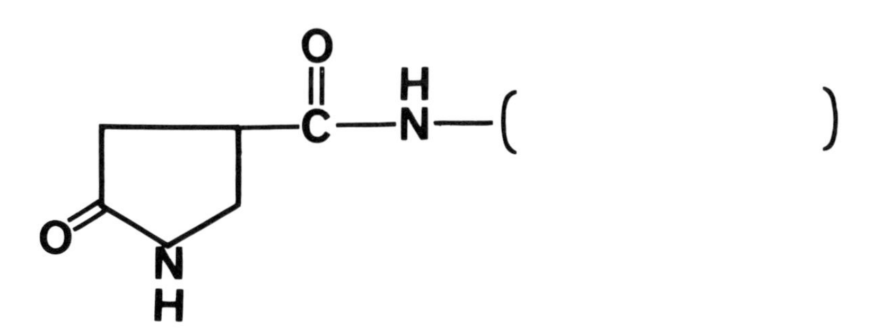

Figure 15. Chemical structure of pyroglutamyl residue at the amino terminus of the putative lobster molt-inhibiting hormone. The rest of the molecule is represented by the parentheses.

weight estimate of at least 8,300 was obtained.

Initial attempts to sequence the MIH peptide by means of a gas-phase sequencer were unsuccessful. From data obtained from insects (Stone *et al.*, 1976; Scarborough *et al.*, 1984; Hayes *et al.*, 1986) and crustacean neuropeptides (Fernlund and Josefsson, 1972; Fernlund, 1976; Rao *et al.* , 1985), it was hypothe-

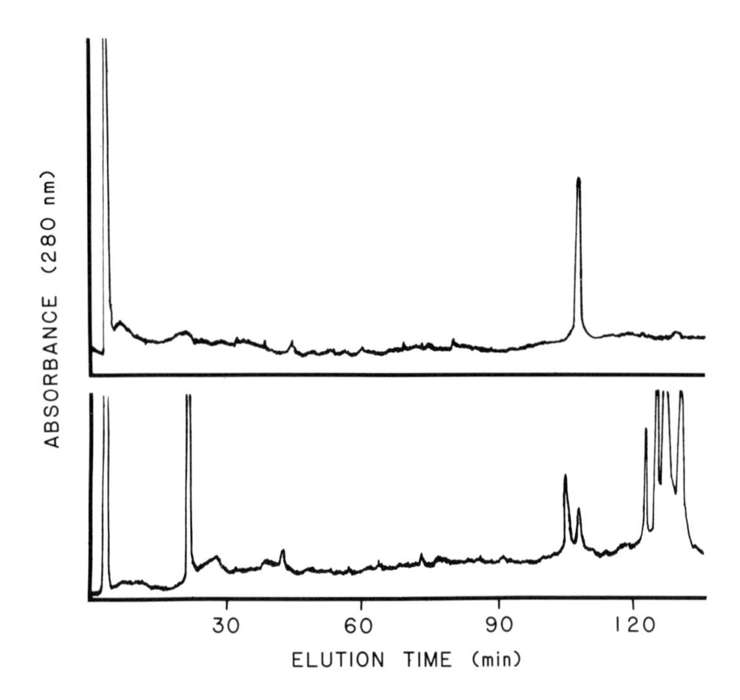

Figure 16. High-performance liquid chromatogram of the putative lobster molt-inhibiting hormone prior to (top) and following (bottom) incubation with pyroglutamyl aminopeptidase. The top chromatogram shows the elution of a single peak at 110 min. The collected MIH (1.7 nmol) was then incubated with the enzyme (170 units) in 120 ul of water. After 16 min, the entire mixture was injected onto the column. The HPLC column was eluted with a linear gradient of 0-80% solvent B (methanol) over 2 hr. Solvent A was 0.1% trifluoroacetic acid in water. Flow rate was 1.0 ml/min. Detection was by absorbance at 280 nm. The bottom chromatogram shows that most of the MIH has been deblocked as evidenced by the appearance of a peak eluting at 108 min. The other additional peaks in the bottom chromatogram are due to the enzyme (Chang and Bruce, unpublished).

sized that the amino terminus of the peptide was blocked by a pyroglutamyl residue (Figure 15). Upon incubation with the enzyme pyroglutamyl aminopeptidase, a distinct shift in the elution profile of MIH resulted (Figure 16). This was evidence that the amino terminus of this peptide was in fact a pyroglutamyl residue.

The deblocked MIH was then subjected to microsequencing techniques. A partial sequence of the first 53 amino acid residues was obtained. We used these data to generate DNA probes to screen a genomic DNA library of the lobster (contained in the phage lambda).

A number of positive clones have been isolated that hybridize with our probes. We hope to isolate the MIH gene from these clones which will then permit a detailed analysis of the molecular organization and control of this important regulatory gene.

Conclusions

It is becoming increasingly apparent that culture techniques will be important for the supplementation of traditional fishery methods for a number of marine species. This is especially true of the decapod crustaceans. Economical manipulation of the growth rate will be a key to this enterprise. In order for these applications to come to fruition, however, a great deal of basic information about the endocrine regulation of molting in crustaceans must first be obtained.

In recent years this basic understanding of the hormonal control of molting in the Crustacea has been greatly advanced since the pioneering classical endocrinological experiments involving glandular ablation and reimplantation. This advancement has been due primarily to the development of new assay (RIA), separation (HPLC), and detection (mass spectrometric) methods. In addition, the cooperation between basic scientists and applied culturists has been mutually beneficial. The use of cultured stocks, that have been raised under controlled temperature, photoperiod, and nutritional regimes, has made possible some of the studies detailed in this paper.

Future research will likely examine the molecular basis of regulation of the Y-organ by MIH and the mediation of differential gene expression by ecdysteroids. This research will be aided by recombinant DNA, monoclonal anti-body, and peptide microsequencing techniques.

Acknlowledgments

I thank the following students and colleagues in my laboratory whose work has contributed to the progress of crustacean endocrinology: D. Aronstein, M. Brody, M. Bruce, W. Hertz, R. Okazaki, and M. Snyder. Much of the work from my laboratory was generously supported by NOAA, National Sea Grant College Program, Department of Commerce, under grant number NA85AA-D-SG140, project number R/A-68, through the California Sea Grant College Program. The U.S. Government is authorized to reproduce and distribute this publication for governmental purposes.

Literature Cited

Baldaia, L., P. Porcheron, J. Coimbra, and P. Cassier. 1984. Ecdysteroids in the shrimp *Palaemon serratus:* Relations with molt cycle. Gen. Comp. Endocrinol. 55, 437-443.

Bliss, D. 1951. Metabolic effects of sinus gland or eyestalk removal in the land crab, *Gecarcinus lateralis*. Anat. Rec. 11, 502-503.

Bliss, D. E., and J. H. Welsh. 1952. The neurosecretory system of brachyuran Crustacea. Biol. Bull. 103, 169-175.

Brody, M. D., and E.S. Chang. 1989. Development and utilization of crustacean long-term primary cell cultures: Ecdysteroid effects in vitro. Internat. J. Invert. Reprod. Develop. In press.

Brown, F., and O. Cunningham. 1939. Influence of the sinus gland of crustaceans on normal viability and ecdysis. Biol. Bull. 77, 104-114.

Bruce, M. J., and E.S. Chang. 1984. Demonstration of a molt-inhibiting hormone from the sinus gland of the lobster *Homarus americanus*. Comp. Biochem. Physiol. 79A, 421-424.

Chaix, J.-C., and M. De Reggi. 1982. Ecdysteroid levels during ovarian development and embryogenesis in the spider crab *Acanthonyx lunulatus*. Gen. Comp. Endocrinol. 47, 7-14.

Chang, E. S. 1984. Ecdysteroids in Crustacea: Role in reproduction, molting, and larval development. In. Advances in Invertebrate Reproduction (W. Engels, W. H. Clark, A. Fischer, P. J. W. Olive, and D. F. Went, eds.), vol. 3, pp. 223-230. Elsevier Science Publishers, Amsterdam.

Chang, E. S. 1989. Endocrine regulation of molting in Crustacea. Rev. Aquatic Sci. 1, 131-157.

Chang, E. S., and M. D. Brody. 1989. Crustacean organ and cell culture. Adv. Cell Cult. 7, 19-86.

Chang, E. S., and M. J. Bruce. 1980. Ecdysteroid titers of juvenile lobsters following molt induction. J. Exp. Zool. 214, 157-160.

Chang, E. S., and D.E. Conklin. 1983. Lobster *(Homarus)* hatchery techniques. In. CRC Handbook of Mariculture (J. P. McVey, ed.), vol. 1, pp. 271-275. CRC Press, Boca Raton.

Chang, E. S., and J. D. O'Connor. 1977. Secretion of a-ecdysone by crab Y-organs *in vitro*. Proc. Nat. Acad. Sci. U.S.A. 74, 615-618.

Chang, E. S., and J. D. O'Connor. 1978. *In vitro* secretion and hydroxylation of a-ecdysone as a function of the crustacean molt cycle. Gen. Comp. Endocrinol. 36, 151-160.

Chang, E. S., and J. D. O'Connor. 1979. Arthropod molting hormones. In. Methods of Hormone Radioimmunoassay (B. M. Jaffe and H. R. Behrman, eds.), 2nd ed., pp. 797-814. Academic Press, New York.

Chang, E. S., and J. D. O'Connor. 1988. Crustacea: Molting. In. Endocrinology of Selected Invertebrate Types (H. Laufer and R. G. H. Downer, eds.), pp. 259-278. Alan R. Liss, New York.

Chang, E. S., B. A. Sage and J. D. O'Connor. 1976. The qualitative and quantitative determinations of ecdysones in tissues of the crab, *Pachygrapsus crassipes,* following molt induction. Gen. Comp. Endocrinol. 30, 21-33.

Chang, E. S., M. J. Bruce, and R. W. Newcomb 1987. Purification and amino acid composition of a peptide with molt-inhibiting activity from the lobster, *Homarus americanus*. Gen. Comp. Endocrinol. 65, 56-64.

Conklin, D. E., and E. S. Chang. 1983. Grow-out techniques for the American lobster *Homarus americanus*. In. CRC Handbook of Mariculture (J. P. McVey, ed.), vol. 1, pp. 277-286. CRC Press, Boca Raton.

Fernlund, P. 1976. Structure of a light-adapting hormone from the shrimp, *Pandalus borealis*. Biochim. Biophys. Acta 439, 17-25.

Fernlund, P., and L. Josefsson. 1972. Crustacean color change hormone: Amino acid sequence and chemical synthesis. Science 177, 173-175.

Freeman, J., and C. Bartell. 1976. Some effects of the molt-inhibiting hormone and 20-hydroxyecdysone upon molting in the grass shrimp, *Palaemonetes pugio*. Gen. Comp. Endocrinol. 28, 131-142.

Hampshire, R., and D. H. S.Horn. 1966. Structure of crustecdysone, a crustacean moulting hormone. Chem. Commun. 1966, 37-38.

Hayes, T. K., L. L. Keeley, and D. W. Knight. 1986. Insect hypertrehalosemic hormone: Isolation and primary structure from *Blaberus discoidalis* cockroaches. Biochem. Biophys. Res. Commun. 140, 674-678.

Hedgecock, D. (1983). Maturation and spawning of the American lobster *Homarus americanus*. In. CRC Handbook of Mariculture (J. P. McVey, ed.), vol. 1, pp. 261-270. CRC Press, Boca Raton.

Hiatt, R. W. 1948. The biology of the lined shore crab, *Pachygrapsus crassipes* Randall. Pacific Sci. 2, 134-213

Horn, D. H. S., E.J. Middleton, J. A. Wunderlich. and F. Hampshire. 1966. Identity of the moulting hormones of insects and crustaceans. Chem. Commun. 1966, 339-340.

Jegla, T. C., C. Ruland, G. Kegel and R. Keller. 1983. The role of the Y-organ and cephalic gland in ecdysteroid production and the control of molting in the crayfish, *Orconectes limosus*. J. Comp. Physiol. 152, 91-95.

Keller, R., and J. D. O'Connor. 1982. Neuroendocrine regulation of ecdysteroid production in the crab *Pachygrapsus crassipes*. Gen. Comp. Endocrinol. 46, 384.

Keller, R., and E. Schmid. 1979. *In vitro* secretion of ecdysteroids by Y-organs and lack of secretion by mandibular organs of the crayfish following molt induction. J. Comp. Physiol. 130, 347-353.

Kleinholz, L. H. 1985. Biochemistry of crustacean hormones. In. The Biology of Crustacea (D. E. Bliss and L. H. Mantel, eds.), vol. 9, pp. 463-522.

Academic Press, New York.
Kleinholz, L. H., and R. Keller. 1979. Endocrine regulation in Crustacea. In. Hormones and Evolution (E. J. W. Barrington, ed.), vol. 1, pp. 159-213. Academic Press, New York.
Kurata, H. 1968. Induction of molting in a prawn *Penaeus japonicus,* by inokosterone injection. Bull. Jpn. Soc. Sci. Fish. 34, 909-914.
Lachaise, F., M. Goudeau, C. Hetru, C. Kappler and J. A. Hoffmann. 1981. Ecdysteroids and ovarian development in the shore crab, *Carcinus maenas.* Hoppe-Seyler's Z. Physiol. Chem. 362, 521-529.
Mattson, M. P. 1986. New insights into neuroendocrine regulation of the crustacean molt cycle. Zool. Sci. 3, 733-744.
Mellon, D., Jr., and E. Greer. 1987. Induction of precocious molting and claw transformation in alpheid shrimps by exogenous 20-hydroxyecdysone. Biol. Bull. 172, 350-356.
Newcomb, R. W. 1983. Peptides in the sinus gland of *Cardisoma carnifex:* Isolation and amino acid analysis. J. Comp. Physiol. 153, 207-221.
Okazaki, R. K., and E. S. Chang. 1990. Ecydysteroids in the embryos and sera of the crabs, *Cancer magister* and *C. anthonyi.* Gen. Comp. Endocrinol. In press.
Passano, L. M. 1953. Neurosecretory control of molting in crabs by the X-organ sinus gland complex. Physiol. Comp. Oecol. 3, 155-189.
Rao, K. R., J. P. Riehm, C. A. Zahnow, L. H. Kleinholz, G. B. Tarr, L. ohnson Norton, S., Landau, M., Semmes, O. J., Sattelberg, R. M., Jorenby, W. H., and M. F. Hintz. 1985. Characterization of a pigment-dispersing hormone in eyestalks of the fiddler crab *Uca pugilator.* Proc. Nat. Acad. Sci. U.S.A. 82, 5319-5322.
Scarborough, R. M., G. C. Jamieson, F. Kalish, S. J. Kramer, G. A. McEnroe, C. A. Miller and D. A. Schooley. 1984. Isolation and primary structure of two peptides with cardioacceleratory and hyperglycemic activity from the corpora cardiaca of *Periplaneta americana.* Proc. Nat. Acad. Sci. U.S.A. 81, 5575-5579.
Skinner, D. M. 1985. Molting and regeneration. In. The Biology of Crustacea (D. E. Bliss and L. H. Mantel, eds.), vol. 9, pp. 43-146. Academic Press, New York.
Spindler, K.-D., R. Keller andJ. D. O'Connor. 1980. The role of ecdysteroids in the crustacean molt cycle. In. Progress In Ecdysone Research (J. A. Hoffmann, ed.), pp. 247-280. Elsevier/North-Holland Biomedical Press, Amsterdam.
Spindler, K.-D., A. Van Wormhoudt, D. Sellos and M. Spindler-Barth. 1987. Ecdysteroid levels during embryogenesis in the shrimp, *Palaemon serratus* (Crustacea Decapoda): Quantitative and qualitative changes. Gen. Comp. Endocrinol. 66, 116-122.
Stone, J. V., W. Mordue, K. E. Batley and H. R. Morris. 1976. Structure of locust adipokinetic hormone, a neurohormone that regulates lipid utilisation during flight. Nature 263, 207-211.
Warren, J. T., S. Sakurai, D. B. Rountree, L. I. Gilbert, S.-S. Lee and K. Nakanishi. 1988. Regulation of the ecdysteroid titer of *Manduca sexta:* Reappraisal of the role of the prothoracic glands. Proc. Nat. Acad. Sci. U.S.A. 85, 958-962.

Webster, S. G. 1983. Effects of exogenous ecdysterone upon moulting, proecdysial development, and limb regeneration in the prawn *Palaemon elegans*. Gen. Comp. Endocrinol. 49, 459-469.
Zeleny, C. 1905. Compensatory regulation. J. Exp. Zool. 2, 1-102.

Structure-Activity Relations of Crustacean Neuropeptides

K. Ranga Rao and John P. Riehm
Department of Biology, The University of West Florida
Pensacola, FL 32514

Abstract

This paper focuses on the chemistry and structure-activity relationships of pigmentary-effector hormones, pigment-dispersing hormones (PDHs) and pigment-concentrating hormones (PCHs). The PDHs occur in multiple forms, each of which elicits chromatophoral pigment dispersion and ommatidial light adaptation. The major forms of PDH in several species of Crustacea and related peptides from two species of insects have been characterized and recognized as an authentic family of neuropeptides. This peptide family shows conservation of chain length (18 residues), common termini (N-terminal Asn, C-terminal Ala-NH_2), and 50% sequence homology. The influence of residue deletion, modification, or replacement on the biological activity of PDHs has been investigated; tentative identification of message sequence and preparation of superpotent analogs have been accomplished. Antisera have been raised against PDHs, immunoreactive soma and fibers have been localized in crustacean and insect nervous tissues, and an ELISA has been developed. In the case of PCHs, the structure of only RPCH is known. RPCH elicits pigment concentration in one or more of the chromatophores and is also able to cause ommatidial dark adaptation. RPCH is structurally related to insect adipokinetic hormones (AKHs). The RPCH/AKH family of peptides can trigger pigment concentration in crustacean chromatophores, but with widely differing potencies related to structure. Evidence has been presented for extra-pigmentary actions of PDH and RPCH; these peptides seem to have inhibitory and stimulatory effect, respectively, on the secretion of methylfarnesoate from crustacean mandibular organs. RPCH is also known to stimulate pyloric rhythms by acting on the stomatogastric gangion. Further analysis of extra-pigmentary functions, development of sensitive bioassays, and characterization of hormones from commercially important species would be fruitful ventures.

Introduction

Crustaceans utilize neuropeptides for humoral regulation of a wide variety of physiological processes (Fingerman, 1987; Laufer and Downer, 1988). This report focuses on the chemistry and structure-activity relationships of neuropeptide hormones involved in the control of pigmentary effectors, epithelial chromatophores and certain eye pigment cells. Intracellular pigment movements, dispersion or concentration (aggregation) of pigment granules, within chromatophores lead to rapidly reversible color changes. Somewhat less conspicuous pigment movements occur in certain ommatidial cells, as part of photomechanical adaptations of the compound eye. Such light/dark-adaptive pigment movements may be displayed exclusively by the retinular cells (photoreceptor cells) or may also be evident in certain extra-retinular pigment cells in the eye (Rao, 1985).

Studies on decapod Crustacea indicate that, whereas pigment movements within retinular cells are triggered mainly by a direct action of light, the extra-retinular eye pigment cells and epithelial chromatophores are under hormonal control (Koller, 1927; Perkins, 1928; Kleinholz, 1934; 1936; Rao, 1985). The pigmentary-effector hormones are present in the eyestalks and in extra-eyestalk nervous tissues. Extracts of any of these tissues may elicit a complex set of responses, light or dark adaptation of the eye as well as dispersion or concentration of chromatophoral pigments, depending on the initial states of the effectors in the assay system. This is attributable to the widespread distribution of mutually antagonistic hormones in various parts of the nervous system. The total number of pigmentary-effector hormones present in any given species remains unknown, but they can be clearly differentiated into two sets. The hormones triggering chromatophoral pigment dispersion and ommatidial light adaptation belong to one set, and these are distinct from the hormones responsible for chromatophoral pigment concentration and ommatidial dark adaptation (Kleinholz and Kimball, 1965; Fingerman and Bartell, 1969; Rao, 1985; Fingerman, 1987; Rao and Riehm, 1988a).

Pigment-Dispersing Hormones (PDHs)

Heterogeneity and Characterization of Crustacean PDHs

Several forms of PDH can be separated from extracts of crustacean eyestalks by means of ion-exchange chromatography (Kleinholz, 1970; 1972; Fernlund, 1971). Bioassays of the multiple hormone forms isolated from the shrimp *Pandalus* revealed that each form of PDH elicits not only chromatophoral dispersion, but also ommatidial light adaptation (Kleinholz, 1970). These findings and subsequent tests (Kleinholz, 1972; 1975; Rao *et al.*, 1983) indicate that PDH and light-adapting hormone (LAH; also known as DRPH, light adapting distal retinal pigment hormone) are not distinct molecules, but that their diverse pigmentary actions are triggered by any one of several forms of a single hormone.

The structural relationships (compositional and sequence homology) among the multiple forms of PDH/LAH present in any given species remain unknown. The amino acid sequences are known for only the major forms of PDH in representative species. These identified major forms are: α-PDH (first described as DRPH) in the shrimp *Pandalus borealis* (Fernlund, 1971; 1976); β-PDH in the crabs *Uca pugilator* (Rao *et al.*, 1985) and *Cancer magister* (Kleinholz *et al.*, 1986); and an analog of β-PDH in the brown shrimp *Penaeus aztecus* (Phillips *et al.*, 1988). Their sequence relationships are shown below.

	1 6 12 18
Uca/Cancer ß-PDH:	N S E L I N S I L G L P K V M N D A amide
Penaeus PDH:	N S E L I N S *L* L G *I* P K V M N D A amide
Pandalus α-PDH:	N S *G M* I N S I L G *I* P *R* V M *T E* A amide

The identified PDHs are octadecapeptides with considerable sequence homology. *Penaeus* PDH differs from β-PDH at only two positions (8 and 11), whereas β-PDH differs from α-PDH at six positions (3, 4, 11, 13, 16, and 17). The latter two peptides share a common hexapeptide core sequence (residues 5-10). Within this core region, a highly conservative substitution (Leu for Ile at position 8) can be seen in *Penaeus* PDH.

Because of the substitution of Glu^3 in place of Gly^3, β-PDH and *Penaeus* PDH are more acidic than α-PDH. Thus, when subjected to cation-exchange chromatography in low ionic strength buffers at pH 4.8-4.9, β-PDH and *Penaeus* PDH elute relatively fast (within one column volume) whereas α-PDH elutes later at about three column volumes. When the cation-exchange chromatography profiles are compared, α-PDH could not be detected in extracts of eyestalks from *Uca*, *Cancer* and *Penaeus*. This is in contrast to the situation in *Pandalus*, in which α-PDH is the identified major form; two faster eluting PDHs are also evident in *Pandalus* (Fernlund, 1971), one of which is related to β-PDH (Rao *et al.*, 1989).

Pigment-Dispersing Factors from Insects

In view of the finding that insect head extracts can trigger melanophore pigment dispersion in *Uca* (Dores and Herman, 1981), we proceeded to purify and characterize the pigment-dispersing factors from representative insect species. The first peptide characterized in this effort was from the lubber grasshopper *Romalea microptera* (Rao *et al.*, 1987). Subsequently, 17 of the 18 residues in a PDF from the cricket *Acheta domesticus* have been identified (Rao and Riehm, 1988b); the unidentified residue was presumed to be Arg or Lys at position 13. More recently, by comparison with HPLC (high-performance liquid chromatography) elution profiles of synthetic peptides, we concluded that Lys is present at position 13 in *Acheta*. The sequence relationships of *Romalea* and *Acheta* PDFs to crustacean PDHs are shown below.

	1 6 12 18
Uca/Cancer ß-PDH:	N S E L I N S I L G L P K V M N D A amide
Penaeus PDH:	N S E L I N S *L* L G *I* P K V M N D A amide
Acheta PDF:	N S E *I* I N S *L* L G L P K V *L* N D A amide
Romalea PDF:	N S E *I* I N S *L* L G L P K *L L* N D A amide
Pandalus α-PDH:	N S *G M* I N S I L G *I* P *R* V M *T E* A amide

The sequence homology between insect PDFs and crustacean PDHs clearly establish them as an authentic family of neuropeptides common to arthropods. This peptide family shows conservation of chain length (18 residues), conservation of termini (amino-terminal Asn and carboxyl-terminal Ala-amide), and at least 50% sequence homology. On the basis of identity of net charge (including the presence of Glu^3) and the degree of sequence homology, the insect PDFs as well as *Penaeus* PDH are more closely related to β-PDH than to α-PDH. In assays for melanophore pigment dispersion in *Uca*, insect PDFs are about 50% as potent as β-PDH (Rao *et al.*, 1987). *Penaeus* PDH and β-PDH are nearly equipotent (Phillips *et al.*, 1988), and the latter peptide is known to be 21-fold more potent than α-PDH (Rao *et al.*, 1985).

Synthetic Octadecapeptide Analogs of -PDH

We prepared several analogs of α-PDH that contain substitutions related to β-PDH (Hintz *et al.*, 1989) or were designed for altered stability (Semmes *et al.*, 1985; Jorenby *et al.*, 1987) for evaluating the contributions of specific residues to the potency of PDHs (Table 1). As noted earlier, α-PDH and β-PDH differ at six positions. β-PDH-related substitutions at four positions (Leu^{11,} Lys^{13}, Asn^{16,} or Asp$^{17)}$ did not markedly alter the potency of α-PDH. Substitutions at positions 3 and 4, Glu^{3} and Leu^{4} , were each effective in increasing the potency of α-PDH (2.4 and 3.3 fold, respectively), but these did not account for the net increase in potency of ß-PDH. These results suggest that the 21-fold increase in potency of β-PDH over α-PDH is an interactive effect of multiple substitutions rather than from the product of the effects resulting from individual substitutions.

Table 1. Relative pigment-dispersing potencies of α-PDH and its Analogs

Peptide	Relative Potency
NSGMINSILGIPRVMTEA amide*	1.0
NSG$INSILGIPRVMTEA amide	3.29
NSGMINSILGIPRV$TEA amide	2.94
NSG$INSILGIPRV$TEA amide	6.22
NSGLINSILGIPRVMTEA amide	2.44
NSGMINSILGLPRVMTEA amide	1.20
NSGMINSILGIPKVMTEA amide	0.72
NSGMINSILGIPRVMNEA amide	0.89
NSGMINSILGIPRVMTDA amide	1.12
NSEMINSILGIPRVMTDA amide	3.25
NSDMINSILGIPRVMTEA amide	0.58
NSNMINSILGIPRVMTEA amide	0.26
NSQMINSILGIPRVMTEA amide	0.11
NSAMINSILGIPRVMTEA amide	0.08
NKIMINSILGIPRVMTEA amide	0.0073
NSIMINSILGIPRVMIEA amide	0.0066
NSELINSILGLPKVMNPA amide**	21.0

*α-PDH; **β-PDH; $ Norleucine; Potencies of the various analogs were calculated relative to α-PDH and were derived from assays on melanophores of *Uca*.

A, Alanine; D, Aspartic acid, E, Glutamic acid; G, Glycine; I, Isoleucine; K, Lysine; L, Leucine; M, Methionine; N, Asparagine, P. Proline; Q, Glutamine; R. Arginine; S. Serine; T, Threonine; V, Valine.

Among the residue substitutions made at position 3 (Glu, Asp, Asn, Gln, Ala, Lys, or Ile), only Glu^3 caused an increase in potency. This is thought to result from a more stable ligand-receptor interaction: i.e., the position 3 side chain carboxylate anion of $[Glu^3]$- α -PDH stabilizes the receptor-bound conformer through a charge-charge interaction (Jorenby *et al.*, 1987).

The increased potency of $[Leu^4]$- α-PDH, like that of $[Nle^4]$- α-PDH, is attributable to partial protection from oxidation (Met^{15}, which is oxidizable is present in these analogs). A greater increase in potency, 6-fold, is displayed by the fully oxidation-resistant analog, $[Nle^{4,15}]$-α -PDH (Semmes *et al.*, 1985). Similarly, in β-PDH which contains a single methionine residue (Met^{15}) the substitution of Nle imparts superpotency (Riehm and Rao, 1987).

<u>Message Sequence of PDH</u>

As reviewed recently (Rao and Riehm, 1988b), we have synthesized and assayed a number of truncated analogs of α-PDH with a hope of identifying the minimal sequence required for PDH action. These studies indicated that the sequences with residues 1-9-NH_2 and 6-18-NH_2 are the shortest analogs showing biological activity among those subjected to C-terminal and N-terminal truncation, respectively. Since the overlapping region of these two peptides contains residues 6 to 9, this part of the sequence seemed to be the possible message sequence for PDH (Riehm *et al.*, 1985).

Recent work by Zahnow (1987) in our laboratory dealt with the synthesis and assay of several peptides covering the core sequence (residues 5-10: -Ile-Asn-Ser-Ile-Leu-Gly-) common to α- and β-PDH. This study showed that peptides 5-9-NH_2, 5-10-NH_2, 6-9-NH_2, and 6-10-NH_2 were each able to elicit pigment dispersion in melanophores. These peptides were, however, ineffective in dispersing leucophore pigments. This indicates that, although residues 6-9 may serve as the minimal sequence required for melanophore activation, some other component(s) of the octadecapeptide sequence may be required to exert the full complement of the multiple actions of the PDHs. The octadecapeptide PDHs are known to trigger dispersion of pigments in melanophores, leucophores, and erythrophores (Kleinholz, 1975; Rao and Riehm, 1988b) in addition to causing ommatidial light adaptation. The critical sequence and structural requirements for activation of different effector cell types by

PDH merit further exploration.

Immunocytochemistry

Antisera have been raised against α-PDH (Schueler *et al.*, 1986) and β-PDH (Dircksen *et al.*, 1987). The application of α -PDH antiserum showed immunoreactive soma in the brain and the various optic ganglia of the crayfish *Orconectes immunis* (Schueler *et al.*, 1986). The β-PDH antiserum has been applied to study the distribution of reactive elements in the eyestalk ganglia of *Orconectes limosus*, *Carcinus maenas* (Mangerich *et al.*, 1987) and *Penaeus aztecus* (Phillips *et al.*, 1987). These studies indicate that the PDH immunoreactivity is not predominantly associated with the X-organ-sinus gland system. Most of the PDH tracts are not connected to the sinus gland, but appear as fiber networks in the neuropil of various optic ganglia; this suggests that, in addition to its established hormonal function, PDH may also serve as a neuromodulator or neurotransmitter. Unlike the distribution of PDH, the immunoreactivities of crustacean molt-inhibiting hormone (MIH) and hyperglycemic hormone (CHH) are found exclusively in the X-organ-sinus gland system (Dircksen *et al.*, 1988).

In eyestalk ganglia of *Orconectes limosus*, *Carcinus maenas* (Mangerich *et al.*, 1987) and *Penaeus aztecus* (Phillips *et al.*, 1987), some of the PDH-positive soma and nerve tracts were also reactive to antiserum against the molluscan neuropeptide FMRFamide; in addition, there were soma and tracts that reacted to only one of the two antisera. Similar studies in the lubber grasshopper *Romalea microptera* (Zahnow *et al.*, 1987) indicated that PDH-positive cells are restricted to the optic lobes, and some of these cells showed co-localization of FMRFamide immunoreactivity; FMRFamide-positive cells are more widely distributed and are found in the brain as well as in the optic lobes of the grasshopper. In the tobacco hawkmoth *Manduca sexta* (Homberg *et al.*, 1987), PDH immunoreactivity and gastrin/cholecystokinin-like immunoreactivity are co-localized in a number of cells distributed widely in the brain, optic lobes, and subesophageal ganglion. The significance of these co-localizations of different immunoreactivities remains unknown, but the dissimilar distribution of PDH in the two insect species studied is expected to stimulate additional comparative studies and analysis of function of PDHs in insects and other arthropods.

Immunoassay

Quackenbush and Fingerman (1985) developed an enzyme-linked immunosorbant assay (ELISA) for a black pigment-dispersing hormone (β-PDH) from *Uca*. The antiserum used in their assay was raised against β-PDH of unknown purity, and the antiserum specificity has not been assessed with synthetic PDHs.

Recently, utilizing the antiserum raised against synthetic β-PDH (Dircksen *et al.*, 1987), an ELISA has been developed in our laboratory and applied for evaluating antibody specificity (Bonomelli *et al.*, 1988). This ELISA involves the usage of rabbit anti- β-PDH immune serum which is identified after combination with an alkaline phosphatase- conjugated goat anti-rabbit IgG and color development with p-nitrophenyl phosphate. In competitive tests the antiserum (at a dilution of 1:100,000) recognizes β-PDH with an IC_{50} of 160 fmol/well, but has little affinity for α-PDH (Figure 1).

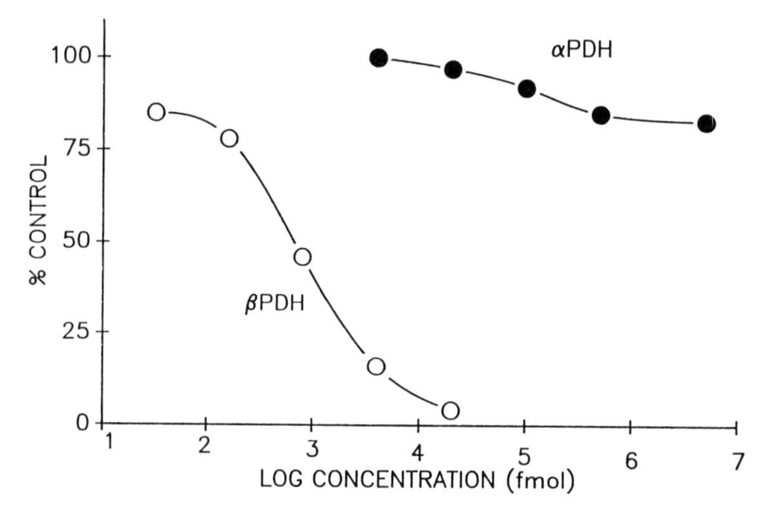

Figure 1. Dose-response antigen inhibition curves derived from ELISA (enzyme-linked immunosorbant assay). Antiserum dilution: 1:100,000. Peptide concentration given is fbmol/ml (it should be noted that each well received 200ul of this solution). When no β-PDH is present in the assay, the obtained value is defined as 100%

Tests with substituted analogs of α-PDH and truncated analogs of β-PDH indicate that the antiserum requires several of the residues closer to the C-terminus in β-PDH for antigen recognition. Consequently, in this ELISA it is possible to detect and quantify β-PDH and several of the closely-related octadecapeptides such as *Penaeus* PDH and the insect PDFs.

Pigment-Concentrating Hormones

Heterogeneity and Characterization

The number of pigment-concentrating hormones present in any given species remains unknown. In species such as *Crangon crangon* (Skorkowski, 1971), *Rhithropanopeus harrisi* (Skorkowski, 1972), and *Crangon septenspinosa* (Fingerman, 1973), it is possible to differentiate two distinct zones of pigment-concentrating biological activity by subjecting eyestalk extracts to gel filtration chromatography (Sephadex G-25). Knowledge of the structure of these hormones, however, is limited to the red pigment concentrating hormone (RPCH) isolated from the shrimp *Pandalus borealis*. This RPCH is an octapeptide with the sequence: pGlu-Leu-Asn-Phe-Ser-Pro-Gly-Trp-NH_2(Fernlund and Josefsson, 1972).
On the basis of compositional analysis, RPCH isolated from eyestalks of *Palaemon squilla* (Carlsen *et al.*, 1976) and RPCH from the sinus glands of *Cardisoma carnifex* (Newcomb, 1983; Newcomb *et al.*, 1985) are thought to be identical to *Pandalus* RPCH. Much remains to be known of the chemistry and diversity of pigment-concentrating hormones in Crustacea.

Pigmentary Actions of RPCH

As reviewed recently (Rao, 1985; Rao and Riehm, 1988a), synthetic RPCH causes pigment concentration in one or more of the chromatophore types depending on the decapod species tested. Thus RPCH seems to act on erythrophores alone (*e.g.*, *Uca pugilator*, *Cambarellus shufeldtii*); on leucophores as well as erythrophores (*e.g.*, *Palaemon squilla*, *Penaeus aztecus*); or on melanophores, leucophores, and erythrophores (*e.g.*, *Crangon crangon*). RPCH failed to act on the chromatophores of the stomatopod *Squilla empusa* and the isopod *Ligia occidentalis*.

Synthetic RPCH has been shown to cause dark-adaptational movement of the distal retinal screening pigment in the eye of the fiddler crab, *Uca pugilator*

(Kulkarni and Fingerman, 1986). This indicates that: (a) RPCH and dark-adapting hormone (DAH) are not distinct, their actions being exerted by a single molecule; (b) RPCH and DAH are structurally related (not necessarily identical) molecules. Further resolution depends on the characterization of this set of hormones and more extensive testing of synthetic hormone preparations.

Structure-Activity Relationships

Relatively little is known of the structure-activity relationships of RPCH. Earlier work (Christensen *et al.*, 1978; 1979) indicated that any modifications to the carboxyl-terminus results in a marked loss of activity, Trp^8 is essential for the activity, and Trp^8 must be amidated to provide maximum potency for RPCH action on the erythrophores of *Palaemon squilla*. The C-terminal Trp-NH_2 requirement for maximum potency was not evident in tests on melanophores of *Crangon*; in this case, locust adipokinetic hormone (AKH; a C-terminally extended RPCH analog) with the sequence pGlu-Leu-Asn-Phe-Thr- Pro-Asn-Trp-Gly-Thr-NH_2 proved to be as effective as RPCH in causing melanophore pigment concentration (Mordue and Stone, 1977).

In recent years, crustacean RPCH and insect AKHs have been recognized as a family of peptides common to arthropods; the number of identified peptides, exclusively from insects, has been growing steadily. Members of this peptide family share several features: N-terminal pGlu; Phe at position 4; Trp at position 8; and amidated C-terminus. The chain length varies from 8 to 10 residues and there are a variety of intramolecular substitutions. In insect assay systems, these peptides elicit hyperlipemia, hyperglycemia, hypertrehalosemia, or cardioacceleration, or a combination of some of these responses. The relative potency of the RPCH/AKH peptides in insect assay systems has been examined (Goldsworthy and Wheeler, 1986), and more recently our laboratory has evaluated their pigment-concentrating effects on erythrophores of the dwarf crayfish, *Cambarellus shufeldtii* (Mohrherr *et al.*, 1987).

As shown in Table 2, RPCH proved to be the most potent peptide in inducing pigment concentration. The most pronounced reduction in potency seems to result from substitution of Val^2 in place of Leu^2. This substitution is present in three of the least active peptides. The Val^2 substitution alone accounts for 1300-fold reduction in

Table 2. Relative Pigment-Concentrating Potencies of Peptides of the RPCH/AKH Family

Peptide	Structure 1 4 8 10	Relative Potency
RPCH	pQ L N F S P G W amide	1.0
AKH-2S	pQ L N F S T G W amide	0.81
CC-2/M-2	pQ L T F T P N W amide	0.17
CHTF-2	pQ L T F T P N W G T amide	0.14
AKH-2L	pQ L N F S A G W amide	0.08
AKH	pQ L N F T P N W G T amide	0.05
BN-HTF	pQ V N F S P G W G T amide	0.002
MH-AKH	pQ L T F T S S W G amide	0.0009
GA-AKH	pQ V N F S T G W amide	0.0006
CC-1/M-1	pQ V N F S P N W amide	0.0005

Potencies for various peptides were calculated relative to RPCH and were derived from assays on erythrophores of the dwarf crayfish, *Cambarellus shufeldtii*. pQ, Pyroglumatic Acid; F, Phenylalanine; W, Tryptophan; for others, see legend to Table 1. RPCH, *Pandalus* red pigment-concentrating hormone; AKH-2S, *Schistocerca* adipokinetic hormone-2; CC-2/M-2, *Periplaneta* cardioactive/myotropic peptide-2: CHTF-2, *Carausius* hypertrehalosemic factor-2: AKH-2L, *Locusta* adipokinetic hormone-2; AKH, locust adipokinetic hormone; BN-HTF, *Blaberus/Nauphoeta* hypertrehalosemic factor; MH-AKH, *Manduca/Heliothis* adipokinetic hormone; GA-AKH (first identified as *Gryllus* AKH, Gade and Rinehart, 1987; recent work in our laboratory revealed that *Acheta* AKH has an identical sequence); CC-1/M-1, *Periplaneta* cardioactive/myotropic peptide-1. GA-AKH was synthesized in our laboratory; BN-HTF was a generous gift from Dr. Tim Hayes (Texas A & M University); all other peptides were purchased from Peninsula Laboratories.

potency (as revealed by comparing *Schistocerca* AKH-2 and *Gryllus/Acheta* AKH). The marked reduction of potency displayed by *Manduca/Heliothis* AKH may result from the presence of Ser^6 Ser^7, unique for this peptide. Substitution of Thr^3 for Asn^3, Thr^5 for Ser^5 , and Ala^6 for Pro^6 seem to partially account for some of the reduction in potency noted in other peptides. Comparison of *Periplane*ta CC-2 with *Carausius* HTF-2 indicates that C - terminal extension causes about 20%

reduction in potency.

Whereas crayfish erythrophores reveal a broad range of potency differences (up to 2000-fold) among the RPCH/AKH peptides, the insect assay systems (*e.g.*, adipokinetic responses in locusts) show only a 12-fold difference in potency (Goldsworthy and Wheeler, 1986). Furthermore, in insect assay systems, *Periplaneta* CC-1 and CC-2 are equipotent (Scarborough *et al.*, 1984); in contrast, CC-2 is 340-fold more potent than CC-1 in crayfish erythrophore assays (Table 2). This shows that the structural requirements for ligand-receptor interaction in crayfish erythrophores are more rigid than those for corresponding receptors in insect assay systems.

Extra-Pigmentary Effects of RPCH and PDH in Crustacea

Stomatogastric System

Nusbaum and Marder (1988) have shown that synthetic RPCH increases cycle frequency in isolated stomatogastric ganglion (STG) displaying slow pyloric rhythms, and is able to initiate rhythmic activity in silent STG of *Cancer borealis*. These data, along with evidence for the presence of RPCH or a related peptide in input fibers to STG, indicate that RPCH serves transmitter/modulator functions in the central nervous system, in addition to the well known hormonal functions. Nusbaum and Marder (1988) suggest that blood-borne RPCH may aid in the coordination of the animal's feeding behavior by directly stimulating rhythmic activity in STG or by acting as a primer for neurally released RPCH.

As noted in the assays on crayfish erythrophores (previous section), RPCH proved to be more effective than AKHs in stimulating the STG of *Cancer* (Nusbaum and Marder, 1988). Furthermore, *Periplaneta* CC-1 seemed to be one of the least potent of the AKH peptides tested on STG, as was also the case in tests with crayfish erythrophores (Mohrherr *et al.*, 1987). This provides preliminary evidence for some similarity in structural requirments for peptide recognition by RPCH receptors in the erythrophores and stomatogastric ganglion.

Mandibular Organs

Laufer *et al.* (1987) showed that synthetic RPCH and β-PDH have stimulatory and inhibitory effects, respectively, on the secretion of methylfarnesoate by crustacean mandibular organs. Since methylfarnesoate is implicated in the regulation of reproduction in Crustacea (as discussed by Laufer in this volume), the functions of PDHs and RPCH in reproductive physiology merit detailed exploration.

Perspectives

Although a variety of physiological processes are known to be under hormonal control in Crustacea, relatively few of the hormones have been identified and chemically synthesized. Substantial progress has been made in elucidating the chemistry and structure-activity relations of pigmentary-effector hormones. Several of these hormones have been isolated from shrimp such as *Pandalus borealis* and *Penaeus aztecus*, and the hormones are available in synthetic form for detailed analysis of function.

In view of recent findings of RPCH actions on stomatogastric system (with implied role in coordination of feeding behavior) and PDH/RPCH actions on mandibular organs (with implied role in reproductive physiology), a more detailed analysis of these functions as well as search for other physiological roles for pigmentary effector hormones would be fertile areas of future research. This depends on successful development of sensitive and appropriate *in vitro* and *in vivo* bioassays. When such assays are available it would be possible to isolate, differentiate, and characterize hormones with specific or broad-range functions.

It should be noted that the chemistry of pigmentary-effector hormones is derived from studies with relatively few species. The hormonal systems of most of the commercial species remain unexplored. Since hormones are key regulators, management of a given species can be more effectively accomplished when the functional biology (including hormonal regulation) is well known. Considering the protein/peptide nature of neurosecretory hormones and the contemporary advances in gene manipulation, the prospects for biotechnology applications of hormonal manipulation in crustacean culture are not remote.

Acknowledgements

This work was supported by National Science Foundation Grant DCB-8711403.

The original research reviewed here was conducted in collaboration with several of our students and associates: Carl Mohrherr, Cynthia Zahnow, Jessica Phillips, Sherman Bonomelli, Mary Hintz, and William Morgan. We are grateful to Alan Tomlinson and Philip Conklin for assistance with instrumentation, and to Paula Fitzgibbons for assisting in manuscript preparation.

Literature Cited

Bonomelli, S. L., K. R. Rao, and J. P. Riehm. 1988. Development and application of an ELISA for crustacean B-PDH. American Zoologist 28: 117A

Carlsen, J., M. Christensen, and L. Josefsson. 1976. Purification and chemical structure of the red pigment-concentrating hormone of the prawn *Leander adspersus*. General and Comparative Endocrinology 30:327-331.

Christensen, M., J. Carlsen, and L. Josefsson. 1978. Structure-function studies on red pigment-concentrating hormone: The significance of the terminal residues. Hoppe-Seyler's Zeitschrift fur Physiologische Chemie 359:813-818.

Christensen, M., J. Carlsen, and L. Josefsson. 1979. Structure-function studies on red pigment-concentrating hormone, II. The significance of the C-terminal tryptophan amide. Hoppe-Seyler's Zeitschrift fur Physiologische Chemie. 360: 1051-1060.

Dircksen, H., S. G. Webster, and R. Keller. 1988. Immunocytochemical demonstration of the neurosecretory systems containing putative moult-inhibiting hormone and hyperglycemic hormone in the eyestalk of brachyuran crustaceans. Cell and Tissue Research 251: 3-12.

Dircksen, H., C. A. Zahnow, G. Gaus, R. Keller, K. R. Rao, and J. P. Riehm. 1987. The ultrastructure of nerve endings containing pigment-dispersing hormone (PDH) in crustacean sinus glands: Identification by an antiserum against synthetic PDH. Cell and Tissue Research 250: 377-387.

Dores, R. M., and W. S. Herman. 1981. Insect chromatophorotropic factors: the isolation of polypeptides from *Periplaneta americana* and *Apis mellifera* with melanophore dispersing activity in the crustacean, *Uca pugilator*. General and Comparative Endocrinology 43: 76-84.

Fernlund, P. 1971. Chromactivating hormones of *Pandalus borealis*: Isolation and purification of a light-adapting hormone. Biochimica Biophysica Acta 237: 519-529.

Fernlund, P. 1976. Structure of a light-adapting hormone from the shrimp, *Pandalus borealis*. Biochimica Biophysica Acta 439: 17-25.

Fernlund, P., and L. Josefsson. 1972. Crustacean color change hormone: amino acid sequence and chemical synthesis. Science 177: 173-175.

Fingerman, M., 1973. Comparison of the effects of partially purified eyestalk extracts of the shrimp *Crangon septenspinosa* on its black, red, and white chromatophores. Physioloigical Zoology 46: 173-179.

Fingerman, M. 1987. The endocrine mechanisms of crustaceans. Journal of

Crustacean Biology 7: 1-24.
Fingerman, M., and C. K. Bartell. 1969. Gel filtration of chromatophorotropins from eyestalk extracts of the fiddler crab, *Uca pugliator,* and the prawn, *Palaimonetes vulgaris*. Biological Bulletin (Woods Hole, Mass.) 137: 399.
Gade, G., and K. L. Rinehart, Jr. 1987. Primary sequence analysis by fast atom bombardment mass spectrometry of a peptide with adipokinetic activity from the corpora cardiaca of the cricket *Gryllus bimaculatus*. Biochemical Biophysical Research Communications 149: 908-914.
Goldsworthy, G. J., and C. H. Wheeler. 1986. Structure-activity relationships in the adipokinetic hormone/red pigment concentrating hormone family. In. A.B. Borkovec and D.B. Gelman (ed.) Insect Neurochemistry and Neurophysiology. pp. 183-186. Humana Press, Clifton.
Hintz, M. F., J. P. Riehm, and K.R. Rao. 1989. Synthesis and assay of structural intermediates of crustacean pigment-dispersing hormones (α- and β-PDH). International Journal of Invertebrate Reproduction and Development, 16: 135-139..
Homberg, U., T. G. Kingan, and J. G. Hildebrand. 1987. Gastrin/CCK-like peptides in the brain of the tobacco hawkmoth *Manduca sexta*. Society of Neuroscience Abstracts 13: 235.
Jorenby, W. H., J. P. Riehm, and K.R. Rao. 1987. Position 3 analogues of a crustacean pigment-dispersing hormone: Synthesis and assay. Biochemical Biophysical Research Communications 143: 652-657.
Kleinholz, L.H. 1934. Eye-stalk hormone and the movement of the distal retinal pigment in *Palaemonetes*. Proceedings of the National Academy of Sciences U.S.A. 20:659-661.
Kleinholz, L. H. 1936. Crustacean eye-stalk hormone and retinal pigment migration. Biological Bulletin (Woods Hole) 70: 159-184.
Kleinholz, L. H. 1970. A progress report on the separation and purification of crustacean neurosecretory pigmentary-effector hormones. General and Comparative Endocrinology 14: 578-588.
Kleinholz, L. H. 1972. Comparative studies of crustacean melanophore-stimulating hormones. General and Comparative Endocrinology 19: 473-483.
Kleinholz, L. H. 1975. Purified hormones from the crustacean eyestalk and their physiological specificity. Nature 258: 256-257.
Kleinholz, L.H., and F. Kimball. 1965. Separation of neurosecretory pigmentary-effectors hormones of the crustacean eyestalk. General and Comparative Endocrinology 5: 336-341.
Kleinholz, L. H., K. R. Rao, J.P. Riehm, G. E. Tarr, L. Johnson, and S. Norton. 1986. Isolation and sequence analysis of pigment-dispersing hormone from eyestalks of the crab *Cancer magister*. Biological Bulletin (Woods Hole) 170: 135-143.
Koller, G. 1927. Uber Chromatophoren-system, Farbensinn und Farbwechsel bei *Crangon vulgaris*. Zeitschrift fur Vergleichende Physiologie 5: 191-246.
Kulkarni, G. D., and M. Fingerman. 1986. Distal retinal pigment of the fiddler crab, *Uca pugliator:* Evidence for stimulation of release of light adapting and dark adapting hormones by neurotransmitters. Comparative Biochemistry and Physiology 84C:219-224.
Laufer, H., and R. G. H. Downer (eds.) 1988. Invertebrate Endocrinology Vol. 2,

Endocrinology of selected invertebrate types. Alan R. Liss, New York. pp 1-500.

Laufer, H., E. Homola, and M. Landau. 1987. Control of methyl farnesoate in crustacean mandibular organs. American Zoologist 27:69A.

Mangerich, S., R. Keller, H. Dircksen, K. R. Rao, and J. P. Riehm. 1987. Localization of pigment-dispersing hormone (PDH) and coexistence with FMRFamide immunoreactivity in the eyestalks of two decapod crustaceans. Cell and Tissue Research 250: 265-375.

Mohrherr, C. J., K. R. Rao, and J. P. Riehm. 1987. Erythrophorotropic activity of the AKH/RPCH peptide family in the dwarf crayfish *Cambarellus shufeldti.* American Zoologist 27: 70A.

Mordue, W., and J. V.Stone. 1977. Relative potencies of locust adipokinetic hormone and prawn red pigment concentrating hormone in insect and crustacean systems. General and Comparative Endocrinology 33: 103-108.

Newcomb, R. W. 1983. Peptides in the sinus gland of *Cardisoma carnifex.* Isolation and amino acid analysis. Journal of Comparative Physiology 153: 207-221.

Newcomb, R. W., E. Stunkel, and I. M. Cooke. 1985. Characterization, biosynthesis, and release of neuropeptides from the X-organ-sinus gland system of the crab *Cardisoma carnifex.* American Zoologist 25: 157-171.

Nusbaum, M. P., and E. Marder. 1988. A neuronal role for a crustacean red pigment concentrating hormone-like peptide: neuromodulation of the pyloric rhythm in the crab, *Cancer borealis.* Journal of Experimental Biology 135: 165-181.

Perkins, E .B. 1928. Color changes in crustaceans, especially in *Palaemonetes.* Journal of Experimental Zoology 50: 71-105.

Phillips, J. M., C. A. Zahnow, and K. R. Rao. 1987. An immunocytochemical study of the eyestalk of *Penaeus aztecus* utilizing antisera for synthetic B-PDH and FMRFamide. American Zoologist 27: 69A.

Phillips, J. M., K. R. Rao, J. P. Riehm, and W. T. Morgan. Isolation and characterization of a pigment dispersing hormone from the shrimp, *Penaeus aztecus.* Society of Neuroscience Abstracts 14: 534.

Quackenbush, L. S., and M. Fingerman. 1985. Enzyme linked immunosorbant assay of black pigment dispersing hormone from the fiddler crab, *Uca pugilator.* General and Comparative Endocrinology 57: 438-444.

Rao, K. R. 1985. Pigmentary effectors. In. D.E. Bliss and L.H. Mantel (eds.) The biology of Crustacea. Vol. 9. Integument, pigments, and hormonal processes. pp. 395-462. Academic Press, Orlando.

Rao, K. R., D. G. Doughtie, P. A. Robinson, and J .P. Riehm. 1983. Effects of synthetic *Pandalus* DRPH on the ommatidial pigmentary effectors in *Palaemonetes pugio.* American Zoologist 23: 951.

Rao, K. R., L. H. Kleinholz, and J. P. Riehm. 1989. Characterization of three forms of pigment-dispersing hormone from the shrimp *Pandalus jordani.* Society of Neuroscience Abstracts 15: 26.

Rao, K .R., C. J. Mohrherr, J. P. Riehm, C. A. Zahnow, S. Norton, L. Johnson, and G.E. Tarr. 1987. Primary structure of an analog of crustacean pigment-dispersing hormone from the lubber grasshopper, *Romalea microptera.*

Journal of Biological Chemistry 262: 2672-2675.
Rao, K. R., and J. P. Riehm. 1988a. Chemistry of crustacean chromatophorotropins. In. J.T. Bagnara, (ed.), Advances in pigment cell research. pp. 407-422. Alan R. Liss, New York.
Rao, K. R., and J.P. Riehm. 1988b. Pigment-dispersing hormones: a novel family of neuropeptides from arthropods. Peptides 9, Supplement 1:153-159.
Rao, K. R., J. P. Riehm, C. A. Zahnow, L. H. Kleinholz, G. E. Tarr, L. Johnson, S. Norton, M. Landau, O.J. Semmes, R.M. Sattelberg, W.H. Jorenby, and M.F. Hintz. 1985. Characterization of a pigment-dispersing hormone in eyestalks of the fiddler crab, *Uca pugilator*. Proceedings of the National Academy of Sciences U.S.A. 82: 5319-5322.
Riehm, J. P., and K. R. Rao. 1987. Synthesis and assay of tyrosinated analogs of a crustacean pigment-dispersing neuropeptide hormone. International Journal of Peptide and Protein Research 29: 415-420.
Riehm, J .P., K. R. Rao, O. J.. Semmes, W. H. Jorenby, M. F. Hintz, and C. A. Zahnow. 1985. C-terminal deletion analogs of a crustacean pigment-dispersing hormone. Peptides 6: 1051-1056.
Scarborough, R .M., G. C. Jamieson, F. Kallish, S. J. Kramer, G. A. McEnroe, C. A. Miller, and D. A. Schooley. 1984. Isolation and primary structure of two peptides with cardioacceleratory and hyperglycemic activity from the corpora cardiaca of *Periplaneta americana*. Proceedings of the National Academy of Sciences U.S.A. 81:5575-5579.
Schueler, P. A., A. J. Madsen, W. S. Herman, and R. Elde. 1986. Immunohistochemical mapping of distal retinal pigment hormone in the crayfish central nervous system. Society of Neuroscience Abstracts 12: 242.
Semmes, O. J., J. P. Riehm, and K. R. Rao. 1985. Substitution of norleucine for methionine residues in a crustacean pigment dispersing hormone. Peptides 6: 491-494.
Skorkowski, E. F. 1971. Isolation of three chromatophorotropic hormones from the eyestalk of the shrimp *Crangon crangon*. Marine Biology 8: 220-223.
Skorkowski, E .F. 1972. Separation of three chromatophorotropic hormones from the eyestalk of the crab *Rhithropanopeus harrisi* (Gould). General and Comparative Endocrinology 18: 329-334.
Zahnow, C. A. 1987. Synthesis and bioassays of N-terminal deletion peptides and certain "core" analogs of a crustacean pigment-dispersing hormone. Masters Thesis, The University of West Florida, Pensacola.
Zahnow, C. A., K. R. Rao, C. J. Mohrherr, and J. P. Riehm. 1987. Immunocytochemistry of neuropeptides in the cephalic neuroendocrine system of the lubber grasshopper, *Romalea microptera*. Society of Neuroscience Abstracts 13: 993.

Regulation Of Vitellogenesis In Penaeid Shrimp

L.S. Quackenbush
Department of Biological Sciences
Florida International University
Miami, Florida, 33199

Introduction

A general characteristic of crustacean reproduction is the production of yolk laden eggs. The yolk is a combination of proteins, lipids, sugars, and some steroid hormones (Adiyodi, 1985; Couch and Hagino, 1983; Wallace *et al.*,1967). The purpose of the yolk is primarily nutrition for the developing embryo. Yolk may serve as the sole source of nutrition until the functional mouthparts are developed (Anderson *et al.*,1949; Cook and Murphy, 1969). Crustacean eggs are dominated by a single group of proteins called vitellins, which may constitute 60% to 90% of the total protein within an egg prior to oviposition (Eastman-Reks and Fingerman, 1985). As development progresses, the embryo consumes the yolk, incorporating the protein into tissues. The diversity of crustacean life histories is demonstrated in the various patterns of reproduction. For example, Palinurid lobsters may produce 500,000 or more small (350 um) eggs per brood annually, whereas talitrid amphipods usually produce only 20 large (500-700 um) eggs per brood, but they may have multiple broods per year (Aiken and Waddy, 1980 ; Charniaux-Cotton, 1978). The coordination of both egg release and larval release to variations in the environment further constrains crustacean reproduction (Bergin, 1981 ; Christy 1982). Controlled larval release may optimize larval survival by decreasing relative predation, and by taking advantage of beneficial currents for larval dispersal (Christy, 1982).

Studies of crustacean egg yolk protein synthesis or vitellogenesis have focused on representatives from only a few of the many families of crustaceans (Charniaux-Cotton, 1985). The fiddler crab, *Uca pugilator,* has a pattern of reproduction that is common to many small brachyuran crabs. Females (13-22 mm carapace width) produce a large brood (10,000) of small eggs (250 um). During the summer reproductive season, crabs are capable of multiple broods. Fertilized eggs are maintained on the pleopods for about 14 days, while further development proceeds (Christy and Salmon, 1984). Larval release is timed to the occurrence of locally high tides (Bergin, 1981; Christy and Salmon, 1984). The investment of yolk by the female is a considerable drain on stored resources. An ovary that has completed vitellogenesis may have a wet weight that is 6% to 10% of the entire body weight of the crab (Webb, 1977). After the larvae are released the female may molt to a larger size. Like many decapods, the fiddler crab with an increased size may also increase the number of eggs it can maintain on its pleopods. Even though clutch size does vary between species, not all fiddler crabs maintain the same number of eggs in their broods (Christy and Salmon, 1984). Larger decapods may increase the total number of eggs brooded per spawn, while decreasing the number of annual spawns. For example, *Homarus americanus*, the american lobster, spawns on alternate years a brood of 20,000 to 50,000 large (1400-1600 um) eggs (Aiken and Waddy, 1980).

A completely different pattern of crustacean reproduction is demonstrated by the talitrid amphipod, *Orchestia gammarella* (Charniaux-Cotton, 1978) (Table 1). Female amphipods are reproductively active for 9 months of their annual cycle.

Table 1. Reproduction in Crustaceans

	Crab *Uca pugilator*	Amphipod *Orchestia gammarella*	Shrimp *Penaeus vannamei*	Lobster *Homarus americanus*
Size of female	22mm/20g	14mm/5g	18cm/40g	90cm/700g
Number of Eggs/spawn	10,000	20	100,000	50,000
Egg size				
1) Primary oocyte	50 um	100 um	50 um	100 um
2) Secondary oocyte	250 um	700 um	250 um	1,600 um
Post Fertilization				
Maternal Care	14 days	3 - 6 days	0 days	3 - 6 months
Spawns per year	2 - 3	7 - 9	5 - 6	1/2

These small (14 mm total length) animals produce about 20 large (500-700 um) eggs per spawn. In contrast to the decapods, yolk protein production in amphipods occurs during the premolt phase of their molt cycle. The eggs are spawned, externally fertilized, and then maintained for several days on specialized modifications of the pereopods called oostegites. The oostegites are secondary sexual characters which form a brooding pouch called a marsupium. After larval release the female finishes the molt cycle that was stopped in premolt. Females can produce and release a brood of eggs every 25 days during the reproductive period. These crustaceans have developed a reproductive system that is characterized by repetitive small broods of large eggs during a long reproductive period (Charniaux-Cotton, 1978).

Penaeid shrimp have a reproductive pattern that is uncommon among the crustaceans. These animals are called broadcast spawners because they release fertilized eggs into the ocean with no parental maintenance. Egg release generally occurs during the night in this group, in an effort to avoid predators on the helpless eggs. The shrimp, *Penaeus vannamei,* presents a typical penaeid pattern of broadcast spawnners. Mature females (18-22 cm total length, 30-40 grams total weight) will release from 20,000 to 100,000 small (250 um) eggs each spawn (Table 2; Aquacop, 1983).

Table 2. Penaeid Fecundity

Species	Female Size (grams)	Number Eggs/Spawn (thousands)	Size Eggs (um)	Spawns per Month	Mature Time (Day)
P. indicus	25	25-80	300	2	3-10
P. japonicus	50-80	200--700	280-300	1	5-10
P. monodon	60	175	280-300	1	7-10
P merguiensis	6	20-300	220-300	1	6
P. stylorostris	40	100-300	200-250	3	6-10
P. vannamei	35	60-200	280-300	1	7-10

Data obtained from Chamberlain *et al.*, 1985; McVey and Fox, 1983; Primavera, 1985; Yang, 1975.

Multiple spawns during a single molt cycle have been observed. The duration between spawns varies according to conditions from 5 to 30 days (Primavera, 1985). As the female grows after molting, they may produce an incrementally larger ovary capable of producing more eggs. A mature ovary may weigh from 3% to 10% of the total body weight of the animal. During the first 72 hours to 100 hours of larval life, the larvae are completely dependent on the accumulated yolk for nutrition. Functional mouth parts do not develop until the fourth naupliar stage (Cook and Murphy, 1969). This pattern of reproduction and larval life is considerably different from most other crustacean groups. The most significant difference is the release of the fertilized egg with no parental care. This pattern was of course a major factor in the success of the mass culture of these animals.

Egg Yolk Protein Production

Vitellogenin (Vg) is the precursor to the egg yolk protein, vitellin (Vn). Vg can be produced in crustacean ovaries, but it may also be produced in extra-ovarian sites (Adiyodi, 1985 ; Charniaux-Cotton, 1985; Lui and O'Connor, 1977; Junera and Meusy, 1982). Depending on the species examined, Vg may be exclusively produced in the ovary (*i.e.*, crayfish) or by both ovarian and extra-ovarian tissues (*i.e.*, crabs and lobsters) or exclusively by extra-ovarian tissues (*i.e.*, isopods and amphipods) (Adiyodi, 1985). In the insects, Vg is exclusively produced in the adipocytes of the fat body (Downer and Laufer, 1983). In the decapod crustaceans, several sites for yolk protein production have been proposed. These sites include the hemocytes of the hemolymph, the hepatopancreas, and subepidermal adipose tissues (Adiyodi and Subramoniam, 1983; Charniaux-Cotton, 1978; Kerr, 1969; Paulus and Laufer, 1987; Tom *et al.,* 1987). Histological evidence suggests that decapod eggs can take up extra-ovarian proteins via micropinocytosis. This is the method proposed for extra-ovarian yolk incorporation in the developing eggs (Dhainaut and Leersnyder, 1976 ; Talbot, 1981 a and b; Schade and Schivers, 1980 ; Wolin *et al.*, 1973). Vg has been found in the hemolymph of several crustaceans during the reproductive periods using many different methods (Adiyodi, 1985; Fyffe and O'Connor, 1974 ;

Kerr, 1969). These direct observations support the extra-ovarian Vg synthesis model for some authors (Adiyodi, 1985). However, others suggest that the Vg found in the hemolymph is a result of oocyte lysis, which refutes the role of extra-ovarian Vg synthesis (Lui and O'Connor, 1976).

The decapod hepatopancreas was proposed as a logical site for extra-ovarian Vg synthesis. However, there has been no published demonstration that the hepatopancreas of any decapod actually makes Vg in any quantity (Charniaux-Cotton, 1985; Quackenbush and Keeley, 1988). The gross morphology of the hepatopancreas varies among the decapods, but its role in digestion and absorption of food appears to be a common function to all decapods (Caceci *et al.*, 1988; Johnson, 1980 ; McLaughlin, 1983). The histology of the cells of the hepatopancreas demonstrates that this tissue is capable of both synthesizing proteins and storing fats, a role that is similar to the insect fat body (Al-Mohanna and Nott, 1987 ; Caceci *et al.*, 1988, Gibson and Barker, 1979 ; Miyawaki *et al.*, 1985). The hepatopancreas can be maintained in vitro and it is capable of synthesizing proteins (Gorrell and Gilbert, 1971 ; Malcoste *et al.,* 1983). The hepatopancreas makes the respiratory pigment, hemocyanin in the american lobster, *Homarus americanus* (Senkbeil and Wriston, 1981). The hepatopancreas of both the crayfish and shrimp also makes several digestive enzymes (Durliat and Vranckx, 1982 a and b ; Malcoste *et al.*, 1983). Dehn *et al.* (1983) tested the hepatopancreas of *Homarus americanus in vitro* and found it did not make vitellogenin. Immunocytochemical demonstrations of yolk proteins in cells and tissues of several crustaceans support the possibility of yolk protein synthesis in the hepatopancreas (Paulus and Laufer, 1987 ; Tom *et al.*, 1987).

In insects and some isopods the subepidermal adipose tissue can produce egg yolk proteins (Downer and Laufer, 1983 ; Souty and Picaud, 1984). Insect Vg is exclusively produced in these tissues. In the amphipod, *Orchestia gammarella*, the subepidermal adipose tissue produces several egg yolk proteins (Charniaux-Cotton, 1985). Tom *et al.* (1987) demonstrated the presence of an egg yolk protein in the adipose tissue of a penaeid shrimp . Crustaceans are not restricted to a single tissue in the production of egg yolk proteins. The relative roles and contributions of each tissue to the overall egg yolk investment has not yet been demonstrated.

Hemocytes of the blue crab, *Callinectes sapidus*, can produce Vg. However, the rate and amount of Vg produced is not enough to account for complete vitellogenesis (Kerr, 1969). The rate of Vg synthesis of the crayfish ovary was also judged too slow and too low to account for complete vitellogenesis (Lui and O'Connor, 1976). During a reproductive period the mature crustacean ovary may weigh as much as 10% of the total body weight (Quackenbush and Herrnkind, 1981;1983). During the secondary vitellogenesis the decapod ovary usually increases in weight by 300% or more depending on the species (Adiyodi, 1985 ; Quackenbush, 1986). This massive protein production may require contributions from ovary, hepatopancreas, hemocytes, and subepidermal adipose tissue. Coordination of all these tissues, and timing to environmental variations is essential to optimize larval success.

Patterns of Egg Yolk Protein Production

The overall goal of maturation in crustaceans is the investment of the oocyte with egg yolk proteins. Additionally cortical specializations are also developed to help seal off the developing embryo from the external environment. Egg maturation has traditionally been divided into two phases : primary and secondary vitellogenesis. Details of these two basic stages vary among the diverse groups of crustaceans. In general, primary vitellogenesis is characterized by little growth in egg diameter, but rather development of cytological features that prepare the cell for protein synthesis. Secondary vitellogenesis is characterized by a massive increase in oocyte size and weight, and the development of cortical granules or a chorion.

Primary oocytes are derived from oogonia in the germanative zone of the ovary. In crustaceans, the production of primary oocytes continues throughout adult life, unlike many vertebrates. During primary vitellogenesis, ribosomes are developed as is an extensive rough endoplasmic reticulum. An immunologically distinct primary yolk, a glycoprotein, develops in the eggs of *Orchestia gammarella* during primary vitellogenesis. Follicle cells surround the primary oocytes during this initial stage. In the penaeid shrimp, *Sicyonia ingentis* the oocyte develops reticular elements during a cisternal phase of primary vitellogenesis (Anderson, *et al.*, 1984; Duronsolet et *al.*, 1975). In penaeid shrimp, the next stage, called the platelet stage, bridges the transition to secondary vitellogenesis. During the platelet stage the oocytes increase in size and yolk spheres appear in the cytoplasm for the first time, micropinocytosis activity also increases during this transition (Anderson *et al.*, 1984).

Secondary vitellogenesis is the phase of maturation where the majority of the egg yolk protein accumulates within the oocyte. Contributions from the oocyte are added to exogenous egg yolk obtained by micropinocytosis from the surrounding extracellular fluid. In penaeid shrimp the origin of the exogenous yolk seems to be quite variable. In *Penaeus japonicus*, the ovary appears to be the sole source of egg yolk proteins. The follicular cells which surround the oocyte are suggested as the site of exogenous egg yolk protein production (Yano and Chinzei, 1987). In *Parapenaeus longirostris*, egg yolk proteins were found in the ovary as well as a subepidermal adipose tissue. The subepidermal adipose tissue was suggested as a potential site for yolk protein synthesis (Tom *et al.*, 1987). In *Penaeus vannamei*, egg yolk protein synthesis was demonstrated in both ovarian and hepatopancreatic tissue, and the hepatopancreas was suggested as a source for exogenous egg yolk proteins (Quackenbush and Keeley, 1987; Quackenbush, 1989). Despite these specific variations, it is clear that the oocytes do receive egg yolk proteins from exogenous sources in penaeid shrimps.

In the homarid lobster, the primary oocyte grows from 100 um to 1200 um or larger during secondary vitellogenesis. In the amphipod, *Orchestia gammarella*, the primary oocyte (150 um) grows to 500 um during secondary vitellogenesis. In penaeid shrimp, the mature oocyte of 250 um has grown from the primary oocyte of 50-70 um. The majority of this growth is the accumulation of egg yolk proteins. Included within the egg of *Orchestia gammarella* is a distinct tubular network of electron dense material. The function of this network has been speculated as the transport of cellular material during secondary vitellogenesis (Charniaux-Cotton, 1985). In the oocytes of penaeid shrimp distinct cortical granule precursors (2-9 um) begin to develop at the end of secondary vitellogenesis. These cortical granules precursors will eventually form the characteristic penaeid cortical crypts (Clark *et al.*, 1984.) The cortical granules eventually fuse with the egg oolemma, and contribute to an egg jelly which surrounds a fertilized egg (Anderson *et al.*, 1984, Clark *et al.*, 1984). In *Orchestia gammarella*, some cortical granules do form in secondary vitellogenesis, but they are smaller and less numerous than the penaeid cortical inclusions (Charniaux-Cotton, 1978). In the homarid lobster a chorion forms around the oocyte after vitellogenesis is completed. The chorion is derived from the follicular cells and appears to perform the same function as the penaeid cortical granules (Talbot, 1981b).

The next easily distinguished cytological event is germinal vesicle breakdown. In *Orchestia gammarella*, the germinal vesicle breaks down antecedent to ovulation (Charniaux-Cotton, 1978). In several penaeid shrimp the germinal vesicle also breaks down antecedent to ovulation. This event is considered the end of secondary vitellogenesis, and it usually occurs very soon before ovulation (24 hours in penaeids, Anderson *et al.*, 1984). At ovulation the shrimp follicle cells degenerate, though this is not the case in the amphipod, *Orchestia gammarella* (Anderson *et al.*, 1984; Charniaux-Cotton, 1978; 1985).

The cortical crypts of penaeid shrimp represent a demand on oocyte protein synthesis in addition to the demand for egg yolk proteins. These specializations are clearly of oocyte origin in contrast to the egg chorion of the lobsters which is derived from the follicle cells. Extensive cortical inclusions are also seen in several free spawning echinoderms (Wasserman, 1987). The role of the cortical granules and their egg jelly product is speculated to be protective of the fragile egg. Animals which maintain or brood their eggs after fertilization may not require such an extensive production of cortical granules. The penaeid oocyte must produce both egg yolk proteins as well as the egg jelly proteins. This probably provides a need for extra-ovarian protein synthesis in this group of crustaceans.

Endocrine Regulation of Gonadal Development

Panouse (1943) first demonstrated the eyestalk neurohormonal control of gonadal development in crustaceans. Depending on the time of year, all crustaceans begin a rapid vitellogenesis after eyestalk ablation (Fingerman, 1987; Quackenbush, 1986). The eyestalk neuroendocrine complex produces an inhibitory hormone. When this is removed, precocious gonadal development ensues. In most crustaceans, only secondary vitellogenesis is under eyestalk hormone control, almost nothing is known about the endocrine regulation of primary vitellogenesis. In the penaeid shrimp farms, unilateral eyestalk ablation is used to induce maturation in captive shrimp. Unfortunately, this procedure results in continuous gonadal development. Eventually the female shrimps become exhausted, and in time they produce eggs which are no longer viable (Primavera, 1985). The peptide neurohormone which affects gonadal development has been called gonad inhibiting hormone or GIH.

In the shrimp, *Penaeus vannamei* both the ovary and the hepatopancreas produce egg yolk proteins (Quackenbush, 1989). In other shrimp the adipose tissue contains egg yolk proteins, but whether this tissue actually produces these proteins has not yet been established (Tom *et al.*, 1987). Just after eyestalk ablation in *P. vannamei*, the rate of protein synthesis in both the ovary and hepatopancreas increases. During this induced vitellogenesis these tissues both produce egg yolk proteins, but only the ovary seems to retain yolk proteins in any significant quantity. In the shrimp, *Penaeus japonicus*, the follicle cells and not the hepatopancreas appear to produce egg yolk proteins after eyestalk ablation (Yano and Chinzei, 1987). Crude extracts from shrimp eyestalks can stop or inhibit the induced vitellogenesis in the isolated tissues of *P. vannamei*. This suggests that the target tissues of GIH include both the ovary and the hepatopancreas. GIH may have other target tissues within the shrimp, for example the adipose tissue or other endocrine centers. Little is known about the interactions of the eyestalk peptide GIH and other endocrine processes , though there is speculation that this peptide may also affect steroid endocrine centers.

The chemical nature of GIH has not been completely characterized. From several different crustacean sources GIH is a small water soluble peptide (5,000 daltons or less; Charniaux-Cotton, 1985; Fingerman, 1987 ; Quackenbush, 1986). With the advent of new assays based on *in vitro* protein synthesis of ovaries, characterization of this hormone should proceed faster. The mechanism of action of this inhibitory peptide has not yet been determined. The assays used to identify its presence suggest that GIH can directly affect target tissues and slow or stop protein synthesis. GIH results in a cessation of oocyte growth or a cessation in eyestalk ablated gonadal growth in various different crustaceans (Charniaux-Cotton, 1985). This activity appears to be rapid, since partially purified GIH can slow protein synthesis in ovarian tissue *in vitro* in only 4 hours. Since all eyestalk peptide hormones must circulate within the hemolymph to reach target tissues, it is also possible that GIH has targets other than ovaries or hepatopancreas.

Peptide hormones which are capable of stimulating gonadal development in crustaceans have been postulated based on several observations in various different crustacean groups. Thoracic ganglion implants, or treatments with thoracic ganglion extracts have produced increases in gonadal growth in shrimps, crabs and lobsters (Fingerman, 1987; Quackenbush, 1986). In the amphipod, *Orchestia gammarella*, the ovary produces a putative peptide hormone which results in a stimulation of egg yolk protein synthesis in the adipose tissues (Blanchet-Tournier, 1982). This hormone has activity on several target tissues , it affects secondary sexual characteristics, and may also regulate the molt cycle in the amphipods. To date, the characteristics of the putative gonad stimulating factors have not been determined. This is an area for continued research effort.

Steroid hormones have been localized by several methods to many different crustacean tissues (Fingerman, 1987; Skinner, 1985). The steroid hormone, ecdysone, regulates the molt cycle of crustaceans, its target tissues and its pattern of release have been well documented (Chang, 1989; Skinner, 1985). Steroid hormones other than ecdysone have been found in crustacean eggs, ovarian tissue, and the mandibular gland (Adiyodi, 1985 ; Couch and Hagino, 1983). The location of these steroid hormones, progesterone and estradiol, suggest that they may have a role in the regulation of reproduction. In *Orchestia gammarella*, ecdysone stimulates the adipose tissue to produce egg yolk proteins. In insects the fat body is a target tissue for ecdysone, this tissue produces vitellogenin in the presence of ecdysone (Downer and Laufer, 1983). Vertebrate steroid hormones, progesterone and 17 alpha hydroxyprogesterone can stimulate yolk protein synthesis in both isopods and penaeid shrimp (Kulkarni *et al.*, 1979 ; Souty and Picaud, 1984). Tsukimura and Kamemoto (1988) have shown that progesterone and progesterone-like steroid hormones have a direct effect on the development of immature penaeid ovaries. In summary, vertebrate steroid hormones seem to be present in crustacean tissues. Exogenous applications of these hormones produces effects consistent with a role promoting ovarian development. A direct action of the steroid hormones on a target tissue has also been observed. In order to fully characterize the role of steroid hormones in the regulation of reproduction several areas must be explored. First, the circulating hemolymph levels of the putative hormone should be determined during either natural or induced maturation. Second, a specific response in the target tissue to the physiological level of the putative hormone must be measured. In the future, the use and manipulation of steroid hormones in the regulation of penaeid maturation seems a real possibility.

Summary

Current research on shrimp maturation has focused on understanding the basic mechanism of egg yolk protein production. Within the family Penaeidae, there is a significant variation in the pathways of vitellogenesis. Before adequate means of optimizing reproduction can be achieved, the processes which regulate yolk protein synthesis need to be characterized. The tissues which contain yolk proteins have been identified in the penaeid shrimps as : ovary, hepatopancreas and adipose tissue. The relative contributions of these tissues to the overall yolk production needs to be determined. The timing of exogenous yolk protein synthesis also needs to be measured. Endocrine regulation of yolk synthesis has been assumed based on several observations and experiments.The nature of the interaction of the inhibitory peptides and steroid hormones clearly needs to be addressed. Optimization of reproduction will require the full characterization of the roles of both peptide and steroid hormones. This work is currently in progress in several laboratories.

Predictable induced reproduction in captive penaeids without the use of eyestalk ablation is a long term goal for shrimp mariculture. Progress in the technology of maintenance of large broodstock animals has achieved about as much as can be expected. Analysis of diet and feeding requirements may improve fecundity and larval survival. However, real breakthroughs will only be achieved when it is no longer necessary to use eyestalk ablation to induce reproduction. The potential for endocrine manipulation of broodstock shrimp opens new opportunities for shrimp mariculture. Selective breeding programs require the evaluation of the progeny during the lifetime of the parents. Using current techniques, by the time the larval growth can be evaluated the female has long since ceased to produce viable eggs. A challenge for endocrinologists is to determine which hormones can induce reproduction and which hormones can inhibit reproduction so that selective breeding programs can become feasible.

Endocrine research has recently focused on the identification of GIH and the characterization of the roles of steroid and terpenoid hormones in the regulation of vitellogenesis. These efforts should produce significant results in the next five years. Additional new research should be directed towards the elucidation of a putative gonad stimulating peptide hormone. The importance of the females' egg yolk contribution to the survival of shrimp larvae has been taken for granted. Characteristics of the quality and quantity of egg yolk proteins that the females incorporate into the eggs may have a direct and measurable effect on larval survival. In coordination with ongoing studies of maturation, the composition of the yolk needs to analyzed for the potential of improvements in larval survival.

Acknowledgments

This work was supported in part through an institutional grant NA 86AA-D-SG068 to Florida International University by the National Oceanic and Atmospheric Administration's Sea Grant program, project R/LR A-11 to L.S.Q. . The author would also like to thank W.J. Dougherty and Benjamin P. Quackenbush for their comments and assistance.

Literature Cited

Adiyodi, R. 1985. Reproduction and its control. In. *The Biology of Crustacea* , D.E. Bliss and L.H. Mantel, editors,Vol. 9, pp. 147-217. Academic Press, New York.

Adiyodi, R. and T. Subramoniam. 1983. Arthropoda-Crustacea. In. *Reproductive Biology of Invertebrates*, K.G. Adiyodi and R.G. Adiyodi, editors, pp. 443-496. John Wiley and Sons, New York.

Al-Mohanna, S. Y. and J. A. Nott. 1987. R-cells and the digestive cycle in *Penaeus semisulcatus*. Marine Biology 95 : 129-137.

Aiken, D. E. and S. L. Waddy. 1980. Reproductive biology, In. *The Biology and Management of Lobsters*, J.S. Cobb and B.F. Phillips (editors). pp. 215-276. Academic Press, New York.

Anderson, W. W., J. E. King,, and M. J. Lindner. 1949. Early stages in the life history of the common marine shrimp, *Penaeus setiferus*. Biol. Bull. 96 : 168-172

Anderson, S. L., Chang, E. S., and W. H. Clark. 1984. Timing of postvitellogenic ovarian changes in the ridgeback prawn, *Sicyonia ingentis* (Penaeidae) determined by ovarian biopsy. Aquaculture 42: 257-271.

Aquacop. 1983. Constitution of broodstock, maturation, spawning and hatchery system for penaeid shrimps in the Centre Oceanologique du Pacifique . In. *CRC Handbook of mariculture*, volume 1, crustacean aquaculture, J.P. McVey (editor). pp. 105-122. CRC Press, Boca Raton, FL.

Bergin, M., 1981. Hatching rhythms in *Uca pugilator*. Marine Biology 63: 151-158.

Blanchet-Tournier, M. F. 1982. Quelque aspects des interactions hormonales entre la mue et al vitellogenese chez le Crustace Amphipode, *Orchestia gammarella*. Reprod. Nutr. Develop. 22: 325-344.

Caceci, T, Neck, K. F., Lewis, D. H., and R. F. Sis. 1988. Ultrastructure of the hepatopancreas of the pacific white shrimp, *Penaeus vannamei* (Crustacea: Decapoda). J. Mar. Biol. Ass. U.K. 68 : 323-337.

Chamberlain, G. W., M. G. Haby, and R. J. Miget. 1985. Texas shrimp Farming Manuel. pp. 1-158. Texas Agricultural Extension Service, Corpus Christi, Tx.

Chang, E. S. 1989. Regulation of crustacean molting: role of peptides and steroid hormones. this volume.

Charniaux-Cotton, H. 1978. L'Ovogenese la Vitellogenine et leur controle chez le Crustace Amphipode, *Orchestia gammarella*. Arch. Zool. Exp. Gen. 119 : 365-397.

Charniaux-Cotton, H. 1985. Vitellogenesis and its control in malacostracan crustacea. Amer. Zool. 25: 197-206.

Christy, J. H. 1982. Adaptive significance of semilunar cycles of larval release in fiddler crabs (Genus *Uca*) test of an hypothesis. Biol. Bull. 163 : 251-263.

Christy, J. H. and M. Salmon. 1984. Ecology and evolution of mating systems of fiddler crabs (Genus *Uca*). Biol. Rev. 59: 483-509.

Clark, W. H. Jr., A. I. Yudin, F. J. Griffin, and K. Shigekawa. 1984. The control of gamete activation and fertilization in the marine Penaeidae, *Sicyonia ingentis* In. *Advances in Invertebrate Reproduction,* W. Engels, W. H. Clark, Jr., A. Fischer, P. J. W. Olive, and D. F. Went (editors). Volume 3, pp. 459-472. Elsevier, Amsterdam.

Cook, H. L. and A. M. Murphy. 1969. Culture of larval penaeid shrimp. Trans. Amer. Fish. Soc. 98: 751-755.

Couch, E. F. and N. Hagino. 1983. Correlation of progesterone and estradiol production by the mandibular organ and other tissues with egg development in the american lobster. J. Cell. Biol. 97 : 158A.

Dehn, P. F., D. E. Aiken and S. L. Waddy. 1983. Aspects of vitellogenesis in the lobster, *Homarus americanus*. Can. Tech. Rep. Fish. Aqua. Science 1161: 1-24.

Dhainaut, A. and M. D. Leersnyder. 1976. Etude cytochemique et ultrastructurale de l'Evolution ovocytaire du Crabe Eriocheir sinensis. Arch. Biol. 87: 261-282.

Downer, R. G. H. and H. Laufer. 1983. *Endocrinology of Insects*. Liss Inc., New York.

Durliat, M., and R. Vranckx. 1982a. Proteins of aqueous extracts from the hepatopancreas of *Astacus leptodactylus*. 1. Changes in proteins during the molt cycle. Comp. Biochem. Physiol. 71B : 155-163.

Durliat, M., and R. Vranckx. 1982b. Proteins of aqueous extracts of the hepatopancreas of *Astacus leptodactylus*. 2. Immunological identities of proteins from

hepatopancreas and blood. Comp. Biochem. Physiol. 71B: 165-171.
Duronsolet, M. J., A. I. Yudin, R. S. Wheeler, and W. H. Clark, Jr. 1975. Light and fine structural studies of natural and artificially induced egg growth of penaeid shrimp. Proc. World Maricult. Soc. 6 : 105-122.

Eastman-Reks, S. B. and M. Fingerman. 1985. *In vitro* synthesis of vitellin by the ovary of the fiddler crab, *Uca pugilator*. J. Exp. Zool. 233 : 111-116.

Fingerman, M. 1987. The endocrine mechanisms of crustaceans. J. Crust. Biol. 7 : 1-24.

Fyffe, W. E. and J. D. O'Connor. 1974. Characterization and quantification of a crustacean lipovitellin. Comp. Biochem. Physiol. 47B : 851-867.

Gibson, R. and P. L. Barker. 1979. The decapod hepatopancreas. Oceanog. Mar. Biol. Ann. Rev. 17 : 285-316.

Gorell, T.A. and L. Gilbert. 1971. Protein and RNA synthesis in premolt crayfish, *Orconectes virilis*. Z. vergl. Physiologie. 73 : 345-356.

Johnson, P. T. 1980. *Histology of the Blue Crab, Callinectes sapidus: a model for the Decapoda*. Praeger Press, New York.

Junera, H. and J. J. Meusy. 1982. Vitellogenin and lipovitellin in *Orchestia gammarella*; labeling subunits after *in vivo* administration of 3H-leucine. Experientia 30: 252-253.

Kerr, M. S. 1969. The hemolymph proteins of the blue crab, *Callinectes sapidus*. II. A lipoprotein serologically identical to oocyte lipovitellin. Develop. Biol. 20: 1-17.

Kulkarni, G. K., R. Nagabushanam, and P. K. Joshi. 1979. Effect of progesterone on ovarian maturation in a marine penaeid prawn, *Parapenaeopsis hardwickii* (Miers, 1978). Ind. J. Exp. Biol. 17 : 986-987.

Lui, C. W. and J. D. O'Connor. 1976. Biosynthesis of lipovitellin by the crustacean ovary. II. Characterization of and *in vitro* incorporation of amino acids into purified subunits. J. Exp. Zool. 195 : 41-52.

Lui, C. W., and J. D. O'Connor. 1977. Biosynthesis of lipovitellin. III. The incorporation of amino acids into purified lipovitellin of the crab, *Pachygrapsus crassipes*. J. Exp. Zool. 199 : 105-108.

Malcoste, R., A. Wourmhoudt, and C. Bellon-Humbert. 1983. La characterization de l'hepatopancreas de la crevette *Palaemon seratus* en culture organotypiques. C.R. Acad Science Paris 296 : 597-602.

Mc Laughlin, P. A. 1983. Internal Anatomy. In. *The Biology of Crustacea*, L.H. Mantel (editor). Volume 5, pp. 1-52, Academic Press, New York.

McVey, J. P. and J. M. Fox. 1983. Hatchery techniques for penaeid shrimp utilized by Texas A and M NMFS Galveston laboratory program. In. *Handbook of Mariculture*, J.P. McVey (editor). Vol. 1, pp 129-154 . CRC Press, Boca Raton, FL.

Miyawaki, M., Y. Taketomi. and T. Tsuruda. 1985. Absorption of experimentally administered materials by the hepatopancreas cells of the crayfish, *Procambarus clarkii*. Cell. Biol. Int. Reports.8 : 873-877.

Panouse, J. B. 1943. Influence de l'ablation de peduncle oculaire sur la croissance de l'ovaire chez la crevette, *Leander serratus*. C.R. Acad. Science Paris 217 : 553-555.

Paulus, J. E., and H. Laufer. 1987. Vitellogenocytes in the hepatopancreas of

Carcinus maenus and *Libinia emarginata*. Int. J. Invert. Reprod. Develop. 11 : 29-44.

Primavera, J. H. 1985. A review of maturation and reproduction in closed thelycum penaeids. Proc. First Inter. Conf. Culture Penaeid Prawns. Iloilo City, Philippines.

Quackenbush, L. S. 1986. Crustacean endocrinology; a review. Can. J. Fish. Aqua. Science 43 : 2271-2282.

Quackenbush, L. S. 1989. Vitellogenesis in the shrimp, *Penaeus vannamei: In vitro* studies of the isolated hepatopancreas and ovary. Comp. Biochem. Physiol. 94B: 253-261.

Quackenbush, L. S., and W. F. Herrnkind. 1981. Regulation of molt and gonadal development in the spiny lobster, *Panulirus argus* : effect of eyestalk ablation. Comp. Biochem. Physiol. 69A : 523-527.

Quackenbush, L.S. and W.F. Herrnkind. 1983. Partial characterization of eyestalk hormone controlling molt and gonadal development in the spiny lobster, *Panulirus argus*. J. Crust. Biol. 3: 34-44.

Quackenbush, L. S, and L. L. Keeley. 1987. Vitellogenesis in the shrimp, *Penaeus vannamei*. Amer. Zool. 26 : 810A.

Quackenbush, L. S. and L. L. Keeley. 1988. Regulation of vitellogenesis in the fiddler crab, *Uca pugilator*. Biol. Bull. 175:321-331.

Schade, M. L., and R. R. Schivers. 1980. Structural modulation of the surface and cytoplasm of oocytes during vitellogenesis in the lobster, *Homarus americanus:* an electron microscope protein tracer study. J. Morphol. 163 : 13-26

Senkbeil, E. G. and J. C. Wriston. 1981. Hemocyanin synthesis in the american lobster, *Homarus americanus*. Comp. Biochem. Physiol. 68B : 163-171.

Skinner, D. M. 1985. Molting and regeneration. In. *The Biology of Crustacea*, D.E. Bliss and L.H. Mantel (editors). Volume 9, pp. 43-146, Academic Press, New York.

Souty, C., and J. L. Picaud. 1984. Effect de l'injection d'une gonadotropin humaine sur la synthese et la liberation de la vitellogenine par le tissu adipeaux du Crustace Isopod marin, *Idotea bathica basteri*. Gen. Comp. Endocrinol. 54 : 418-421.

Talbot, P. 1981a. The ovary of the lobster, *Homarus americanus*. I. Architecture of the mature ovary. J. Ultrastruc. Res. 76 : 235-248.

Talbot, P. 1981b. The ovary of the lobster, *Homarus americanus*. II. Structure of the mature follicle and origin of the chorion. J. Ultrastruc. Res. 76 : 249-262.

Tom, M., M. Goren, and M. Ovadia. 1987. Localization of the vitellin and its possible precursors in various organs of *Parapenaeus longirostris*. Int. J. Invert. Reprod. Develop. 12: 1-12.

Tsukimura, B. and F. I. Kamemoto. 1988. Organ culture assay of the effects of putative reproductive hormones on immature penaeid ovaries. World Aqua. Soc. abstract 288.

Wallace, R. A., S. L. Walker and P. V. Hauschka. 1967. Crustacean lipovitellin. Isolation characterization of the major high density lipoprotein from the eggs of decapods. Biochemistry 6 : 1582-1590.

Wasserman, P. 1987. The biology and chemistry of fertilization. Science 235 : 553-559.

Webb, M. 1977. Eyestalk regulation of molt and vitellogenesis in *Uca pugilator*. Biol. Bull. 153 : 630-642.

Wolin, E. M., H. Laufer, and D. F. Albertini. 1973. Uptake of yolk protein, lipovitellin by developing crustacean oocytes. Develop. Biol. 35 : 160-170.

Yang, W. T. 1975. A manual for large tank culture of penaeid shrimp to postlarval stages. University of Miami Sea Grant Technical Publication 31 : 1-94. University of Miami, Coral Gables, FL.

Yano, I., and Y. Chinzei. 1987. Ovary is the site of vitellogenin synthesis in kuruma prawn, *Penaeus japonicus*. Comp. Biochem. Physiol. 86B: 213-218.

Dietary Analysis of Penaeid Shrimp: The Immunoassay Approach

Robert J. Feller
Department of Biology
Marine Science Program
Belle W. Baruch Institute
for Marine Biology and Coastal Research
University of South Carolina
Columbia, SC 29208

Abstract

Penaeid shrimp present special problems to investigators of their feeding biology. They cannot be observed directly while feeding, their gastric mill renders food visually unidentifiable, and they do not exhibit strong periodicity in their feeding behavior. Traditional methods for gut content analysis are not always sufficient for determining diet. Immunoassays based on specific reactions between antisera to soluble proteins of potential prey offer some advantages over traditional methods. Qualitative descriptions of diet are possible using immunodiffusion techniques, whereas immunoelectrophoresis techniques offer opportunities for the quantitative analysis of feeding. Sample results of both qualitative and quantitative approaches show that otherwise invisible prey can be identified and that meals of known size that have been digested for known periods of time can be quantified in the laboratory. Future applications of immunoassays for trophic analysis will have to solve the problem of distinguishing between meal size and time since the meal was ingested before ecologically meaningful data can be gathered on natural populations of penaeid shrimp.

Introduction

The analysis of an organism's diet has been a cornerstone of ecological investigations in all habitats. Trophic studies, both qualitative and quantitative, have resulted in an immense quantity of data that has promulgated a considerable body of testable theory (*e.g.*, size-selective predation, optimal foraging). Fretwell (1987) has proposed elevating the study of food chain dynamics to the central theory of ecology (but see Peters, 1977). The primacy of nutritional studies in ecology, then, assumes that the ability to successfully analyze diets is almost a given in most fields of endeavor. Although this may be the case for large terrestrial fauna whose feeding behavior may be observed in some way, this is decidedly not the case in studies of most aquatic organisms which utilize the benthic habitat as a feeding substrate. Needless to say, penaeid shrimp feeding falls in the same category - only qualitative data are available on its feeding in the natural environment. Immunoassays offer several advantages to traditional methods of dietary analysis which will be illustrated with a few examples from work on post-larval and juvenile white shrimp (*Penaeus setiferus*) collected from tidal creeks in North Inlet, South Carolina, and from laboratory experiments conducted at the Baruch Institute's facilities in Columbia, SC. Only highlights of this research are presented here, as detailed results will be published elsewhere.

Typical objectives and traditional methods of dietary analysis have been reviewed by several authors. Interestingly, most of these reviews concern the analysis of fish stomach contents (Hynes, 1950; Berg, 1979; Hyslop, 1980). The study of insect feeding comprises an even larger body of literature than that for fishes, but very similar questions are asked and similar methods used in insect nutritional ecology (Slansky, 1982; Calver, 1984; Calver and Porter, 1986; Luck *et al.*, 1988; Sunderland, 1988). Just what are some typical questions related to feeding in aquatic invertebrates such as shrimp ? What is eaten, when is it eaten, how much of each food type is eaten (daily ration), what are the ontogenetic shifts in diet, what are digestion rates, is diet catholic or does selective ingestion occur ? These are among many questions that might be asked by investigators of any organism's feeding ecology. All of these questions require information about what is ingested, with knowledge gained either by direct observation of the feeding process or by some other less direct method. These include gut contents analysis, fecal analysis, or may even include indirect methods utilizing information about depletion of a prey population or manipulation of predator and prey in some type of exclusion or caging study. Direct observation is the method of choice but, unfortunately, is not feasible in many instances (Feller *et al.*, 1979).

Penaeid shrimp present several special, though not unique, problems to researchers of their feeding ecology. First, they cannot be reliably observed feeding in their natural habitat. Turbid water, rapid movement of the feeding apparatus, nocturnal feeding, and their tendency to seek cover are a few of the difficulties with observation. Second, the presence of a gastric mill renders ingested food visually unidentifiable by standard gut contents analysis in most instances (typically a large portion of the ingesta's volume is unrecognizable mush). Third, shrimp are not known to be strongly periodic in their feeding, thus making it difficult to sample them at a time when their guts might be most full (McTigue and Feller, 1989). Fourth, the presence of a midgut gland, the hepatopancreas, in penaeids complicates analysis of digestive processes (Caceci *et al.*, 1988). Lastly, shrimp appear to be omnivorous, including both plant and animal food in their diet along with detritus, and show little consistency in dietary content within different habitats, although common diet items include polychaete worms, bivalves, crustaceans, gastropods, and detritus (Cockroft and McLachlan, 1986; Wassenberg and Hill, 1987; Stoner, 1988). Some may view these adaptively advantageous traits as aproblematic for specific questions, but in view of the common questions asked, any, several, or all of these five general complicating factors may come into play.

How one approaches these problems will, of course, depend upon the specific goals of the researcher, but indirect methods of dietary analysis seem to be utilized more frequently at present. For example, Nelson (1981) found that pink shrimp (*P. duorarum*) exerted a significant predatory impact upon macrobenthic fauna in cage enclosure experiments, both in the lab and in the field - no gut contents analyses were performed, however. Lobster feeding patterns have been studied with variable success using pattern recognition algorithms for the analysis of high resolution gas chromatography residues from animals fed known diets (Knutsen and Vogt, 1985). Natural tracers which integrate an organism's past feeding history (*e.g.*, stable isotopes of carbon, oxygen, and sulfur) allow broad descriptions of food consumed by various predators, including penaeid shrimp (Fry *et al.*, 1982; Peterson *et al.*, 1985; Gleason, 1986; Gleason and Wellington, 1988). Natural radioactive tracers are also being measured in shrimp to see if they reflect the source of dietary accumulation of these radionuclides (Heyraud *et al.*, 1988). Indirect methods, whether they analyze material from an organism's gut contents or from its body tissues, share the common problem of relatively low resolution - it is extraordinarily difficult to describe the diet with any detail other than to major taxonomic group at best. This is an inherent limitation of the so-called natural tracer methods that is not shared by immunoassays.

Immunoassays

Immunoassays utilize the highly specific binding and recognition capabilities of antibodies to identify immunogenic moieties present in the stomach contents of predators. The antibodies are usually first prepared to soluble proteins of whole-organism extracts of animals, plants, bacteria, fungi, detritus, or other targeted species suspected as being potential prey. The antibodies are usually prepared by injecting the extract into a rabbit, goat, rat, or chicken. Our laboratory has used rabbits exclusively because of their wide availability, low cost, and capacity to produce antibodies to just about any mixture of non-toxic foreign proteins with which they were injected. Once the antiserum is produced, it must be tested for specificity (its ability to react with the same antigenic substance that was injected into the host), sensitivity, and for its tendency to give false positive reactions due to cross-reactions with non-self antigens. That is, an antiserum may contain antibodies that recognize not only homologous (self) antigens but also heterologous (non-self) antigens. Such an antiserum is said to be poly- or oligo-specific (as opposed to monospecific), because it will recognize non-self antigens in addition to those that it was designed to recognize. Such broad specificity or cross-reactivity is sometimes a problem when assaying for the presence of specific prey, but often the ability of the antiserum to recognize very similar types of prey is advantageous and can be exploited. Clearly, the single most important aspect determining the utility of immunoassays is the quality of the antibodies produced for use in the assay. They must be specific enough to recognize target prey at whatever taxonomic level is desired. Fortunately, antisera can be produced or modified so they will detect prey proteins at nearly any taxonomic level desired, whether that be order, family, genus, species, or even life stages within a species (Feller, 1986).

Dietary immunoassays are performed by reacting native gut contents material collected from the predator with antisera to the target prey taxa. The reactions occur within a support matrix of agarose by simple diffusion or electrophoresis. Immunodiffusion reactions are considerably less quantitative than electrophoretic, but both are capable of some level of quantification (Clausen, 1981; Wang, 1982; Gross and Marz, 1988).

Given that it is possible to produce an antiserum that will recognize almost any antigenic substance that might possibly be ingested by a shrimp, why do we still have so much to learn about shrimp feeding ecology ? Immunoassays also have some inherent limitations that diminish their utility. I view the immunoassay approach as one to be taken in conjunction with other methods only when other more direct methods fail. Specific questions can be answered satisfactorily with the methodology. The technical aspects are not difficult to learn, and antisera can be prepared commercially, but it does take a considerable amount of time to develop a large selection of antibodies whose specificities and sensitivities (titers) are known.

Although most immunoassays are methodologically quantitative for measuring quantities of protein, they are not necessarily interpretable in an ecologically quantitative or meaningful way (Feller and Ferguson, 1988). For example, shrimp do not all take the same size bites, nor do they eat the same sizes of prey. The amount of time that has elapsed since a shrimp's last meal is generally unknown for field collections, hence ingesta may be in varying degrees of digestion when analyzed. Digestion rates are also dependent upon many environmental variables (temperature, food type, activity, etc.), further complicating the analyses. For example, detection of a quantity of protein in a shrimp's stomach says nothing about how much was actually ingested, because the amount remaining in the stomach at the time of collection could easily have resulted from the shrimp's having eaten a small quantity of the prey or from the meal having already been digested for a period of time. Furthermore, it is entirely possible that the shrimp ate prey at different times, so that its various ingesta have experienced widely different residence times in the gut. It is unknown whether all prey ingested are digested at the same rate in shrimp, but it is likely that this factor could also introduce some variability. Another limitation of immunoassay methods is that of contamination of a gut content sample by secondarily ingested prey, *i.e.*, the stomach contents of the prey. Finally, it is difficult to detect a cannibalistic meal because of the possibility of contamination of a gut content sample by the gut lining of the predator.

Regardless of these shortcomings, immunoassays have been used successfully in dietary analyses of a number of different organisms, including euphausiid crustaceans (Theilacker *et al.*, 1986), birds (Walter *et al.*, 1986), insects (Calver *et al.*, 1986), various marine benthic invertebrates (Feller *et al.*, 1979), freshwater invertebrates (Young *et al.*, 1964; Pickavance, 1970), cephalopod molluscs (Grisley and Boyle, 1985), and penaeid shrimp (Hunter and Feller, 1987).

Methodology

Immunoassays are conducted on field-collected shrimp which have been frozen immediately after collection to stop the digestive process. Freezing also serves as a temporary preservation method which does not cause the denaturation of prey proteins in the digestive tract as would occur when formalin is used. Freezing or freeze-drying are both acceptable preservation methods for samples to be analyzed using immunoassays. After thawing, contents of shrimp digestive tracts (usually the entire tract but sometimes only the proventriculus or only the hepatopancreas region) are examined visually for identifiable prey at magnifications of 10-50X and then ground in a buffer solution, centrifuged to remove particulates, and the supernatant assayed directly or refrozen for later analysis.

Qualitative immunoassays utilize the double-diffusion micro-Ouchterlony technique (Wang, 1982) wherein 10-15μl of gut contents material are placed in the central well of a Plexiglass template (2.5 x 2.5 cm^2) resting upon an agarose matrix on a standard 25 x 75 mm glass microscope slide. Four other wells surrounding and equidistant from the template's central well are loaded with 15 μl of antiserum to four different potential prey. The antigenic material (most often proteins) in the shrimp's gut contents drains through a small (1 mm dia) hole in the bottom of the central well and diffuses radially through the agarose until it comes in contact with antisera diffusing towards the central well. Diffusion occurs horizontally at room temperature for 48 h in a humidified chamber. If prey antigens are recognized by antibodies in any of the antisera, precipitin lines form between the central well and the peripheral well containing that particular antiserum. The slide is washed in a buffer solution for 24 h to rid the agarose gel of unprecipitated proteins, washed in distilled water for 4 h to remove salts, dried at 37°C overnight, and stained with Coomassie Brilliant

Blue R (a protein stain) to highlight the precipitin lines. We usually place two four-well templates on a single microscope slide. Depending upon how many potential prey we wish to test for and the volume of the shrimp's stomach contents material, several slides may be run simultaneously. A testing battery of 32 antisera would require four slides and enough buffer-solubilized gut contents material to fill the central well in each of 8 templates (80-120 µl). An algorithm based upon the number of precipitin lines observed minus the number that might result from cross-reactions with other prey that might be present is employed to ascertain whether the presence of particular prey is confirmed or artifactual due to a false-positive reading. Final confirmation is achieved with tests for identity of the antigenic material in a gut content sample with native antigenic material used to create the antiserum (see Hunter and Feller, 1987, for details).

For quantitative immunoassays we use the technique of rocket immunoelectrophoresis (Axelsen and Bock, 1983). This technique is also performed using horizontal slabs of agarose. Gut contents samples are applied to wells punched out of the agarose and immunoprecipitates are generated when the negatively charged antigens in the sample migrate towards the cathode (the positive electrode) through antibody-containing agarose. The precipitin lines have a leptokurtic (pointed at the tip like a rocket) shape that gives the technique its name. The gels are run under constant current at about 3V/cm for periods of about 6 h at room temperature. These running conditions were derived by trial and error. The areas beneath the rocket-shaped precipitin peaks and the peak heights themselves are proportional to the concentration of antigens in the sample, and this technique is capable of detecting as little as 10 ng of antigen under optimal conditions. It works best for soluble protein concentrations above approximately 1 mg/ml. Coomassie stain is also used to highlight the precipitin lines for analysis. Rocket immunoelectrophoresis has an advantage over double-immuno-diffusion in terms of sensitivity and speed, taking only hours to perform instead of days. A variation of the rocket technique called rocket-line immunoelectrophoresis (Kroll, 1983) allows one to simultaneously identify and quantify precipitin lines from 20-30 antigen-antibody complexes in a number of samples from a single run. A small trough cut into the gel is filled with antigen that migrates cathodically and creates precipitin lines that are perpendicular to the direction of the flow field. Wells containing other antigens or gut content samples are located on the anodic side of the trough, and the precipitin lines that form from these wells are rocket-shaped and merge with corresponding lines of identity from the trough. The technique is especially useful for comparing antigens or for evaluating which components of two or more antigens are causing cross-reactions with the antiserum in question (Feller and Gallagher, 1982).

Sample Results Using Immunoassays

Qualitative answers to the question of what do shrimp eat arise from a combination of visual analysis and double-diffusion immunoassays (Table 1). These representative data for white shrimp (*P. setiferus)* were derived for animals of various sizes collected with seines in tidal creeks on several different dates and at various times of day. Small shrimp (< 25 mm CL) were assayed by combining the stomach contents of 5 individuals, whereas shrimp larger than 25 mm carapace length were analyzed individually. What becomes immediately obvious is that the visually identifiable remains are the hard parts of prey that are most resistent to digestion. They may actually be over-represented in the guts of white shrimp because some of these pieces are retained and used to assist in the process of grinding newly ingested prey (Suthers, 1984). The other prey that are identifiable immunologically in many cases confirm what was seen visually in a given sample, but more often prey are identified that would otherwise remain invisible. There are also, however, instances when visually identifiable prey are not seen in the immunoassays. Since we have no control over the feeding behavior of field-collected specimens, the time of feeding is always unknown. Hence, the inability of an antiserum to detect a prey type known to be present in a gut content sample may simply reflect that the prey was eaten a long time ago and its soluble proteins have been completely digested or modified so as to be no longer detectable with an antibody probe.

An interesting aspect of the immunoassay results in Table 1 concerns the mechanisms by which shrimp might have obtained various prey. For instance, do white shrimp capture living fiddler crabs (*Uca pugnax*) or do they simply feed upon remnants left by other predators of these crabs (*e.g.*, blue crabs)? It is not possible to answer these kinds of behavioral questions with the immunoassay technique, but

Table 1. Representative data on feeding by *Penaeus setiferus* as determined by visual analysis (microscopic examination at 25-50X) and by immunoassay (double-immunodiffusion). There is no left-to-right correspondence intended between results of the two assay methods performed on the same sample. Gut contents of 5 shrimp (20-29 mm carapace length) were pooled for analysis on each date.

Date (Time)	Visual	Immunoassay*
20 Aug 85 (0630)	polychaete worm setae sand grains plant material red pigmented material yellow-brown fluids	turbellarians ostracod crustaceans saltmarsh cordgrass ciliates polychaete worms
20 Aug 85 (1500)	polychaete setae & jaws red pigmented material yellow-brown fluids	fish muscle tissue saltmarsh cordgrass nematode worms turbellarians polychaete worms
15 Sep 85 (0600)	polychaete worm setae bivalve shell fragments red pigmented material yellow-brown fluids	fiddler crabs polychaete worms turbellarians ostracod crustaceans fish muscle tissue saltmarsh cordgrass
15 Sep 85 (1500)	polychaete worm setae sand grains ostracod carapaces yellow-brown fluids	ostracod crustaceans turbellarians fish muscle tissue ciliates

* Antisera to the following 27 potential prey taxa were used in the assay: Crustacea: *Penaeus setiferus* (white shrimp), *Neomysis americana* (opossum shrimp), *Uca pugnax* (fiddler crab), *Palaemonetes pugio* (grass shrimp), *Menippe mercenaria* (stone crab), harpacticoid copepod spp., isopod spp., gammarid amphipod spp., ostracod spp., *Lepas* sp. (barnacle); Molluscs: *Littorina irrorata* (periwinkle snail), *Ilyanassa obsoleta* (mud snail), *Crassostrea virginica* (American oyster), *Mercenaria mercenaria* (hard clam), *Geukensia demissa* (horse mussel), *Urosalpinx cinerea* (oyster drill); Annelids: *Diopatra cuprea* (polychaete worm), *Capitella* spp. (polychaete worm), *Glycera dibranchiata* (polychaete worm), oligochaete spp.; *Renilla reniformis* (sea pansey); nematode worm spp.; *Leiostomus xanthurus* (fish); gram-negative marine bacteria; ciliate protozoans; turbellarian spp.; *Spartina alterniflora* (saltmarsh cordgrass); *S. alterniflora* detritus.

simply knowing that a particular type of prey is present in a sample allows one to devise experiments or develop other methods to answer such questions. Quantitative approaches to the feeding behavior of white shrimp have been performed in the laboratory under controlled conditions. Because of the problem involved with not knowing when shrimp feed in the field, we felt it appropriate to examine the gut evacuation time of animals fed known meals. We also wanted to determine how long these known meals could be detected and how well our antisera would detect and quantify meals of a known size after a known time since ingestion. We initially chose to examine single prey meals with just one antiserum before attempting to probe multiple-prey meals with several antisera, the latter being a more realistic situation in nature. This work is now in progress. To enhance our ability to observe the shrimp feeding, we laced their food with fluorescent latex beads (2-5 um diam., available from Radiant Color, 2800 Radiant Ave., Richmond, CA 94804). Shrimp were starved for at least 24 h and then presented a meal which had been pre-weighed. The meal we selected was macerated oyster meat (*Crassostrea virginica*), a food substance eaten readily by shrimp in trial feedings and one which occurs naturally in their diet. The presence of the fluorescent beads had no apparent effect on the feeding response, no matter what color was used (*e.g.*, pink, orange, chartreuse, magenta, green). The feeding arena was a small aquarium filled with seawater of 30 ^{0}/oo salinity at room temperature of about 24°C. A window screen mesh raised off the bottom prevented shrimp from picking up debris or feces from the bottom during the post-ingestion phase. The meal was presented to individual shrimp for 5 min during which time most shrimp ate continuously. If the entire morsel presented was not eaten, the remaining food was removed and weighed to estimate how much of the total had actually been eaten. Shrimp were then simply maintained with aeration for varying lengths of time while they digested their meal and were removed and frozen for later immunoassay. The contents of the proventriculus were removed, weighed wet, and solubilized with a constant amount of buffer solution. If no food remained in the proventriculus, fluid from the organ was removed for analysis.

Meals were detectable for various periods of time depending upon the method of analysis and from where in the gut the samples were collected. For instance, the fluorescent meal could be seen throughout the digestive tract and in fecal strands as they were produced. Feces production began in some cases only 20 min after the meal was eaten but usually required about an hour. Small portions of the meal remained in the proventriculus for about two hours under the experimental conditions and were detectable for up to 5 h post-ingestion with both the immunodiffusion and rocket immunoelectrophoresis methods. Meals were detectable for much longer periods in the hepatopancreas region, usually for 48 h, but often for 5 days. The amount of oyster protein remaining in the proventriculus declined curvilinearly until it disappeared about 2-3 hr post-ingestion (Figure 1). Also, the amount of oyster protein remaining in the pro-ventriculus (measured as area under the peak in the rocket electrophorograms) was linearly correlated with the actual quantity of food ingested at all times post-ingestion.

Discussion

These results have strong implications for shrimp feeding studies. Knowledge of gut evacuation rates is essential for determining daily rations of shrimp (Hill and Wassenberg, 1987). Although our data were gathered under artificial pulse-feeding conditions (as opposed to the "normal" situation in which food is continuously present), we can now begin to make at least minimum estimates of how much food shrimp eat each day. Other gravimetric studies not reported here (McTigue and Feller, 1989) show that shrimp gut contents comprise about 2-5 % of their body weight at all times of day. Since they are capable of emptying their guts every hour or so, they could easily eat half their live body weight or more per day.

The immunoassay results also show great promise for the quantitative analysis of shrimp stomach contents from field collections. Obviously, immunoassays combined with visual analysis of proventricular contents can provide answers to the "what do shrimp eat" question. This is valuable for comparative purposes in terms of habitat differences in diet or for describing ontogenetic changes of diet (*e.g.*, Hunter and Feller, 1987). In addition, since we know how long soluble proteins persist in immunologically recognizable form within the shrimp's digestive tract, interpretation of quantitative immunoassays of field-collected animals becomes somewhat easier. This is especially true with respect to the question of when shrimp feed, as upper and lower bounds on the residence times for food in the proventriculus can

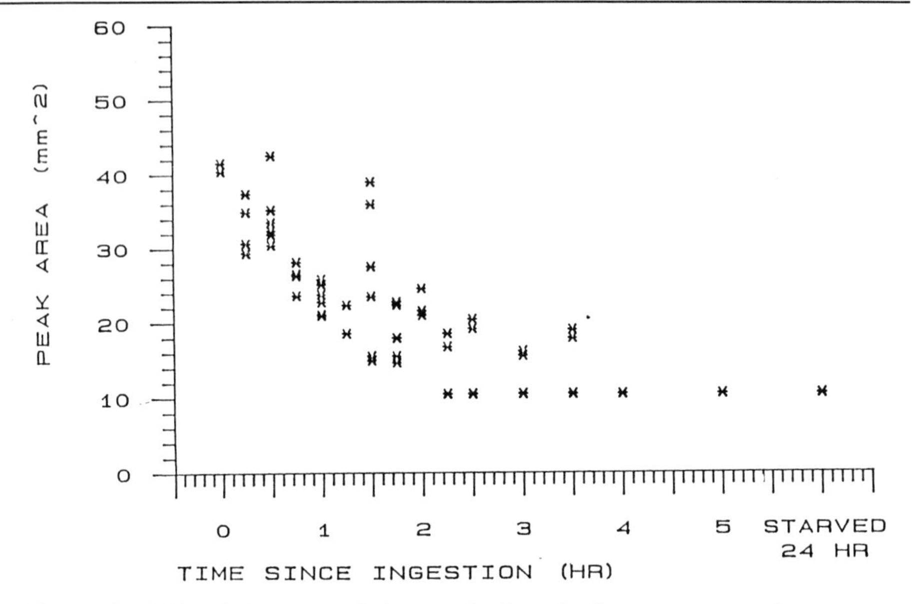

Figure 1. White shrimp were fed a standard meal of oyster meat at time zero and were then serially sacrificed at intervals for up to 5 hours for immunoassay of material remaining in their foreguts. Peak area refers to area beneath the peak of rocket immunoelectrophorograms which is proportional to the amount of oyster protein remaining in the proventriculus. Area under the peak for the 24-hr starved control shrimp represents an empty proventriculus.

now be estimated. How much of a given prey type is present in a shrimp at a particular point in time will always remain a difficult question until we learn more about the rates of digestion of meals eaten under different conditions and the degree to which selective digestion may occur. Our current experiments on the digestive breakdown of multiple-prey meals should provide answers to many of these questions.

Future Directions and Applications

Clearly the single greatest limitation in the use of immunoassays is the quality of the antisera employed. Many investigators suggest that antibody specificity might be magnified immensely by the use of monoclonal antibodies. However, monoclonal antibodies have limitations as well. First, they do not usually precipitate in agarose gels and second, their specificity may actually be too high. A monoclonal antibody recognizes a specific binding site that may exist on a large number of very different proteins or other antigenic substances rather than on a single target protein of interest. When digestion of proteins occurs, protein structures change and reveal binding sites that may not have been present in its native form. A monoclonal antibody thus might recognize and react with portions of proteins completely unrelated to the target protein. Much the same could be said for polyclonal antisera, but with most dietary immunoassays the antisera used have been produced to whole-organism extracts of the target prey. These antisera contain such a diverse mixture of antibodies that no single antigen-antibody complex is responsible for the formation of precipitin products.

An ever-increasing number of quantitative assay methods is available which have a range of sensitivities that can be tailored to the question at hand (Sunderland, 1988). The ELISA (enzyme-linked immunosorbent assay) has great potential in terms of sensitivity and methodological simplicity for use in shrimp dietary research. It is suitable for large-scale testing of field collected specimens and can be used in the field with equipment no more sophisticated than a refrigerator. Gut content sample sizes required for the ELISA procedure are also much smaller than for immunodiffusion or immunoelectrophoretic methods. Our greatest challenge in the quantitative assay of natural shrimp feeding processes is the necessity of distinguishing between meal size and time elapsed since feeding occurred. Until we can state with certainty that a given meal was small because not much was eaten rather than being small because the meal had been digested for a period of time, it will indeed be difficult to measure the impact of shrimp predation on any prey population. Behavioral aspects of shrimp feeding may never be solved without better observational techniques, although some approaches identifying proteins which persist in the gut the longest or whose digestive degradation sequence can be determined have promise (Zagursky and Feller, 1987; Sunderland, 1988).

Ecological investigations of pollutant uptake and transport mediated via ingestion are also possible with immunoassays (*e.g.*, Schwalbe-Fehl, 1986; Aherne, 1984), and immunoassays have already been developed for detecting viral pathogens in penaeids (*e.g.*, Lewis, 1986). Although these techniques are still somewhat new to shrimp biologists, they should soon become available to even more investigators having additional ideas for novel approaches to the study of penaeid shrimp trophic ecology.

Acknowledgements

This work on shrimp feeding was largely due to generous support from the South Carolina Sea Grant Consortium. The National Science Foundation (Biological Oceanography Program) supported the development and testing of antiserum used in the studies. The dedication and assistance of Bob Ferguson, Brian Hentschel, Judy Hunter, Lara Matthews, and Greg Zagursky were indispensible. I am much in their debt. This is contribution number 763 from the Belle W. Baruch Institute.

Literature Cited

Aherne, G. W. 1984. Use and significance of immunoassays in the analysis of water. Anal. Proc. 21:177-179.

Axelsen, N. H., and E. Bock. 1983. Electroimmunoassay (rocket immunoelectrophoresis). Scand. J. Immunol., Vol. 17, Suppl. 10, pp. 103-106.

Berg, J. 1979. Discussion of methods of investigating the food of fishes with reference to a preliminary study of the prey of *Gobiusculus flavescens*. Mar. Biol. 50:263-273.

Caceci, T., K. F. Neck, D. H. Lewis, and R. F. Sis. 1988. Ultrastructure of the hepatopancreas of the Pacific white shrimp, *Penaeus vannamei* (Crustacea: Decapoda). J. mar. biol. Ass. U. K. 68:323-337.

Calver, M. C. 1984. A review of ecological applications of immunological techniques for diet analysis. Aust. J. Ecol. 9:19-25.

Calver, M. C., and B. D. Porter. 1986. Unravelling the food web: dietary analysis in modern ecology. J. Biological Education 20:42-46.

Calver, M. C., J. N. Matthiessen, G. P. Hall, J. S. Bradley, and J. H. Lillywhite. 1986. Immunological determination of predators of the bush fly, *Musca vetustissima* Walker (Diptera: Muscidae), in south-western Australia. Bull. ent. Res. 76:133-139.

Clausen, J. 1981. Immunochemical Techniques for the Identification and Estimation of Macromolecules. 2nd ed. Vol. 1, Part III of the series *Laboratory Techniques in Biochemistry and Molecular Biology*, 387 pp. Elsevier/North Holland Biomedical Press, Amsterdam.

Cockcroft, A., and A. McLachlan. 1986. Food and feeding habits of the surf zone penaeid prawn *Macropetasma africanus* (Balss). P.S.Z.N.I.: Marine Ecology 7:345-357.

Feller, R. J. 1986. Immunological detection of *Mercenaria mercenaria* in a predator and preparation of size-class specific antibodies. The Veliger 28:341-347.

Feller, R. J., and E. D. Gallagher. 1982. Antigenic similarities among estuarine soft-bottom benthic taxa. Oecologia (Berl.) 52:305-310.

Feller, R. J., and R. B. Ferguson. 1988. Quantifying stomach contents using immunoassays: a critique. pp. 295-303, In. Immunochemical Approaches to Estuarine, Coastal and Oceanographic Questions, C. M. Yentsch, F. C. Mague, and P. K. Horan, eds., Springer-Verlag, Coastal Lecture Note Series. No. 25.

Feller, R. J., G. L. Taghon, E. D. Gallagher, G. E. Kenny, and P. A. Jumars. 1979. Immunological methods for food web analysis in a soft-bottom benthic community. Mar. Biol. 54:61-74.

Fretwell, S. D. 1987. Food chain dynamics: the central theory of ecology ? Oikos 50:291-301.

Fry, B., R. Lutes, M. Northam, and P. L. Parker. 1982. A 13C/12C comparison of food webs in Caribbean seagrass meadows and coral reefs. Aquatic Botany 14:389-398.

Gleason, D. F. 1986. Utilization of salt marsh plants by post-larval brown shrimp: comparisons of growth, stable carbon isotopes, and food preferences. Mar. Ecol. - Prog. Ser. 31:151-158.

Gleason, D. F., and G. M. Wellington. 1988. Food resources of postlarval brown shrimp (*Penaeus aztecus*) in a Texas salt marsh. Mar. Biol. 97:329-337.

Grisley, M. S., and P. R. Boyle. 1985. A new application of serological techniques to gut content analysis. J. Exp. Mar. Biol. Ecol. 90:1-9.

Gross, W., and W. Marz. 1988. Immunoelectrophoretic techniques in protein analysis and quantitation. American Biotechnology Laboratory 6:6-19.

Heyraud, M., P. Domanski, R. D. Cherry, and M. J. R. Fasham. 1988. Natural tracers in dietary studies: data for 210Po and 210Pb in decapod shrimp and other pelagic organisms in the Northeast Atlantic Ocean. Mar. Biol. 97:507-519.

Hill, B. J., and T. J. Wassenberg. 1987. Feeding behavior of adult tiger prawns, *Penaeus esculentus*, under laboratory conditions. Aust. J. Mar. Freshw. Res. 38:183-190.

Hunter, J., and R. J. Feller. 1987. Immunological dietary analysis of two penaeid shrimp species from a South Carolina tidal creek. J. Exp. Mar. Biol. Ecol. 107:61-70.

Hynes, H. B. N. 1950. The food of freshwater sticklebacks (*Gasterosteus aculeatus* and *Pygosteus pungitius*) with a review of methods used in studies of the food of fishes. J. Anim. Ecol. 19:36-58.

Hyslop, E. J. 1980. Stomach contents analysis - a review of methods and their application. J. Fish. Biol. 17:411-429.

Knutsen, H., and N. B. Vogt. 1985. An approach to identifying the feeding patterns of lobsters using chemical analysis and pattern recognition by the method of SIMCA. II. Attempts at assigning stomach contents of lobsters, *Homarus gammarus* (L.), to infauna and detritus. J. Exp. Mar. Biol. Ecol. 89:121-134.

Kroll, J. 1983. Rocket-line immunoelectrophoresis. Scand. J. Immunol. Vol. 17, Suppl. 10, pp. 165-169.

Lewis, D. H. 1986. An enzyme-linked immunosorbent assay (ELISA) for detecting penaied baculovirus. J. Fish Diseases 9:519-522.

Luck, R. F., B. M. Shepard, and P. E. Kenmore. 1988. Experimental methods for

evaluating arthropod natural enemies. Ann. Rev. Entomol. 33:367-391.
McTigue, T. A., and R. J. Feller. 1989. Feeding of juvenile white shrimp (*Penaeus setiferus*): periodic or continuous ? Mar. Ecol. Prog. Ser. 52:227-233.
Nelson, W. G. 1981. Experimental studies of decapod and fish predation on seagrass macrobenthos. Mar. Ecol. - Prog. Ser. 5:141-149.
Peters, R. H. 1977. The unpredictable problems of tropho-dynamics. Env. Biol. Fish. 2:97-101.
Peterson, B. J., R. W. Howarth, and R. H. Garritt. 1985. Multiple stable isotopes used to trace the flow of organic matter in estuarine food webs. Science 227:1361-1363.
Pickavance, J.R. 1970. A new approach to the immunological analysis of invertebrate diets. J. Anim. Ecol. 39:715-724.
Schwalbe-Fehl, M. 1986. Immunoassays in environmental analytical chemistry. Intern. J. Environ. Anal. Chem. 26:295-304.
Slansky, F., Jr. 1982. Insect nutrition: an adaptationist's perspective. Florida Entomologist 65:45-71.
Stoner, A. W. 1988. A nursery ground for four tropical *Penaeus* species: Laguna Joyuda, Puerto Rico. Mar. Ecol. - Prog. Ser. 42:133-141.
Sunderland, K. D. 1988. Quantitative methods for detecting invertebrate predation occurring in the field. Ann. appl. Biol. 112:201-224.
Suthers, I. M. 1984. Functional morphology of the mouthparts and gastric mill in *Penaeus plebejus* Hess (Decapoda: Penaeidae). Aust. J. Mar. Freshw. Res. 35:785-792.
Theilacker, G. H., A. S. Kimball, and J. S. Trimmer. 1986. Use of an ELISPOT immunoassay to detect euphausiid predation on larval anchovy. Mar. Ecol. - Prog. Ser. 30:127-131.
Walter, C. B., E. O'Neill, and R. Kirby. 1986. "ELISA" as an aid in the identification of fish and molluscan prey of birds in marine ecosystems. J. Exp. Mar. Biol. Ecol. 96:97-102.
Wang, A.-C. 1982. Methods of immune diffusion, immunoelectrophoresis, precipitation, and agglutination. pp. 139-161, In. Antibody as a Tool, J. J. Marchalonis and G. W. Warr, eds., John Wiley & Sons Ltd.
Wassenberg, T. J., and B. J. Hill. 1987. Natural diet of the tiger prawns *Penaeus esculentus* and *P. semisulcatus*. Aust. J. Mar. Freshw. Res. 38:169-182.
Young, J. O., I. G. Morris, and T. B. Reynoldson. 1964. A serological study of *Asellus* in the diet of lake-dwelling triclads. Arch. Hydrobiol. 60:366-373.
Zagursky, G., and R. J. Feller. 1988. Application of immunoblotting for dietary analysis. pp. 117-128, In. Immunochemical Approaches to Estuarine, Coastal and Oceanographic Questions, C. M. Yentsch, F. C. Mague, and P. K. Horan, eds., Springer-Verlag, Coastal Lecture Note Series. No. 25.

Stable Isotope Methodology for Evaluation of Nutritional Requirements Of Shrimp

Patrick L. Parker, Richard K. Anderson
The University of Texas, Marine Science Institute
Port Aransas, Texas 78373
Addison L. Lawrence
Texas A&M University, Shrimp Mariculture Project
Port Aransas, Texas 78373

Abstract

The isotope tracer approach taken in this study is general so that it may be used in other types of nutrition research. It is especially well suited to aquaculture because it is possible to formulate feeds with a desired $d^{13}C$, $d^{15}N$ or $d^{34}S$ value. The isotopic measurements are very precise yet can be made on a mass production basis. There is reason to expect that carbon and nitrogen assimilation data derived from isotope measurements will be factored into the economic model of large scale aquaculture systems.

Introduction

While it is well established that chemical tracer experiments can be done with both radioactive and enriched stable isotopes, it is just now becoming generally realized that stable isotopes ratio variations at the natural abundance level constitute tracer experiments for many ecological systems. The elements with stable isotope pairs and chemical properties which lead to significant isotope effects include hydrogen, carbon, nitrogen, oxygen and sulfur. These are by good fortune the elements of major interest to biologists and geochemists. Chemical isotope effects in key reactions in the geochemical cycle of these elements have created reservoirs of material which can serve as tracers. For carbon, the isotope effect in carboxylation,

$$\mathrm{RuBP} + {}^{12}\mathrm{CO}_2 \xrightarrow{k_{12}} 2\ \mathrm{PGA}\text{-}{}^{12}\mathrm{C}$$

$$RuBP + {}^{13}CO_2 \xrightarrow{k_{13}} 2\ PGA\text{-}{}^{13}C$$

that is photosynthesis, leads to the reservoirs that are of interest in the context of shrimp nutrition. Thus in the reactions the isotope effect is determined by k_{12}/k_{13}, the ratio of the rate constants. The product 3-P-glyceric acid will have a $^{13}C/^{12}C$ ratio different from that of the ribulose-1,5-bisphosphate (RuBP) as long as k_{12}/k_{13} is not unity. For plants such as soybeans that fix carbon using the RuBP pathway, the ratio of rate constants is approximately 1.026. These are commonly called C_3 plants. Certain other plants including tropical grasses fix carbon by a four-carbon pathway and have different and well resolved values of the rate constant ratio. These plants, of which corn and sorghum are examples, are called C_4 plants. It is the custom to utilize $d^{13}C$ (Defined as: $d^{13}C = (^{13}C/^{12}C)_{sample} - (^{13}C/^{12}C)_{ref}$ X $1000/(^{13}C/^{12}C)_{ref}$ where the reference adopted if the PDB carbonate.) rather than rate constants in this field of research (van der Mewre, 1982). The relationships are summarized as

	k12/k13	d13C
C_4 plants	~1.010	-10
marine algae	~1.020	-20
C_3 plants	~1.026	-26

follows where $d^{13}C$ of the reactant is that of the source inorganic carbon.

Since $d^{13}C$ values can be routinely measured to $\pm$ 0.2°/00, these variations form the basis of the use of $^{13}C/^{12}C$ ratios at the natural abundance level as tracers of carbon flow in terrestrial and aquatic food webs (Fry and Sherr, 1984). It has been confirmed in many studies that the large fractionations in the fixation of carbon by autotrophs are not repeated in metabolism and synthesis by heterotrophs. This relationship of "you are what you eat to $\pm$ 1°/oo" is well illustrated in Figure 1. There are some interesting smaller isotope effects in metabolism which will be noted later. The considerable body of data and relationships which has been discovered in ecological studies form a basis for the animal nutrition studies in an aquaculture setting.

These studies began with an idea for a way to solve the well recognized pond

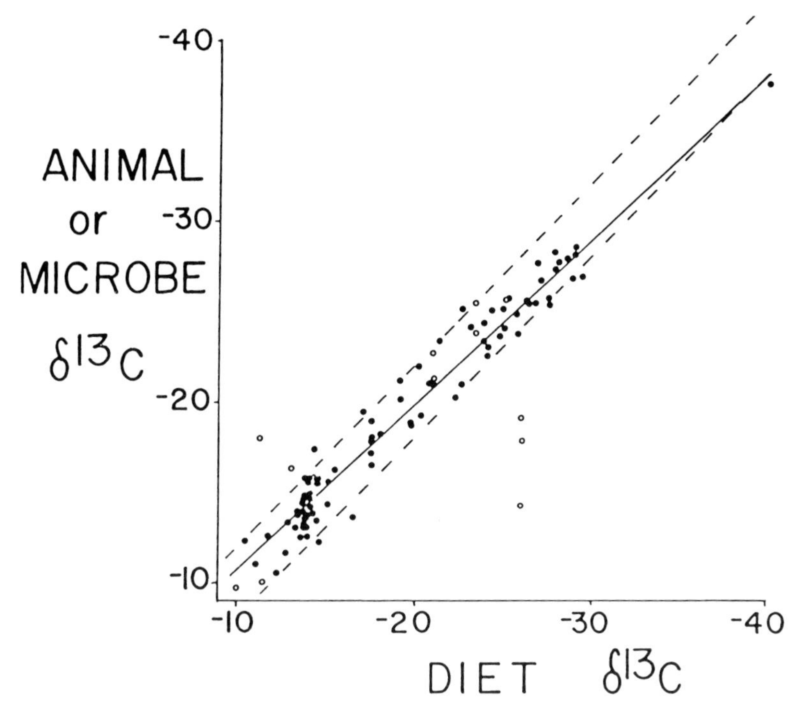

Figure 1. Relationship between organismal and dietary values for animals (.) and microbes (0). (From Fry and Sherr, 1984).

aquaculture problem of the relative contributions of primary productivity and supplied feed rations to cultured animals. Our approach, developed in earlier publications, is described more fully in this paper (Anderson *et al.*, 1987; Parker *et al.*, 1988). The approach is based on using $d^{13}C$ as a tracer of the carbon assimilated by shrimp under various experimental conditions. Based on the $d^{13}C$ values of the potential components of feeds shown in Figure 2, a series of experimental feeds were formulated with overall $d^{13}C$ values ranging from -15.2 to -25.1°/oo. These materials constituted a series of well resolved feeds which were used to examine shrimp nutrition in a unique way. Although our long range goal has been to study nutrition in pond systems, several studies were done in tanks where better experimental control was possible. In the $d^{13}C$ approach, the shrimp is viewed as a mixture of the initial shrimp carbon and carbon assimilated during an experiment. Thus at any time during growth:

$$d_tW_t = d_iW_i + d_gW_g \quad (1)$$

where

d_t, d_i, d_g = the $d^{13}C$ of shrimp tissue at time = t, initial tissue and tissue gained during the experiment, respectively.

and W_t, W_i, W_g = weights of these respective tissues.

If a variety of sources are available with resolved $d^{13}C$ values, then W_g is complex and:

$$W_g = W_1 + W_2 + \ldots\ldots W_n \quad (2)$$

and

$$d_tW_t = d_iW_i + d_1W_1 + d_2W_2 \ldots\ldots d_nW_n \quad (3)$$

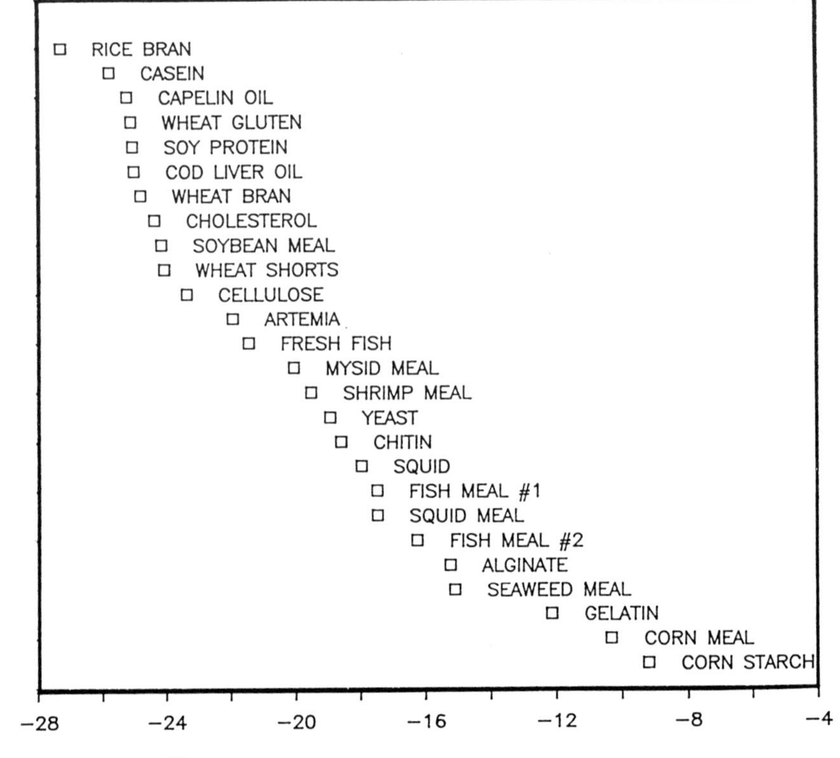

Figure 2. The $d^{13}C$ of the individual components used to formulate the feeds used in tank and pond studies. (from Anderson, Parker and Lawrence, 1987).

where the numerical subscripts refer to different sources of nutrition. Clearly a system wherein the shrimp have only two choices of food would be necessary for unambiguous results. In most cases this has been pond biota and added formulated feeds. However, tank experiments where the shrimp have only one apparent source of food have been very useful for obtaining background data as well as useful rate relationships.

Tank Experiments

Test of the "You are what you eat" Relationship

Tank experiments were used to quantitate the basic assumption that growing shrimp acquire the $d^{13}C$ value of feed. The results shown in Figure 3 are for a 28 day tank experiment wherein feeds of three different $d^{13}C$ values were offered to 1.5g *Penaeus vannamei*. The shrimp appear to track the feed poorly; however, when

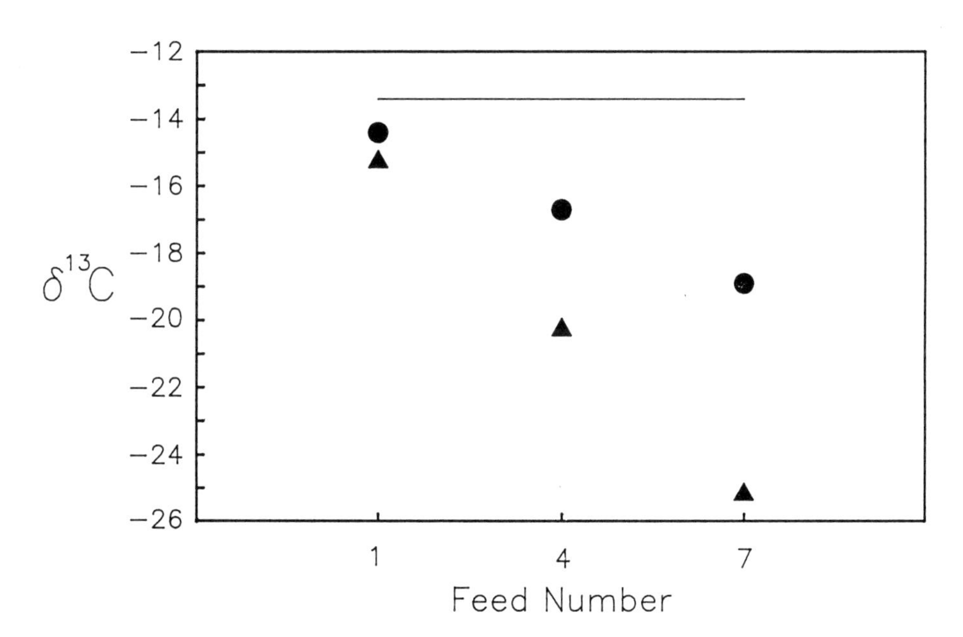

Figure 3. The relationship between the $d^{13}C$ of the shrimp tail muscle and three isotopically distinct offered feeds for a tank study. The (•) values represent the feeds (d_f) and the (Δ) values are the shrimp at the time of harvest (d_i). The line at -13.4 per mil represents the average value of the initial shrimp (d_i). (from Anderson, *et al.*, 1987).

d_g, $d^{13}C$ of the weight gained is calculated using equation 1, the values are within +0.1, +1.4 and +1.3°/oo of the feeds. This ~+1.0°/oo relationship has been found in many systems and may be related to a small isotope effect in metabolism. For numerical modeling of experimental data, this 1°/oo may be used as a correction factor.

Turnover Rates of Animal Tissues.

Tank experiments with small and large *P. setiferus* have provided data on the rate at which various tissues acquire the $d^{13}C$ value of offered feeds. Figure 4a shows the growth rate of the experimental animals and Figure 4b the $d^{13}C$ relationships. The $d^{13}C$ values are for whole tissue, not corrected to dg. The rapid rate at which the digestive gland approached the $d^{13}C$ of the feed is remarkable. The other

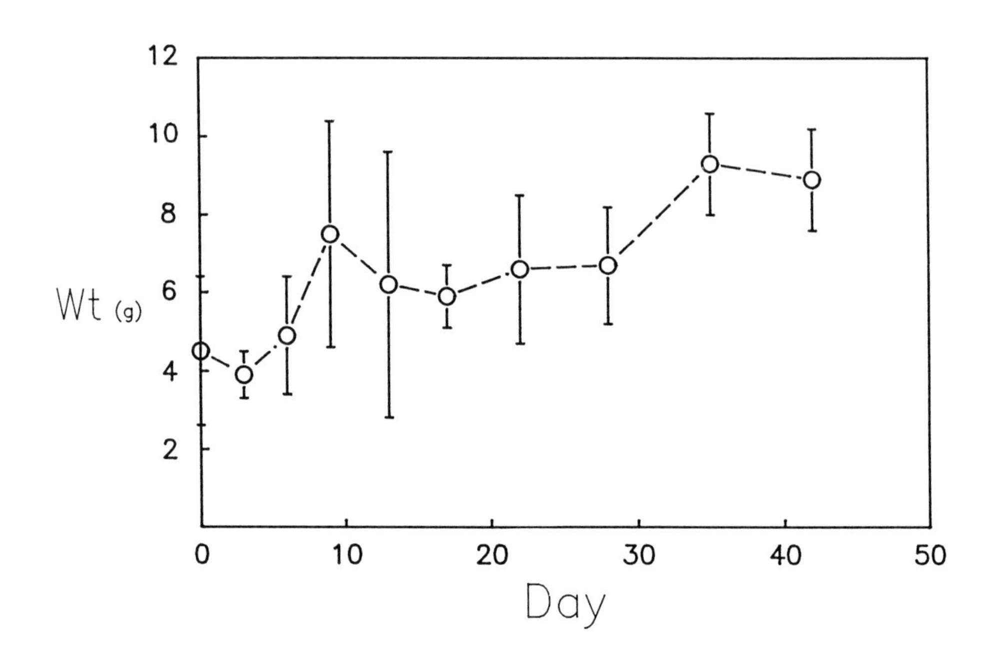

Figure 4a. The weight gain for small *Penaeus setiferus* used in the turnover experiment.

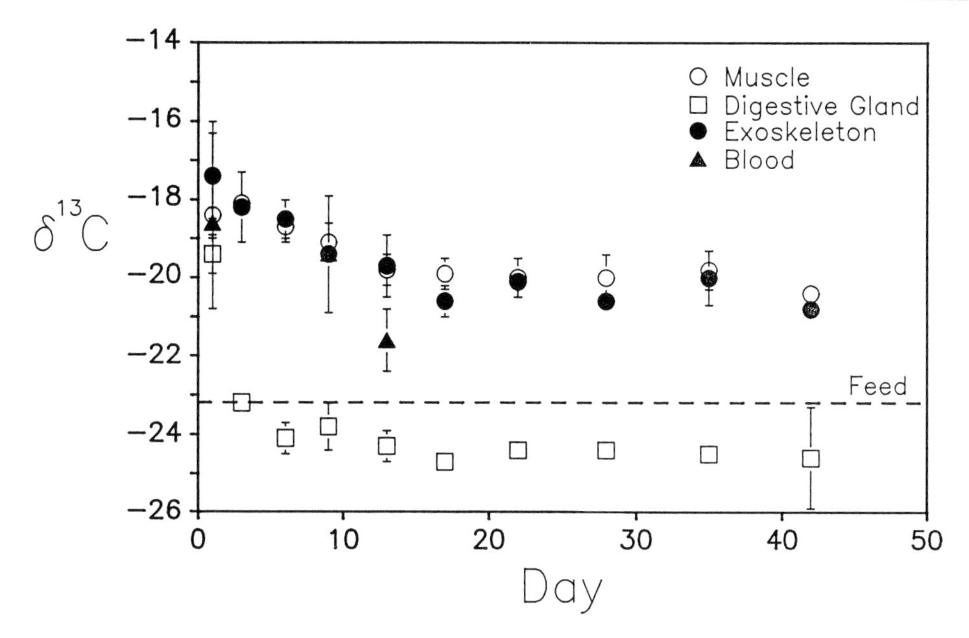

Figure 4b. The change of d^{13}C of various tissues of tank grown *Penaeus setiferus*. Note the rapid change of the digestive gland.

tissues resemble the trend shown in Figure 3 for muscle tissue. Similar results for large *P. setiferus* are shown in Figures 5a and 5b. Again, d^{13}C of the digestive gland quickly approaches that of the feed. These large shrimp show little growth so that the rate of replacement of carbon is being measured rather than growth plus replacement as for the smaller animals. Other experiments using purified feeds have been performed and are being evaluated for more details of selective biochemical assimilation.

Selective Uptake of Fatty Acids.

Tank experiments were used to investigate selective uptake of feed components by shrimp. It is well known that natural biota with a given d^{13}C value can be separated into biochemical fractions with distinct d^{13}C values (Abelson and Hoering, 1961; Parker, 1964). In general, lipids are more negative (^{13}C depleted) and protein more positive (^{13}C enriched) than the total tissue. Carbohydrate falls between these extremes. Certainly the feeds formulated from the components in Figure 2 show this same relationship (whole fish = -20°/oo, while capelin oil = -25°/oo). In order to

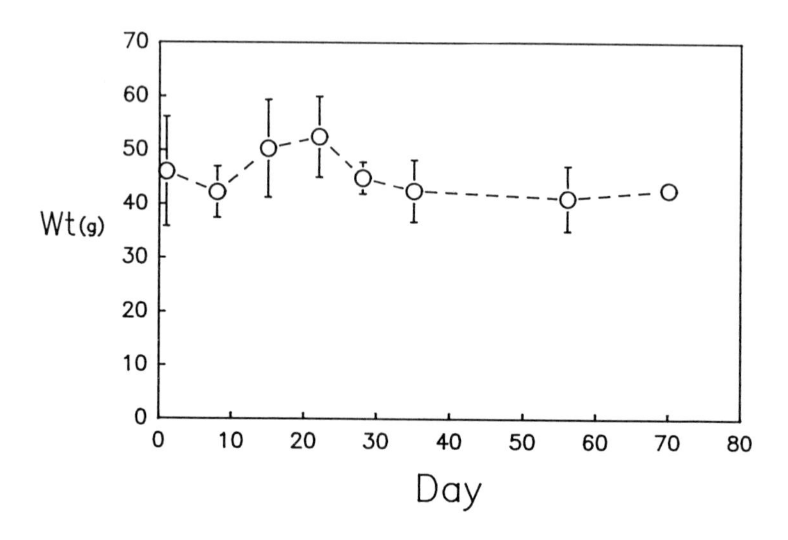

Figure 5a. The weight data for large *Penaeus setiferus* used in the turnover experiment.

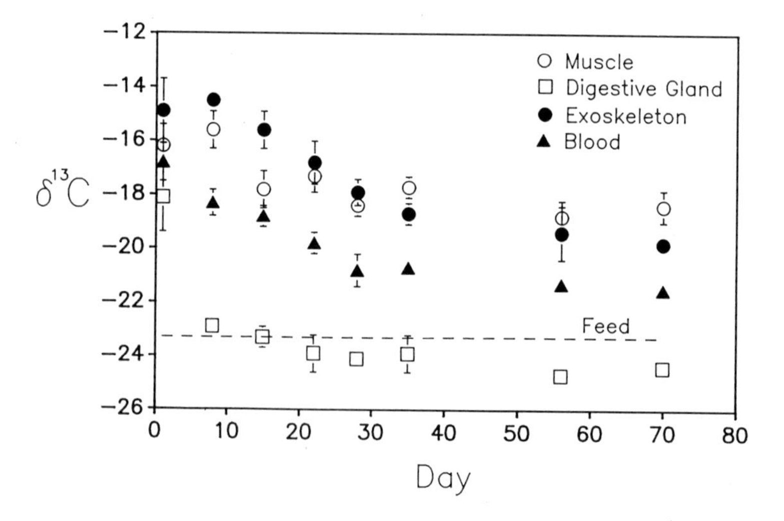

Figure 5b. The change of $d^{13}C$ of various tissues for large *Penaeus setiferus*.

investigate fatty acid nutrition, three feeds were formulated for which d^{13}C was measured for three fractions, the total lipids, the total fatty acids and the non-saponifiables. The lipid contents of the three feeds were based on soybean oil, corn oil and mehaden oil. Controlled feeding experiments using small shrimp were carried out for 28 days. The results are shown in Figures 6a, b and c. The trials with soy

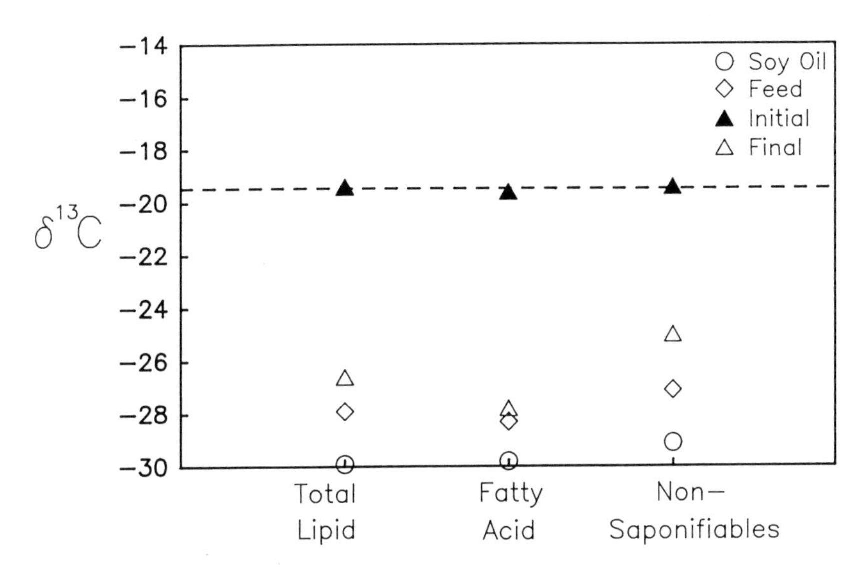

Figure 6a. The d^{13}C relationships of the lipids of *Penaeus setiferus* which were grown in tanks on a feed made with soybean oil. Note that the final shrimp values approach the soy oil values.

and menhaden oil feeds demonstrate efficient assimilation of soy and menhaden lipids but strikingly different results were obtained for the corn oil feed. For the soy oil, the d^{13}C of fatty acids for the pure oil and the acids from the formulated feed were close together at -28°/oo, well resolved from the acids of the initial shrimp. By harvest time d^{13}C of the initial shrimp had moved from -19 to -28°/oo. Total lipid showed the same trend. Non-saponifiables moved toward the feed, but showed considerable lag. In the menhaden oil experiment, all three lipid fractions tracked the d^{13}C of the feed to within +1.0°/oo. The corn oil experiment gave unexpected results. Instead of shifting toward the pure corn oil used in the feed, all three fractions moved in the opposite direction, in fact they moved away from the value of the total feed

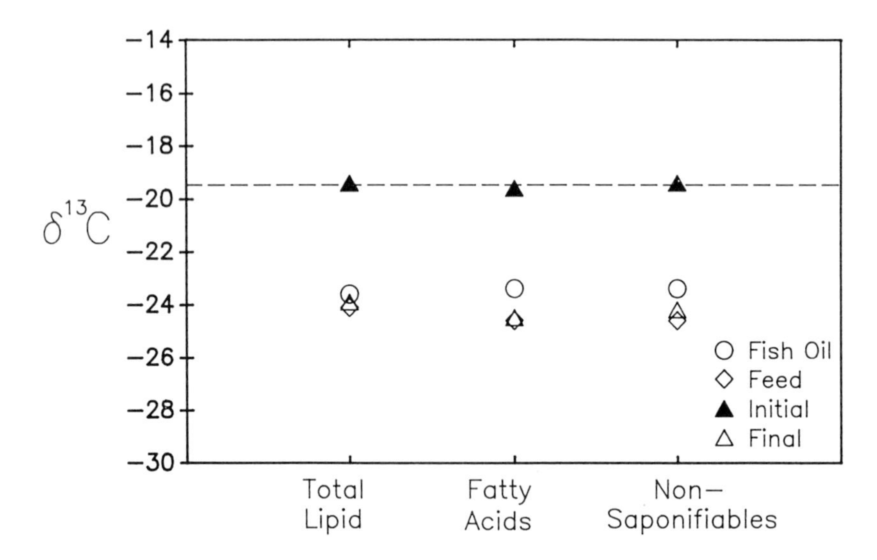

Figure 6b. The d^{13}C relationships of the lipids of *Penaeus setiferus* which were grown in tanks on a feed made with menhaden oil. Note that the shrimp values track the feed oil values.

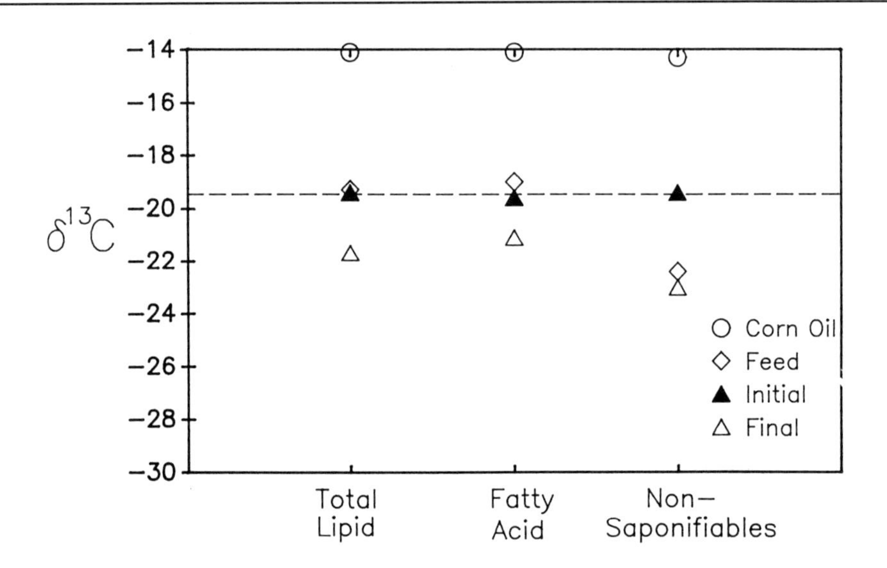

Figure 6c. The d^{13}C relationships of the lipids of *Penaeus setiferus* which were grown in tanks on a feed made with corn oil. The failure of the final shrimp to track the feed oil is due to a trace of fish meal in the feed.

itself. This is only possible if the feed contained lipid with $d^{13}C$ around - 24°/oo or if the shrimp were performing a total synthesis of fatty acids. It was later found that the feed contained small amounts of menhaden oil which were too dilute to shift the $d^{13}C$ of the feed lipids, but which were clearly selectively assimilated by the shrimp. This work suggests that competitive uptake experiments where the shrimp are offered fatty acid or other chemical fractions from various sources (with resolved $d^{13}C$ values) would be useful. In a sense, pond experiments are such multi-source experiments.

Pond Experiments

Feeding Rate and Stocking Density Effects

A series of feeding experiments in ponds were done using the formulated feeds for which $d^{13}C$ of the total carbon is known. The experimental details are published but are essentially as follows (Anderson *et al.,* 1987). A single earth pond with 1 m^2 bottomless cages in contact with sediment was used. Shrimp were also stocked outside the cages, but not offered the formulated feed. Typically pond trials were for 28 days and utilized small (~1.5g) shrimp. In an early experiment designed to measure the influence of feeding rate on cages stocked at three animal densities, it was noted that the pond shrimp did not track $d^{13}C$ of the feed nearly as closely as did the shrimp in the tank experiments (Figure 7). This was not unexpected and in fact the quantitation of this was one of our major project goals. If the shrimp do not assimilate only added feed then they must be utilizing pond biota. The observation that the steady state feed utilization rate is reached at ~5°/oo of body wt/day (Figure 7) suggests that higher feeding rates may be serving only to stimulate the non-shrimp pond biota. It is interesting to note that the low-density stocking experiments showed slightly more pond biota utilization and the high-density stocking slightly more added feed utilization. These trends are probably at the limit of experimental error, but they are consistent for all densities at all feeding rates.

Pond Biota Versus Feed Utilization

Based on these background experiments, a pond experiment can be designed to test the relative assimilation of carbon from formulated feeds and pond biota. In one such experiment seven formulated feeds ranging from -15.2 to -25.1°/oo were used and contrasted to pond biota which was estimated to range between -12 and -15°/oo. The results are shown in Figure 8. First inspection shows that the shrimp did not

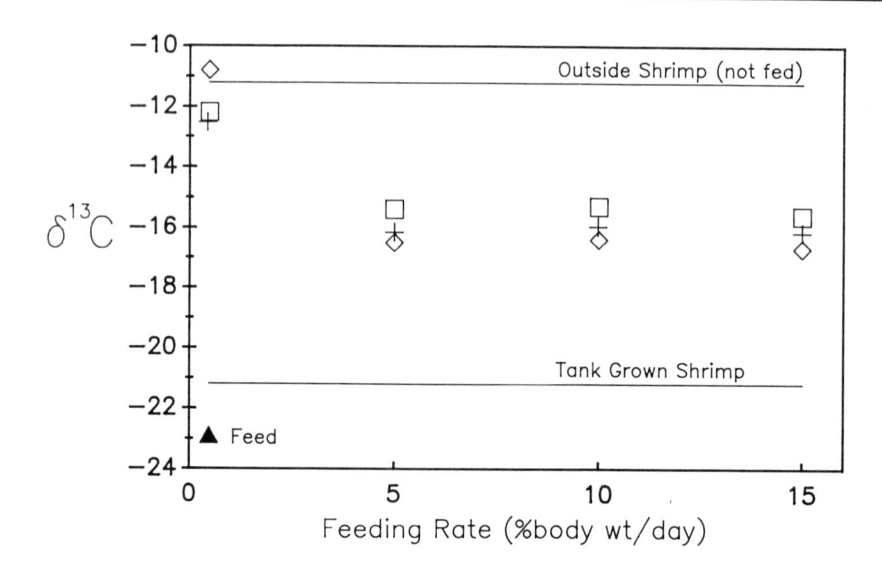

Figure 7. The d13C relationships of Penaeus vannamei grown in ponds at three feeding rates and three densities (, +,) are 10, 20 and 40 shrimp/m^2.

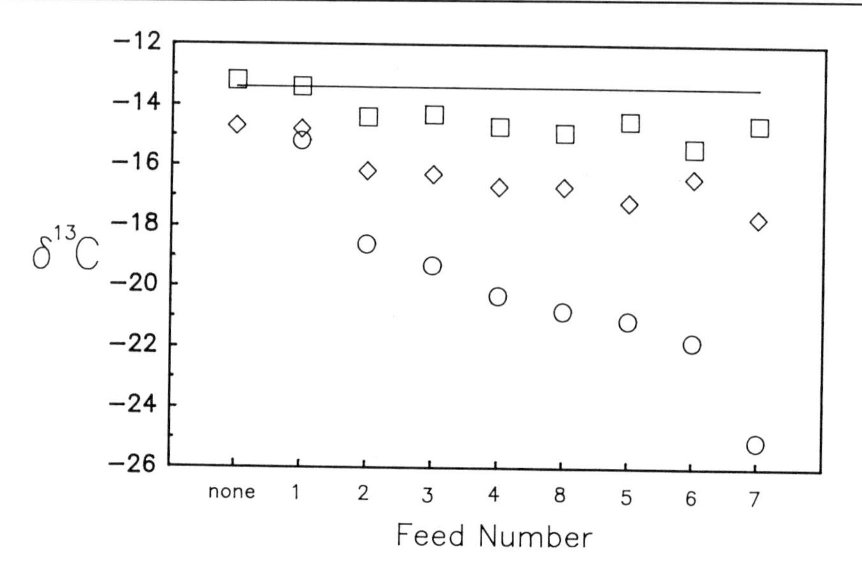

Figure 8. The relations between d^{13}C of shrimp and offered feed for a pond study. The (o) values are the feed (d_f), the ( ) values are the shrimp at harvest time (dt), the (X) values are the carbon gained (d_g). The line at -13.4 represents the initial shrimp (d_i). (from Anderson, Parker and Lawrence, 1987).

track $d^{13}C$ of the added feed to within + 1.0% as they did in the simple tank experiments. Even after the raw data is corrected to dg, the $d^{13}C$ of the weight gained, there is a considerable difference between that and df, the $d^{13}C$ of the feed. There is little doubt that pond biota is a major carbon source for this system. Using the relationship

$$\frac{W_p}{W_f} = \frac{d_f - d_g}{d_g - d_p} \qquad (4)$$

derived from equations (1) and (2), the relative contribution of the feed and pond biota can be calculated. The calculations indicate that pond productivity accounts for 53 - 77% of the carbon assimilated by the pond grown shrimp. Schroeder (1983a;b) used stable isotopes in a similar but less controlled way and reached a similar general conclusion - that pond biota are key elements for aquaculture organisms.

Nitrogen Assimilation in Ponds

Nitrogen is perhaps more important than carbon in aquaculture nutrition because of its cost. The isotope pair $^{15}N/^{14}N$ can be used to do nitrogen tracer experiments at the natural abundance level in a fashion similar to carbon. The parameter $d^{15}N$ is somewhat more complex than $d^{13}C$ because $d^{15}N$ shifts +3 to +4°/oo as organic matter moves to a new trophic level. This is currently regarded as the result of fractionation during nitrogen metabolism. Tracer $d^{15}N$ experiments were in fact done when the $d^{13}C$ runs were made because the feeds had a specific, although unplanned, $d^{15}N$ value. Tank $d^{15}N$ feeding experiments indicated that the trophic shift for *P. vannamei* was +2.4% (Figure 9). Because the pond $d^{15}N$ samples were actually aliquots of the $d^{13}C$ samples, the results may be compared with some confidence. The results in Figure 10 show that the shrimp were assimilating nitrogen with a $d^{15}N$ value of +6.2, a value about midway between that of the pond biota (+3.4) and the formulated food (+9.3). Application of the mixing equation (4) to these data gives a value of about 1, indicating that the feed and pond biota supply equal amounts of the nitrogen assimilated by the pond grown shrimp.

	$d^{15}N$(°/oo)
Initial shrimp (d_0)	+5.2
Feed (d_d)[a]	+9.3
Final Shrimp (d_t)	+9.2

Final weight	3.9g
Initial weight	1.5g
Weight gained	2.4g

$$d_tW_t = d_0W_0 + d_gW_g$$
$$(9.2)\ (3.9) = (5.2)\ (1.5) + d_g(2.4)$$
$$35.9 = 7.8 + d_g(2.4)$$
$$d_g = +11.7°/oo$$

and $$d_g = d_d + \text{trophic shift.}$$

So, $$\text{Trophic shift} = 11.7 - 9.3 = +2.4°/oo$$

[a] In the case of the tank study, the d of the animal's diet is identical to the d of the prepared feed (df).

Figure 9. $d^{15}N$ Data used to calculate the trophic shift and the $d^{15}N$ values of the weight gained (d_g) of shrimp grown in a controlled tank experiment (Parker *et al.*, 1988).

	$d^{15}N$ (°/oo)
Initial shrimp (d0)	+5.2
Final pond shrimp (caged)(dt)	+7.6
Pond biota	+3.4

Caged shrimp:

Final weight	5.0g
Initial weight	1.5g
Weight gained	3.5g

$$d_tW_t = d_0W_0 + d_gW_g$$
$$(7.6)\ (5.0) = (5.2)\ (1.5) + d_g(3.5)$$
$$d_g = +8.6°/oo$$

and since $$d_g = d_d^{\,a} + \text{trophic shift.}$$

Then, $$d_d = 8.6 - 2.4 = +6.2°/oo$$

[a] In the case of the pond study, the diet of the shrimp includes the prepared feed and the pond biota.

Relative nitrogen contributions by pond biota and feed to the diet of the shrimp

$$\frac{W_{feed}}{W_{pond}} = \frac{d^{15}N_{pond} - d^{15}N_{diet}}{d^{15}N_{diet} - d^{15}N_{feed}} = \frac{3.4 - 6.2}{6.2 - 9.3} = \frac{1}{1.1}$$

Figure 10. $d^{15}N$ Data used to calculate $d^{15}N$ of shrimp grown in an earth pond (d_g) and $d^{15}N$ of the total diet of the shrimp (d_d)[a] (Parker *et al.*, 1988).

Literature Cited

Abelson, P.H. and T.C. Hoering. 1961. Carbon isotope fractionation in formation of amino acids by photosynthetic organisms. Proc. Natl. Acad. Sci. USA 47: 623-632.

Anderson, R.K., P.L. Parker and A. Lawrence. 1987. A $^{13}C/^{12}C$ tracer study of the utilization of presented feed by a commercially important shrimp *Penaeus vannamei* in a pond growout system. J. World Aquacult. Soc. 18(3):148-155.

Fry, B. and E. Sherr. 1984. $d^{13}C$ Measurements as indicators of carbon flow in marine and freshwater ecosystems. Contrib. Marine Sci. 27: 13-47.

Parker, P.L. 1964. The biogeochemistry of stable isotopes of carbon in a marine bay. Geochim. Cosmochim. Acta 28:1155-1164.

Parker, P.L., R.K. Anderson and A. Lawrence. 1988. A $d^{13}C$ and $d^{15}N$ tracer study of nutrition in aquaculture: *Penaeus vannamei* in a pond growout system. In. Stable Isotopes in Ecological Research (Rundel, Ehleringer and Nagy, eds.)Springer-Verlag p. 288-303.

Schroeder, G. L. 1983a. Stable isotope ratios as naturally occurring racers in the aquazculture food web. Aquaculture 30: 203-210.

Schroeder, G. L. 1983b. Sources of fish and prawn growth in polyculture ponds as indicated by oC analysis. Aquaculture 35: 29-42.

van der Mewre, N.J. 1982. Carbon isotopes, photosynthesis, and archaeology. Amer. Scientist 70:596-606.

Hosts, Geographic Range and Diagnostic Procedures for the Penaeid Virus Diseases of Concern to Shrimp Culturists in the Americas

D.V. Lightner and R.M. Redman
Environmental Research Laboratory
University of Arizona, 2601 East Airport Drive
Tucson, AZ 85706 USA

Abstract

At least six virus diseases are presently known in cultured penaeid shrimp. All but one of these six have been detected in cultured penaeids in the Americas (North, Central, and South America, the Caribbean Islands, and Hawaii). Each of these six penaeid virus diseases (BP = *Baculovirus penaei*; MBV =*P. monodon*-type baculovirus; BMN = baculoviral midgut gland necrosis; HPV = hepatopancreatic parvo-like virus; IHHNV = infectious hypodermal and hematopoietic necrosis virus; and REO = reo-like virus of the hepatopancreas) may be comprised by a multitude of individual strains, some of which are highly pathogenic to some penaeids, while being of little importance to others.

BP is widespread in its distribution in cultured and wild penaeids in the Americas, and except for Hawaii, it has not been observed elsewhere. MBV-type baculoviruses are analogous to BP in their diverse host range and in their wide distribution on the IndoPacific coasts of Asia, Australia and Africa and in Southern Europe. However unlike BP, MBV has been observed in the Americas in imported stocks and in American penaeids exposed to the virus.

HPV has a geographic range similar to that of MBV in Asia and Australia, and like MBV it has been introduced to the Americas with imported penaeids. IHHNV has a world-wide distribution in cultured penaeid shrimp, but its distribution in wild penaeids remains unknown. The only occurrences of IHHNV (or a similar agent) in shrimp culture facilities using only captive-wild broodstock have been in Southeast Asia. Only BMN has not been detected in penaeids cultured in the Americas.

Three basic diagnostic procedures are used in screening penaeid shrimp stocks for virus infections: 1) direct samples for microscopic (wet-mount) examination or histopathology for signs of virus infection (*e.g.*, polyhedral occlusion bodies); 2) enhancement of infection by severe crowding "stress" followed by microscopic examination or histopathology; and 3) bioassay of a suspect shrimp population with a sensitive indicator species followed by sampling and histopathology. More rapid and sensitive advanced diagnostic procedures based on serological and gene probe technologies are being developed, but are not yet available to the industry.

Introduction

Six virus diseases are presently recognized in the penaeid shrimp (Table 1). These six viruses are: BP = *Baculovirus penaei* (Couch, 1974); MBV =*P. monodon*-type baculovirus (Lightner and Redman, 1981); BMN = baculoviral midgut gland necrosis (Sano *et al.,* 1981); IHHNV = infectious hypodermal and hematopoietic necrosis virus (Lightner *et al.,* 1983a); HPV = hepatopancreatic parvo-like virus (Lightner and Redman, 1985); and REO = reo-like virus (also known as RLV) of the hepatopancreas (Tsing and Bonami, 1987). Each virus may actually be comprised by a multitude of individual strains, some of which are highly pathogenic to some penaeids, while being of little importance to others (Table 2).

The host geographic range of the known penaeid viruses has been updated several times recently (Couch, 1981; Johnson, 1983; Lightner, 1983; and Lighter *et al.*, 1985), and then comprehensively reviewed three years ago, but just recently published (Lightner, 1988). Since then, surveys and investigations of mortality problems undertaken by this laboratory in various shrimp growing areas have provided new data on several of the virus diseases that affect cultured penaeid shrimp. This review of the penaeid viruses emphasizes the current diagnostic procedures for the penaeid viruses, their natural hosts, and their natural and introduced geographic distributions.

Table 1. The known penaeid viruses.

Virus	Approximate Virion Size	Probable Nucleic Acid	Classification
IHHNV	20 nm	ssDNA	Parvovirus
HPV	22 nm	ssDNA	Parvovirus
REO	60 nm	dsRNA(?)	Reo-like virus
BP	~75 x 300 nm	dsDNA	Baculovirus;occluded
MBV	~75 x 300 nm	dsDNA	Baculovirus; occluded
BMN	~75 x 300 nm	dsDNA	Baculovirus; non-occluded

Diagnostic Procedures

Three basic diagnostic procedures are used in screening penaeid shrimp for virus infections: 1) direct samples for microscopic (wet-mount) examination and/or histopathology or electron microscopy; 2) enhancement of infection followed by sampling and histopathology and electron micro-scopy; and 3) bioassay of a suspect shrimp population with a sensitive indicator species combined with direct sampling and examination of the indicator shrimp for signs of infection using wet-mounts or histopathology.

Direct Diagnostic Procedures

Random or (preferably) non-random samples of shrimp, or shrimp feces, are selected in the direct sampling procedure from culture tanks, ponds, or cages and examined directly for signs of BP or MBV in wet-mounts, or they may be preserved in Davidson's AFA or in 10% buffered formalin (Humason, 1967) for histological evaluation. The sensitivity of this procedure is limited, and it will only demonstrate shrimp with viral infections that are acute or subacute in a population with a high incidence rate. All six types of penaeid virus infections (IHHN, REO, BP, MBV, BMN, and HPV) may be diagnosed successfully with direct samples, but such samples have also produced false negative diagnoses on populations later shown by electron microscopy, enhancement, or bioassay diagnostic procedures to be positive for one of these virus diseases.

Enhancement Diagnostic Procedures

A quarantined population in the enhancement procedure is reared under relatively crowded and stressful conditions. Postlarvae are best used for this test, which normally requires 30 to 60 days. Random samples are taken at intervals throughout the test period, or moribund animals are nonrandomly sampled when they are observed. Samples may be prepared for direct wet-mount microscopic examination for occlusion bodies diagnostic of BP and MBV-caused diseases, or for histology or diagnosis of IHHN disease in *P. stylirostris*, *P. vannamei*, and *P. monodon*. Demonstration of inapparent infections due to HPV and REO may also be possible by the enhancement procedure. Enhancement is not a suitable procedure for demonstration of IHHNV

Table 2. The penaeid viruses and their known natural and experimentally infected hosts.

Host Subgenus And Species**	VIRUS* BP	MBV	BMN	IHHNV	HPV	REO
Litopenaeus:						
P. vannamei	+++	+		+	+	
P. stylirostris	++			+++		
P. setiferus	+			+(e)		
P. schmitti	++					
Penaeus:						
P. monodon		++		++	++	++
P. esculentus		+			++	
P. semisulcatus		+		+	+++	
Fenneropenaeus:						
P. merguiensis		++			+++	
P. indicus					++	
P. chinensis (=*orientallis*)					++	
P. penicillatus	++	++			++	
Marsupenaeus:						
P. japonicus			+++	++(e)		+++
Farantepenaeus:						
P. aztecus	+++		+(e)			
P. duorarum	+++		+(e)			
P. brasiliensis	++					
P. paulensis	++					
P. subtilis	++					
Melicertus:						
P. kerathurus		+				
P. marginatus	+++					
P. plebejus		++				

* Abbreviations:

BP = *Baculovirus penaei*
MBV = *P. monodon*-type baculovirus
BMN = Baculoviral midgut gland necrosis
IHHNV = Infectious hypodermal and hematopoietic necrosis virus

"continued"

"Table 2 continued"

HPV = Hepatopancreatic parvo-like virus
REO = Reo-like virus
+ = Infection observed, but without signs of disease
++ = Infection may result in moderate disease and mortality
+++ = Infection usually results in serious disease in some life stages.
e = Experimentally infected; natural infections not yet observed
** Classification according to Holthuis, 1980, FAO Species Catalog.

in asymptomatic carriers (*i.e.*, subadult or adult *P. stylirostris* IHHN epizootic survivors, or in species such as *P. vannamei* which are readily infected by the virus, but seldom show diagnosable infection after the early juvenile stages).

Bioassay Diagnostic Procedures

Carriers of IHHNV may be detected by bioassay with sensitive "indicator" shrimp. Indicator shrimp in this procedure (juvenile *P. stylirostris* of 0.05 to 4 g body weight) may be exposed to samples of suspect carrier shrimp by one or more of three methods: 1) injection with a cell-free filtrate prepared from a homogenate of suspect carrier shrimp (the indicator shrimp will show signs of IHHN disease within 5 to 15 days if the suspect shrimp were infected with IHHNV); 2) rearing in the same tank suspect carrier shrimp with indicator shrimp (the indicator shrimp will show signs of IHHN disease within 30 to 60 days); and 3) feeding minced carcasses of suspect carrier shrimp to indicator shrimp (the indicator shrimp will show signs of IHHN within 15 to 30 days). In actual bioassay tests, the latter technique of exposure (*i.e.*, feeding carcass fragments to the indicator shrimp) has become the method of choice.

Current Disease Specific Procedures

Actual diagnosis of infection by BP, MBV, HPV, IHHNV, and REO is dependent on microscopic or histologic demonstration of the particular cytopathology that is unique to each disease. Gross signs and behavior are usually not sufficiently specific in shrimp with infection by these penaeid viruses to be used reliably in diagnosing these diseases.

BP and MBV:

Patent acute BP and MBV infections may be readily diagnosed by demonstration of their characteristic occlusion bodies in either wet-mounts of feces, hepatopancreas, or midgut, or in histological preparations of the latter two organs. BP occlusions are distinctive tetrahedral bodies (Couch, 1974; 1981) easily detected by bright field or phase microscopy in unstained wet-mounts of feces or tissue squashes (Figures 1a & b), while MBV occlusions are spherical (Figures 2a & b) and therefore difficult to distinguish from lipid droplets, secretory granules, *etc.* The use of a stain, like 0.05% aqueous malachite green, in preparing wet mounts for MBV diagnosis aids in demonstration of the occlusion bodies (Lightner *et al.*, 1983c). Presumably, the protein making up the occlusion absorbs the stain more rapidly than does most other material in the feces or in host tissues, contrasting them relative to other materials present within a few minutes (Figure 2a).

BP and MBV occlusion bodies in histological preparations appear as prominent eosinophilic (with H&E) usually multiple inclusion bodies within the hypertrophied nuclei of hepatopancreatic tubule or midgut epithelial cells (Figures 1c & 2c). Often the affected nuclei have a peripherally displaced compressed nucleolus and marginated chromatin, giving affected nuclei a "signet ring" appearance even before occlusion bodies become well developed. Brown and Brenn histologic Gram stain (Luna, 1968), although not specific for baculovirus occlusion bodies, tends to stain occlusions more intensely (either red or purple, depending upon section thickness, time of decolorization, *etc.*) than the surrounding tissue, aiding in demonstrating their presence in low-grade infections.

Transmission electron microscopy of BP and MBV infected cells show large numbers of rod-shaped baculovirus particles both free and occluded within the proteinogenous crystalline matrix of the occlusion body (Figures 1d & e and 2d &e).

<u>BMN:</u>

BMN affects the same target organs as does BP and MBV, but unlike BP and MBV it does not produce an occlusion body. Hence, its diagnosis is dependent upon history, clinical signs, and wet-mounts and histopathology of the hepatopancreas. Sano and coworkers (1984) in Japan have also reported development of a fluorescent antibody diagnostic technique for BMN. By wet-mount microscopy or histology the principal diagnostic feature of BMN is hypertrophied nuclei within infected hepatopancreatocytes (Sano *et al.*, 1981 and Momoyama, 1983). These enlarged nuclei have marginated chromatin, a laterally displaced or disassociated nucleolus, but lack occlusion bodies (Figure 3).

HPV:

Diagnosis of HPV is dependent upon the histological demonstration of single prominent basophilic (with H&E), Feulgen positive (Luna, 1968) intranuclear inclusion bodies in the hypertrophied nuclei of infected hepatopancreatic tubule epithelial cells (Figures 4a & b). Consequent lateral displacement and compression of the nucleolus and chromatin margination are also prominent features of such infected cell nuclei (Figures 4a & 4c). Early in their development, HPV inclusions are small eosinophilic bodies centrally located within the nucleus and closely associated with the nucleolus.

TEM of HPV-infected hepatopancreatocytes shows the inclusion body to contain virus-like particles of 22 to 24 nm in diameter (Figure 4d). HPV particle size, along with the features of host cell cytopathology, are quite similar to the reported characteristics of the parvovirus group (Kurstak *et al.*, 1977; Longworth, 1978; and Paradiso *et al.*, 1982).

IHHNV:

Diagnosis of infection by IHHNV is dependent upon histological demonstration of prominent eosinophilic (with H&E), Feulgen negative intra-nuclear inclusion bodies (Figures 5a & b) within chromatin marginated, hypertrophied nuclei of cells in tissues of ectodermal (epidermis, hypodermal epithelium of fore and hindgut, nerve cord, and nerve ganglia) and mesodermal origin (hematopoietic organs, antennal gland tubule epithelium, connective tissue, and striated muscle). Usually the midgut, midgut ceca, and the hepatopancreas (endoderm-derived tissues) are unaffected, except in severe cases where hepatopancreatic involvement has been observed (Lightner *et al.*, 1985). These inclusions match closely the characteristics of the type A intranuclear inclusion body class described by Cowdry (1934). Basophilic chromatin strands are occasionally visible by light microscopy within IHHN intranuclear inclusion bodies. These chromatin strands are a prominent feature of IHHN intranuclear inclusion bodies by TEM (Figure 5c).

IHHN intranuclear inclusion bodies are common early in acute infections, later and juvenile *P. monodon* sampled from farms in the Pingtung area of Southern

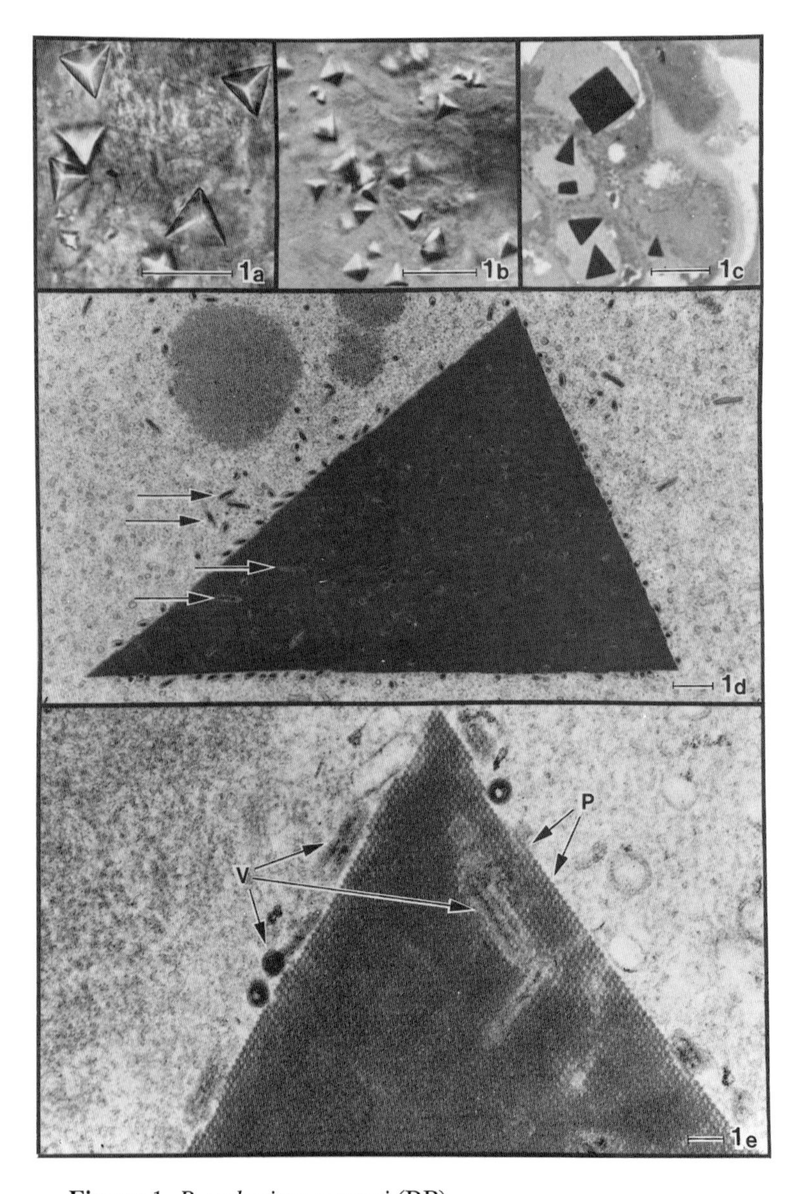

Figure 1. *Baculovirus penaei* (BP).

1a & 1b. BP tetrahedral occlusion bodies as they appear in feces and tissue squashes from *Penaeus vannamei*; no stain. 1a. bright-field microscopy, X1,520; bar

"continued"

"Figure 1 continued"
is 10um; 1b. Hoffman interference contrast, X660; bar = 20 um.

1c. BP occlusions in a histological section of *P. aztecus* hepatopancreas. O-toludine blue, X530; bar = 20 um.

1d. TEM of a BP-infected hepatopancreatocyte from *P. marginatus*. Rod-shaped BP virions (arrows) are present free in the karyoplasm and occluded within the tetrahedral occlusion body. X14,400; bar = 0.5 um.

1e. High magnification view of a BP occlusion body in the hepatopancreas of *P. aztecus*. Virions (V) are being occluded by proteinaceous polyhedrin (P) subunits of the occlusion body. X61,200; bar = 100 nm.

---

decreasing in number, and are followed by necrosis and inflammation of target tissues. Affected cells may also have highly vacuolated cytoplasm with cytoplasmic bodies that range from eosinophilic to basophilic. Although the prominent intranuclear inclusions present in shrimp infected with IHHNV are evidence of nuclear involvement, assembly of the virus occurs in the cytoplasm of affected cells (Figure 5d). The size and morphology of the virus, 17 to 26 nm in sections (Figure 5e) and 20 nm in purified preparations (Figure 5f), and its replication within the cytoplasm support the tentative classification of IHHNV with the picornaviruses.

REO:

REO is the newest of the penaeid viruses and it was discovered by Tsing and Bonami (1987) in juvenile *P. japonicus* in France using electron microscopy, and subsequently, in the same species in Hawaii using the same technique (Lightner *et al.*, 1985). Most recently REO, or a closely related form, has been found associated with a serious disease syndrome in pond-cultured *P. monodon* in Southeast Asia (Nash and Nash, 1987). In both species, other lesions were more apparent by light microscopy, and signs of REO infection were overlooked until found by electron microscopy. The virus was located in the cytoplasm of F-cells and R-cells of the hepatopancreatic tubule epithelium, where it formed large cytoplasmic viral inclusions (Figure 6a). The non-enveloped, icosahedral virions of REO measured about 60 nm and 50 to 70 nm in diameter, respectively, in purified preparations and in tissue sections (Figures 6b & c). Tsing and Bonami (1987) experimentally transferred the disease in juvenile *P. japonicus* by inoculation or by feeding new host shrimp pieces of REO-infected hepatopancreas. Development of the infection was slow, requiring

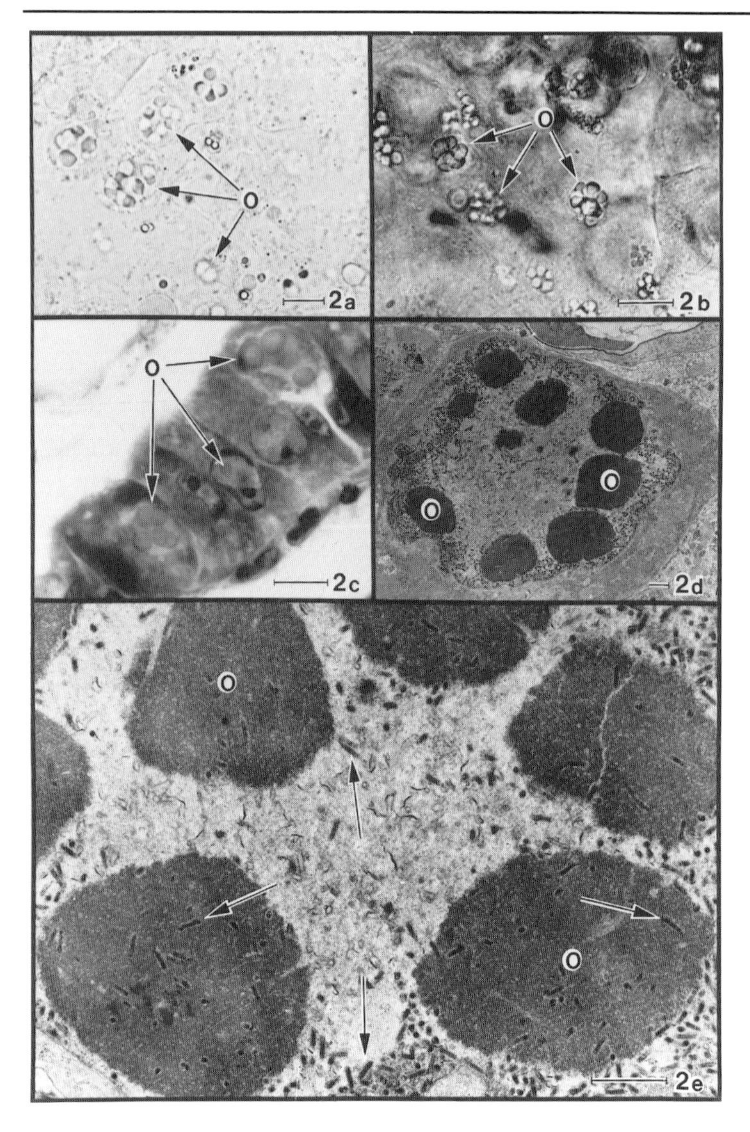

Figure 2. *P. monodon*-type baculovirus (MBV) in *P. monodon*, showing diagnostic spherical occlusion bodies (O) in:

2a. A wet-mount of an MBV-infected hepatopancreas cells. Malachite green stain, X660; bar = 10 um.

2b. In feces from an infected adult. No stain, X250; bar = 10 um.

"continued"

"figure 2 continued"

2c. In hypertrophied nuclei of MBV-infected hepatopancreas. Histological section, H&E stain, X950; bar = 10 um.

2d & 2e. By TEM in MBV-infected hepatopancreatocytes. Rod-shaped MBV virions (arrows) are apparent free and within the occlusion (O) bodies. 2d. X3,630; 2e. X13,300. Bars = 1 um.

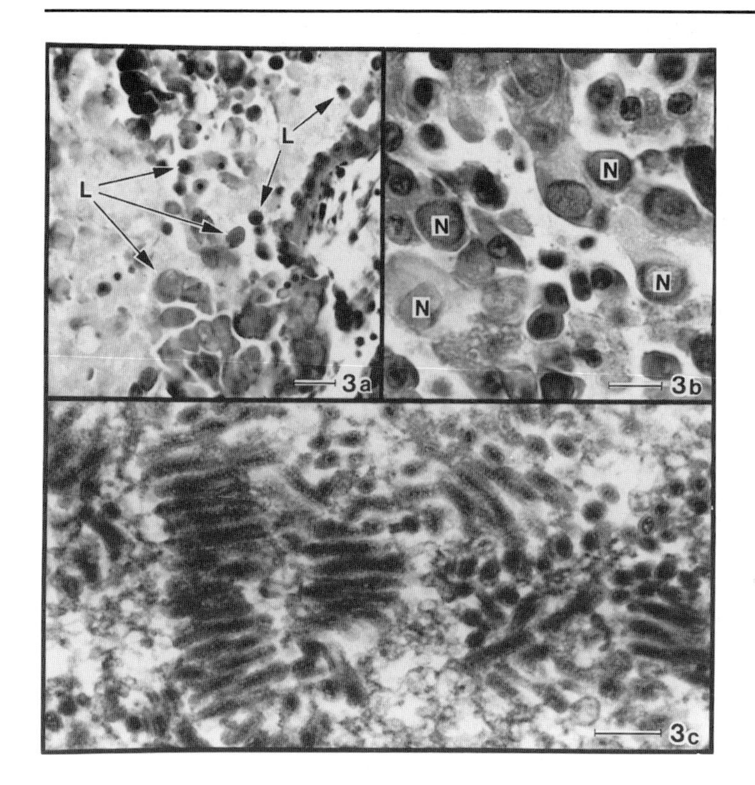

Figure 3. Baculoviral midgut gland necrosis virus (BMN) in *P. japonicus*:

3a. Histological section of necrotic hepatopancreas tubules from a postlarva. Tubule lumens (L) are filled with cell debris and necrotic cells (arrows). H&E stain, X380; bar = 20 um.

3b. Hepatopancreatocytes with diagnostic hypertrophied nuclei (N) that do not contain occlusion bodies. H&E stain, X960; bar = 10 um.

3c. TEM of a BMN-infected nucleus that contains masses of rod-shaped virions. X60,000; bar = 0.2 um.

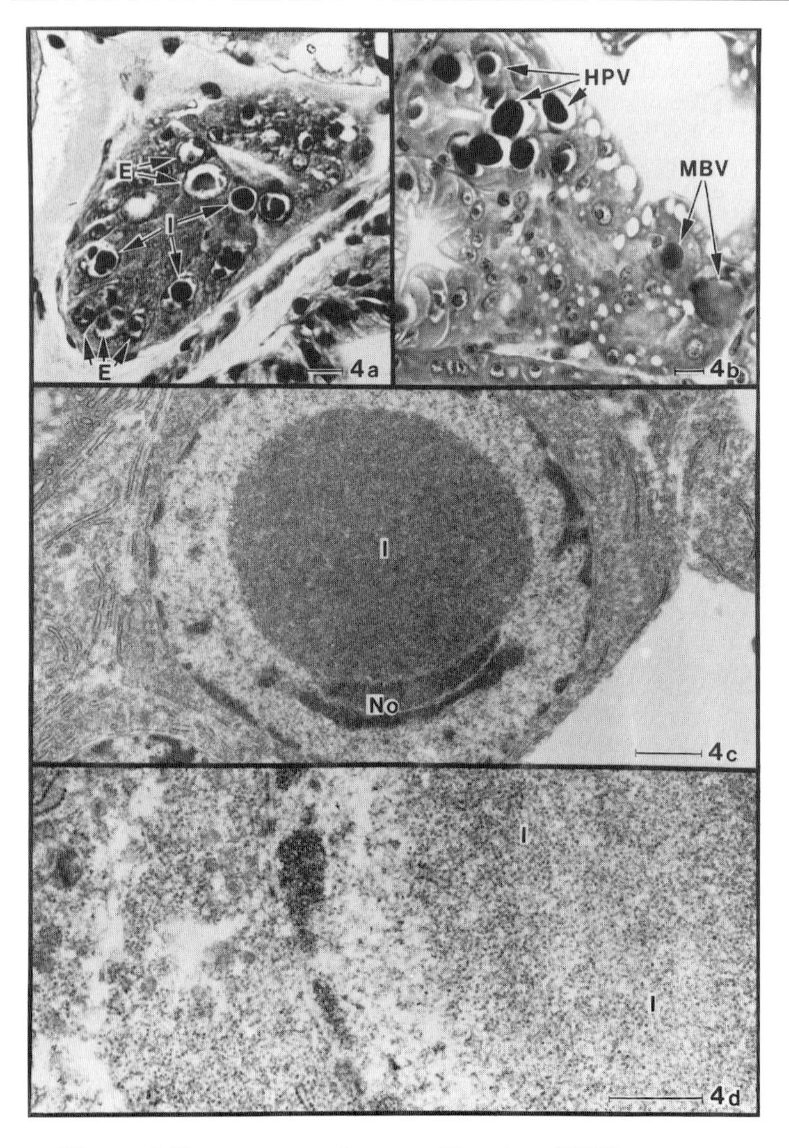

Figure 4. Hepatopancreatic parvo-like virus (HPV):

4a. Histological section of an HPV-infected hepatopancreas from a juvenile *P. monodon*. Infected cells nuclei contain prominent intranuclear inclusion bodies, which are relatively small and eosinophilic early in development (E), but are larger and basophilic when fully developed (I). H&E stain, X660; bar = 10 um.

4b. A dual infection of the hepatopancreas of a juvenile *P. monodon* by HPV and MBV. H&E, X530; bar = 10 um.

"continued"

"Figure 4 Continued"

4c. TEM of a hepatopancreatocyte from *P. orientalis* showing an HPV intranuclear inclusion body (I), which is developing in close association with the laterally displaced host cell nucleolus (No). X11,900; bar = 1 um.

4d. A higher magnification TEM of an HPV-infected hepatopancreatocyte that shows unorganized masses of 22 nm diameter virions developing within the virogenic stroma of the inclusion body (I). X33,600; bar = 0.5 um.

---

about 45 days to develop. Secondary infections by *Fusarium solani* were common in REO-infected shrimp. Some new data (Tsing *et al.*, 1985) suggests a probable link between infection by REO and "gut and nerve syndrome" (GNS), an idiopathic condition found in chronically ill populations of *P. japonicus* cultured in Hawaii (Lightner *et al.*, 1984).

Hosts and Geographic Distribution

BP:

BP is widespread in distribution in cultured and wild penaeids in the Americas, ranging from the Northern Gulf of Mexico south through the Caribbean and reaching at least as far as the State of Bahia in central Brazil. On the Pacific Coast, BP ranges from Peru to Mexico, and it has been observed in wild penaeid shrimp in Hawaii. BP has not yet been observed in wild, cultured or imported (from the Americas) penaeids outside of the Americas. Recent new information on the host and geographic distribution of BP has come from Brazil and Mexico. In South America, BP was found to infect larvae and postlarvae of five penaeid species (in imported *Penaeus vannamei* and *P. penicillatus* and in native *P. schmitti*, *P. paulensis,* and *P. subtilis*), four of which represent new host species for the virus (Table 2).

BP was found for the first time in Mexico in cultured larval and postlarval *P. stylirostris* at a facility near Guaymas, Sonora on the West Coast of Mexico (Lightner *et al.*, 1988). Because the affected facility had no history of stock importations, BP must be assumed to be enzootic in wild penaeids in the region.

MBV:

MBV-type baculoviruses are analogous to BP in their diverse host range and in their wide distribution on the IndoPacific coasts of Asia, Australia, and Africa, and in Southern Europe. However unlike BP, MBV has been observed in the Americas in imported stocks and in an American penaeid exposed to the virus. Although MBV was first discovered in a quarantined population of *P. monodon* that had originated from Taiwan (Lightner and Redman, 1981; Lightner *et al.*, 1983c), it had not actually been demonstrated in Taiwan until it was found to be widely distributed in Taiwanese shrimp farms in a 1986 survey of the country (Lightner *et al.*, 1987). Studies in 1987 linked MBV to serious disease losses in many Taiwanese farms (S.N. Chen and G.H. Kou, unpublished communication, National Taiwan Univ., Taiwan).

Since the information on MBV was last summarized, MBV has been found in Texas, Ecuador, and Brazil in imported stocks of *P. monodon*. Of possible significance was the presence of MBV-like (spherical) occlusion bodies that were found with a heavy BP infection of the hepatopancreas juvenile *P. vannamei* which were being cultured with MBV-infected *P. monodon* in Ecuador.

A similar agent, found first in *P. plebejus* and thus called *Plebejus Baculovirus* (PBV), was found in cultured penaeids in Australia (Lester *et al.*, 1987). Other than its presence in a new host species, the agent of PBV differs little from MBV in host cell cytopathology and in the morphology of the virus, and it probably represents a strain of the MBV-type viruses rather than a separate distinct species.

BMN:

BMN has been reported only in *P. japonicus* cultured in Japan, where it is considered a major problem in the larval and early postlarval stages of that species (Sano *et al.,* 1984; 1985; and Sano and Fukuda, 1987). Despite numerous introductions of *P. japonicus* stocks (larvae, postlarvae, and broodstock) to Hawaii, France, Brazil, and other locations during the past two decades, BMN has not been detected in that species or in other penaeids cultured in the Americas.

HPV:

HPV has a geographic range similar to that of MBV in Asia and Australia, and like MBV it has been introduced to the Americas with imported penaeids. More recently, HPV was found for the first time in Taiwan in dual infections with MBV in postlarval

Taiwan (Figure 4b). This region in 1987 had experienced serious disease losses in its farms due, at least in part, to MBV. The severity of HPV infections in some of the shrimp sampled suggests that HPV, while unrecognized, may have contributed to the 1987 epizootic.

Reports of HPV in captive-wild *P. esculentus* in Australia (Paynter *et al.*, 1985), in *P. monodon* imported to Israel from Kenya (Colorni *et al.*, 1987), and in captive-wild and hatchery reared *P. indicus* and *P. merguiensis* in Singapore (Chong and Loh, 1984) have expanded the known host and geographic distribution of this virus (Tables 2 and 3).

In the Singapore study, of four shrimp farms surveyed, HPV incidence was highest (>50%) in the two farms that reared hatchery-derived postlarvae, and lower (<15%) in the two farms which cultured only feral shrimp collected by tidal entrapment (Chong and Loh, 1984). This suggests that HPV is transmitted either vertically from parent broodstock, or horizontally from shrimp to shrimp with efficiency only during the larval stages.

HPV has been observed in the Americas. In Brazil in 1987, HPV was found in stocks of *P. penicillatus* imported from Taiwan. At the same culture facility, HPV was found in light infections in juvenile *P. vannamei*, which had been exposed to infected *P. penicillatus* indirectly as a result of normal farming practices. The discovery of HPV in cultured shrimp in Brazil represents the first time this pathogen has been documented in the Americas and in an American penaeid (S. Bueno, R. Meyer, and D. Lightner, unpublished observations).

IHHNV:

IHHNV has a world-wide distribution in cultured penaeid shrimp, but its distribution in wild penaeids remains unknown. Infection by the virus causes serious disease in *P. stylirostris*, and acute catastrophic epizootics in intensively cultured juveniles of that species. In other penaeids, IHHNV has been reported to cause infection and disease (Brock *et al.*, 1983; Lightner *et al.*, 1985; and Lightner,1988), but disease severity does not approach that observed in *P. stylirostris*. The natural host(s) and geographic distribution of IHHNV are unknown. However, the occurrence in Southeast Asia (Singapore, Malaysia, and the Philippines) of IHHNV (or a similar agent) in shrimp culture facilities using only captive-wild *P. monodon*

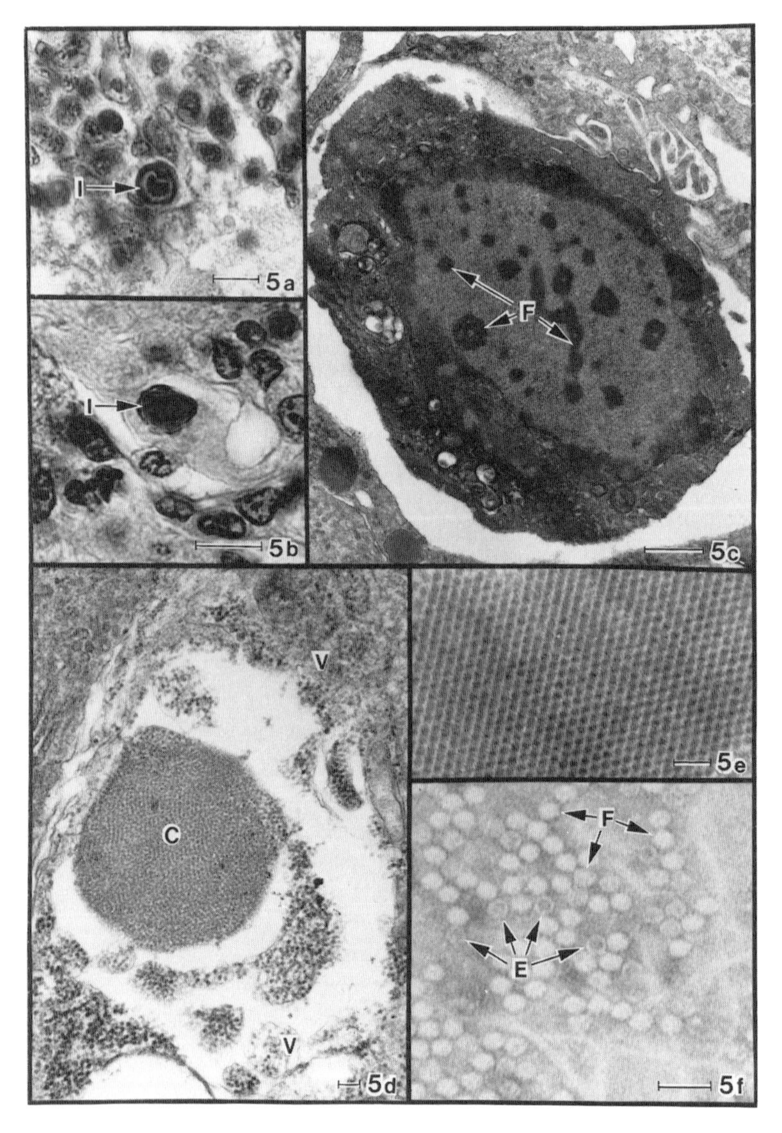

Figure 5. Infectious hypodermal and hematopoietic necrosis virus (IHHNV) in *Penaeus stylirostris*:

5a & 5b. Histological sections of typical, and diagnostic of IHHN, intra-nuclear eosinophilic inclusion bodies (I) in a hematopoietic cell in 5a, and in 5b a gill epidermal cell. H&E stain, 5a is X825; 5b is X1,220; bars = 10 um.

"continued"

"Figure 5 continued"

5c. TEM of a gill podocyte with an IHHN intranuclear inclusion body. The dense fibrillar strands (F) are presumed to be regions of virus-induced nucleic acid embedded in a less dense granular material thought to be mostly ribosomes and protein. No virus particles are evident in this cell's nucleus or cytoplasm. X10,600; bar = 100 nm.

5d. TEM of IHHNV present in loose aggregates (V) and condensed paracrystalline arrays (C) in the cytoplasm of an infected cell in the gills of a juvenile. X37,800; bar = 0.1 um.

5e. A higher magnification TEM of 20 nm average diameter virions of IHHN in a paracrystalline array. X130,000; bar = 50 nm.

5f. IHHN virus from density gradient centrifugation. Full (F) and empty (E) capsids are present. 2% PTA stain; X190,000; bar = 50 nm.

---

broodstock suggests that that region is within the virus' geographic range, and that *P. monodon* may be among its natural host species.

Since 1985, no new hosts for IHHNV have been demonstrated. However, the geographic distribution of the virus in culture facilities has continued to expand. In Mexico in 1987, IHHN was found in an imported population of postlarval *P. vannamei* at a facility in Baja California (Lightner, unpublished data). Likewise, IHHN was found to be present in imported quarantined stocks of *P. vannamei* in a 1986 survey of Taiwanese shrimp culture facilities, but not in cultured stocks of other penaeid species, including *P. monodon*, at the farms surveyed (Lightner *et al.*, 1987).

Discussion and Conclusions

The present diagnostic procedures for the penaeid virus diseases are largely dependent upon history, clinical signs, and histopathology. Electron microscopy is also of importance in some diagnostic applications. Techniques like enhancement and animal bioassays when coupled to histopathology add sensitivity to these diagnostic procedures.

These procedures, however, are very limited. Examination of relatively small samples is one important limitation; the length of time required to carry out routine histopathology and/or electron microscopy is another factor limiting their practical usefulness. Also, the cost of histopathology, electron microscopy, of maintaining

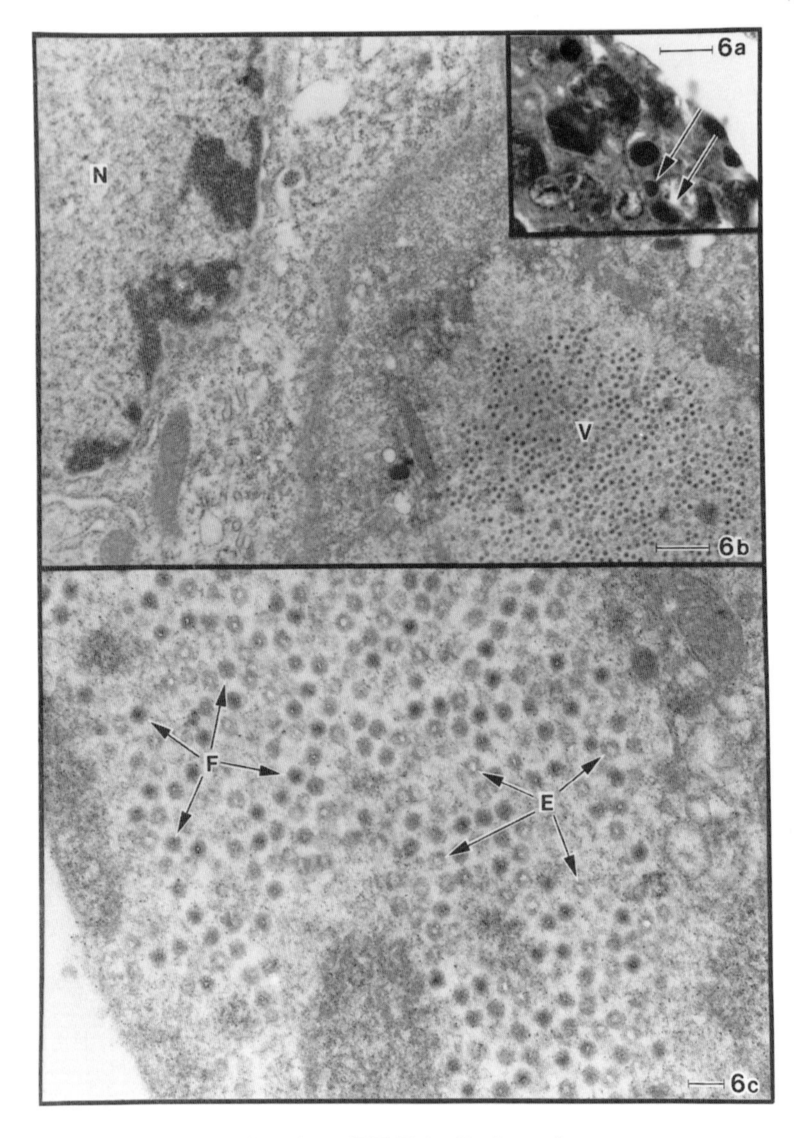

Figure 6. Reo-like virus (REO) in *P. japonicus*:

6a. Histological section of a REO-infected hepatopancreas of a juvenile. Magenta staining cytoplasmic inclusion bodies (arrows) may represent virus masses. H&E stain, X960; bar = 10 um.

"continued"

"Figure 6 continued"

6b. TEM of a REO virus mass (V) in the cytoplasm of an hepatopancreatocyte. The nucleus (N) of the cell appears unaffected. X18,900; bar = 0.5 um.

6c. A higher magnification view of a cytoplasmic mass of REO showing both full (F) and empty (E) 55 nm average diameter virus particles. X190,000; bar = 100 nm

---

enhancement and bioassay areas, and the limited availability of specific pathogen-free indicator shrimp for bioassays, all add to the list of reasons why better, more rapid, more sensitive diagnostic procedures are needed.

Methods using tissue culture, serologic methods, and gene probe diagnostic techniques that have become state-of-the-art in human and veterinary medicine, are being developed for penaeid shrimp. For example, the first documented success at producing primary cell cultures from shrimp was recently reported (Chen *et al.,* 1986). However, while development of these techniques for use with shrimp is underway, none are yet routinely available to the diagnostic labs of the industry. As the new diagnostic methods do become available, they should provide more rapid diagnoses than do current methods; they should be very sensitive and easily standardized among various labs that use them; and they should be inexpensive and simple to run.

The practice of transporting penaeid stock between facilities and between different geographic regions has resulted in the introduction of five of the six known penaeid shrimp viruses to regions where they may not have previously existed. Five of the six known types of penaeid viruses are apparently not native to the Americas, but of these five, four (IHHNV, MBV, HPV, and REO) have been introduced with shrimp intended for aquaculture (Table 4). Whether or not these introduced viruses have escaped the culture facilities to which they have been introduced and have become established in local wild penaeid stocks is not known.

While evaluation of non-native penaeids by the emerging shrimp culture industry is an essential component to the growth and development of that industry, introduction of pathogens like IHHNV to regions where it previously did not occur can have catastrophic consequences to the industry (Lightner *et al.*, 1983a; 1983b; and

Table 3. Observed and reported occurrences of the penaeid viruses in wild and cultured penaeids indicating their probable natural and introduced geographic distributions.

Virus	Region/site where found	Host Status*	Virus Status
IHHNV	Atlantic side: SE U.S., Caribbean, & Brazil	Cul	introduced
	Pacific side: Ecuador, Peru, & Central America		
	Pacific: Hawaii, Guam, Tahiti	Cul	introduced
	Asia: Taiwan	Cul	introduced
	Singapore, Malaysia & Philippines	CW	enzootic?
	Middle East: Israel	Cul	introduced
HPV	IndoPacific: P.R. China, Taiwan, Philippines, Malaysia, Singapore & Australia	Cul, CW, W	enzootic
	Africa: Kenya	W	enzootic
	Middle East: Israel & Kuwait	Cul, CW	enzootic
	Americas: Brazil	Cul	introduced
BP	Atlantic side: SE U.S., Caribbean & Brazil	Cul, CW	enzootic
	Pacific side: Ecuador,		
	Peru & Central America	Cul, CW, W	enzootic
	Mexico	Cul	enzootic?
	Hawaii	W	enzootic
MBV	IndoPacific: P.R. China, Taiwan, Philippines, Malaysia, Singapore & Australia	Cul, CW, W	enzootic
	Africa: S. Africa	W	enzootic
	Middle East: Israel & Kuwait	Cul, CW, W	enzootic
	Mediterranean: Italy	Cul, CW, W	enzootic
	Pacific: Tahiti, Hawaii	Cul	introduced
	Americas: Mexico, Ecuador	Cul	introduced
	Texas, and Brazil	Cul	introduced
BMN	Japan	Cul, CW, W	enzootic
REO	Japan, Malaysia	Cul	enzootic
	Hawaii & France	Cul	introduced

"continued"

"Table 3 continued"

* Cul = "cultured"; from cultured or captive-wild broodstock.

CW = "captive-wild"; from wild-caught seed or from single-spawn wild broodstock.

W = "wild"; from natural sources or commercial fishery.

Table 4. Penaeid viruses in the Americas and their status.

	Virus Status in the Americas and Hawaii
IHHNV	Widely distributed in cultured *P. vannamei* and *P. stylirostris*; not recognized in wild penaeids; enzootic in Southeast Asian wild penaeids; recently introduced into Western Mexico.
HPV	Enzootic in Asia, Australia, and Africa; introduced to one or more sites in South America from Taiwan; will infect the American penaeid *P. vannamei*.
BP	Widely distributed in American penaeids; enzootic in wild penaeids on both Atlantic and Pacific sides of tropical and subtropical America.
MBV	Enzootic in Asia, Australia, Africa, and Mediterranean; introduced to several sites in Hawaii, North, Central, and South America; will infect *P. vannamei*; contaminated stocks eradicated from Hawaii, Mexico, and Texas.
BMN	Enzootic in Japan; not reported outside of Japan.
REO	Enzootic in Japan; introduced to Hawaii from Japan; contaminated stocks eradicated in Hawaii.

Lightner, 1988). Prevention of such introductions of exotic pathogens is dependent upon the use of quarantine, certification and inspection policies, and procedures that are supported by reliable diagnostic procedures. Sindermann (1988) provided a copy of the "Revised Code of Practice to Reduce the Risks of Adverse Effects Arising from Introduction of Non-indigenous Marine Species" as adopted in 1979 by the member countries of the International Council for the Exploration of the Sea. The practices outlined in this code, if practiced, are effective for their intended purpose. Adoption and implementation of such a policy by the penaeid shrimp aquaculture industry of the Americas may be in the best interest of that emerging industry.

Acknowledgements

Grant support for this work was from the U.S.D.A. Cooperative State Research Service; from Sea Grant, U.S. Dept. of Commerce; and from the Hawaii Aquaculture Development Program. Kent Wood's assistance in assembly of the figures is thankfully acknowledged.

Literature Cited

Brock, J. A., D. V. Lightner, and T. A. Bell. 1983. A review of four virus (BP, MBV, BMN and IHHNV) diseases of penaeid shrimp with particular references to clinical significance, diagnosis and control in shrimp aquaculture. International Council for the Exploration of the Sea. C.M. 1983/Gen:10/ Mini-Symposium.

Chen, S. N., S. C. Chi, G. H. Kou, and I. C. Liao. 1986. Cell culture from tissues of grass prawn, *Penaeus monodon*. Fish Pathology 21: 161-166.

Chong, Y. C., and H. Loh. 1984. Hepatopancreas chlamydial and parvoviral infections of farmed marine prawns in Singapore. Singapore Veterinary Journal 9: 51-56.

Colorni, A., T. Samocha, and B. Colorni. 1987. Pathogenic viruses introduced into Israeli mariculture systems by imported penaeid shrimp. Bamidgeh 39: 21-28.

Couch, J. A. 1974. An enzootic nuclear polyhedrosis virus of pink shrimp: ultrastructure, prevalence, and enhancement. Journal of Invertebrate Pathology 24: 311-331.

Couch, J. A. 1981. Viral Diseases of Invertebrates Other Than Insects. pp. 127-160. In. E. W. Davidson (ed.) Pathogenesis of Invertebrate Microbial Diseases. Allanheld, Osmun Publ.,Totowa, N.J.

Cowdry, E. V. 1934. The problem of intranuclear inclusions in virus diseases. Archives of Pathology 18: 527-542.

Humason, G. L. 1967. Animal Tissue Techniques. W.H. Freeman and Co., San Francisco, CA. p. 432.

Johnson, P. T. 1983. Diseases caused by viruses, rickettsiae, bacteria, and fungi. pp. 1-78, In. A.J. Provenzano, Jr.(ed.), The Biology of Crustace*a*, Vol. 6. Academic Press, N.Y.

Kurstak, E., P. Tijssen, and S. Garzon. 1977. Densonucleosis viruses (Parvoviridae). pp. 67-91. In. Karl Maramorosch (ed.), The Atlas of Insect and Plant Viruses. Academic Press, N.Y.

Lester, R. J. G., A. Doubrovsky, J. L. Paynter, S. K. Sambhi, and J. G. Atherton. 1987. Light and electron microscope evidence of baculovirus infection in the prawn *Penaeus plebejus* . Diseases of Aquatic Organisms 3: 217-219.

Lightner, D. V., and R. M. Redman. 1981. A baculovirus-caused disease of the penaeid shrimp, *Penaeus monodon*. Journal of Invertebrate Pathology 38: 299-302.

Lightner, D. V. 1983. Diseases of Cultured Penaeid Shrimp. pp. 289-320. In. J.P. McVey (ed.), CRC Handbook of Mariculture. Vol. 1. Crustacean Aquaculture. CRC Press, Boca Raton, FL.

Lightner, D. V., R. M. Redman, and T. A. Bell. 1983a. Infectious hypodermal and hematopoietic necrosis (IHHN), a newly recognized virus disease of penaeid shrimp. Journal of Invertebrate Pathology 42: 62-70.

Lightner, D. V., R. M. Redman, T. A. Bell, and J. A. Brock. 1983b. Detection of IHHN virus in *Penaeus stylirostris* and *P. vannamei* imported into Hawaii. Journal of the World Mariculture Society 14: 212-225.

Lightner, D. V., R. M. Redman, and T. A. Bell. 1983c. Observations on the geographic distribution, pathogenesis and morphology of the baculovirus from *Penaeus monodon* Fabricius. Aquaculture 32:209-233.

Lightner, D. V., R. M. Redman, T. A. Bell, and J. A. Brock. 1984. An idiopathic proliferative disease syndrome of the midgut and ventral nerve in the Kuruma prawn, *Penaeus japonicus* Bate, cultured in Hawaii. Journal of Fish Diseases 7: 183-191.

Lightner, D. V. and R. M. Redman. 1985. A parvo-like virus disease of penaeid shrimp. Journal of Invertebrate Pathology 45: 47-53.

Lightner, D. V., R. M. Redman, R. R. Williams, L.L. Mohney, J.P.M. Clerx, T.A. Bell, and J.A. Brock. 1985. Recent advances in penaeid virus disease investigatio ns. Journal of the World Mariculture Society 16: 267-274.

Lightner, D.V., R. P. Hedrick, J.L. Fryer, S. N. Chen, I. C. Liao, and G. H. Kou. 1987. A survey of cultured penaeid shrimp in Taiwan for viral and other important diseases. Fish Pathology 22: 127-140.

Lightner, D. V. 1988. Diseases of cultured penaeid shrimp and prawns. pp. 8-127, In. C. J. Sindermann and D. V. Lightner (eds.) Disease Diagnosis and Control in North American Marine Aquaculture. Elsevier, Amsterdam.

Lightner, D. V., R. M. Redman, and E.A. Almada Ruiz. 1988. *Baculovirus penaei* in *Penaeus stylirostris* (Crustacea: Decapoda) cultured in Mexico: unique cytopathology and a new geographic record. Journal of Invertebrate Pathology 51: "in press".

Longworth, J.F. 1978. Small isometric viruses of invertebrates. Advances in Virus

Research 23: 103-157.

Luna L. G. (ed.). 1968. Manual of Histologic Staining Methods of the Armed Forces Institute of Pathology. McGraw-Hill Book Co. N.Y. 259 p.

Momoyama, K. 1983. Studies on baculoviral mid-gut gland necrosis of Kuruma shrimp (*Penaeus japonicus*) - III. Presumptive diagnostic techniques. Fish Pathology 17: 263-268.

Nash, M., and G. Nash. 1987. A reo-like virus observed in the tiger prawn, *Penaeus monodon*, from Malaysia. Journal of Fish Diseases 10: "in press".

Paradiso, P., S. L. Rhode, III, and I.I. Singer. 1982. Canine parvovirus: a biochemical and ultrastructural characterization. Journal of General Virology 62:113-125.

Paynter, J.L., D. V. Lightner, and R. J. G. Lester. 1985. Prawn virus from juvenile *Penaeus esculentus*. p. 61-64, In. P.C. Rothlisberg, B.J. Hill and D.J. Staples (eds.), Second Australian National Prawn Seminar, NPS2, Cleveland, Australia.

Sano, T., T. Nishimura, K. Oguma, K. Momoyama, and N. Takeno. 1981. Baculovirus infection of cultured Kuruma shrimp *Penaeus japonicus* in Japan. Fish Pathology 15: 185-191.

Sano, T., T. Nishimura, H. Fukuda, T. Hayashida, and K. Momoyama.1984. Baculoviral mid-gut gland necrosis (BMN) of kuruma shrimp (*Penaeus japonicus*) larvae in Japanese intensive culture systems. Helgolander Meeresunters. 37: 255-264.

Sano, T., T. Nishimura, H. Fukuda, T. Hayashida, and K. Momoyama. 1985. Baculoviral infectivity trials on kuruma shrimp larvae, *Penaeus japonicus*, of different ages. p. 397-403, In. Fish and Shellfish Pathology, Academic Press, N.Y.

Sano, T., and H. Fukuda 1987. Principal microbial diseases of mariculture in Japan. Aquaculture 67: 59-69.

Sindermann, C. J. 1988. Disease problems caused by introduced species. pp. 394-398, In. C. J. Sindermann and D. V. Lightner (eds.) Disease Diagnosis and Control in North American Marine Aquaculture. Elsevier, Amsterdam.

Tsing, A., and J.R. Bonami. 1987. A new viral disease of the shrimp, *Penaeus japonicus* Bate. Journal of Fish Diseases 10: 139-141.

Tsing, A., D. Lightner, J.R. Bonami, and R. Redman. 1985. Is "gut and nerve syndrome" (GNS) of viral origin in the tiger shrimp Penaeus japonicus Bate? p. 91. In. Programme and Abstracts of the 2nd International Conference of the European Association of Fish Pathologists. Montpellier, France.

Development of Appropriate and Economically Viable Shrimp Pond Growout Technology for The United States

Arlo W. Fast, Ph.D.
Hawaii Institute of Marine Biology
University of Hawaii at Manoa
P.O. Box 1346
Kaneohe, HI 96744

Abstract

The United States now imports more than $1 billion worth of shrimp a year, or more than twice the amount produced domestically. Domestic shrimp production in the U.S. is mostly from the trawler fleet in the Gulf of Mexico, a fishery which is thought to be at maximum sustainable yield. Total U.S. shrimp production is presently about 91,000 MT (heads-off). Shrimp cultured in the U.S. during 1987 totaled 620 MT (heads-off), or 0.7% and 0.2% of total U.S. production and consumption respectively. This situation has motivated concerned interests in the U.S. to increase the production of shrimp in the U.S. through aquaculture. The initial attempts have largely relied on pond culture technology developed elsewhere for shrimp and on technology developed in the U.S. for channel catfish culture. These technologies, while appropriate and profitable for the conditions under which they evolved, have not proven profitable for shrimp culture in the U.S. The U.S. needs to develop a new and appropriate technology for shrimp culture in the U.S. within the climatic, social and legal frameworks which exist in the U.S. Work has begun on this problem, but an appropriate technology has not yet emerged in a readily usable form. Shrimp culture technologies for Ecuador and Taiwan are described and discussed relevant to the U.S. These two countries presently produce perhaps half of the world's cultured shrimp. A simulated transfer of intensive shrimp culture from Taiwan to Hawaii reveals that labor, construction and energy costs are major constraints to shrimp culture profitability in the U.S. using Taiwan technology. Microcomputer applications for pond monitoring and real-time management could reduce labor, energy and risks of crop loss during pond growout of shrimp. These reductions alone, if successfully adapted, could perhaps make shrimp culture profitable in the U.S. The University of Hawaii has concentrated pond research efforts on these microcomputer applications and the development of pond mixing equipment. Other conditions which could improve shrimp culture profitability in the U.S. include: use of alternative shrimp species, especially those species that grow well at cooler temperatures and higher salinities; development of high quality, low cost feeds; reduced seed costs; better farm and farm equipment designs; and use of improved nursery (multi-phasic growout) applications. Greatly increased world production of pond cultured shrimp is expected to reduce shrimp prices. This in turn may force less efficient farms to upgrade techniques, or go out of business.

Introduction

General Motivation

The primary motivation for shrimp culture research in the U.S. has to do with the large negative trade imbalance created by foreign shrimp imports into the U.S. Shrimp imports into the U.S. now exceed $1 billion/yr, while the value of domestically produced shrimp is less than half this amount (Lawrence *et al.*, 1984; Vondruska 1987). There is therefore a strong national interest in reducing our dependence on shrimp imports through increased domestic production.

Domestic shrimp production in the U.S. is largely limited to wild-caught shrimp, mostly from the Gulf of Mexico trawler fleet. The total poundage of these wild caught shrimp (heads-off) has ranged from a low of 108 million lbs (49 thousand metric tons; MT) during 1961, to a high of 288 million lbs (131 thousand MT) during 1977 (Figure 1; Vondruska 1987). Total projected U.S. production of wild caught shrimp from all domestic sources during 1987 was 208.7 million lbs (94.9 thousand MT), or 321 million lbs (146 thousand MT) heads-on. Although there is an upward trend in the catch from 1950 through 1986, it is slight and the catch is probably at or near maximum sustainable yield with landings of 200 million lbs (91 thousand MT) heads-off.

While domestic production from the wild is now static, domestic consumption, per capita consumption, and imports have all continued to increase. Domestic consumption has increased from 151 million lbs in 1950 to more than 700 million pounds heads-off (1,077 million lbs heads-on) during 1986 (Figure 1). This increase is due to population increases in the U.S., and to per capita consumption increases, especially the latter. While the U.S. population increased 160% during this period from 151 to 242 million people between 1950 and 1986, per capita shrimp consumption increased 298% from 1.00 to 2.95 lbs/person/yr during this same time. As a result of this shortfall between U.S. shrimp production and the U.S. demand for shrimp, shrimp imports into the U.S. have increased from 44 million lbs in 1950 to more than 492 million lbs in 1986; thus the negative trade balance for the U.S., and the motivation to increase U.S. domestic shrimp production.

Domestic shrimp production in the U.S. is from wild-caught (trawler) shrimp and

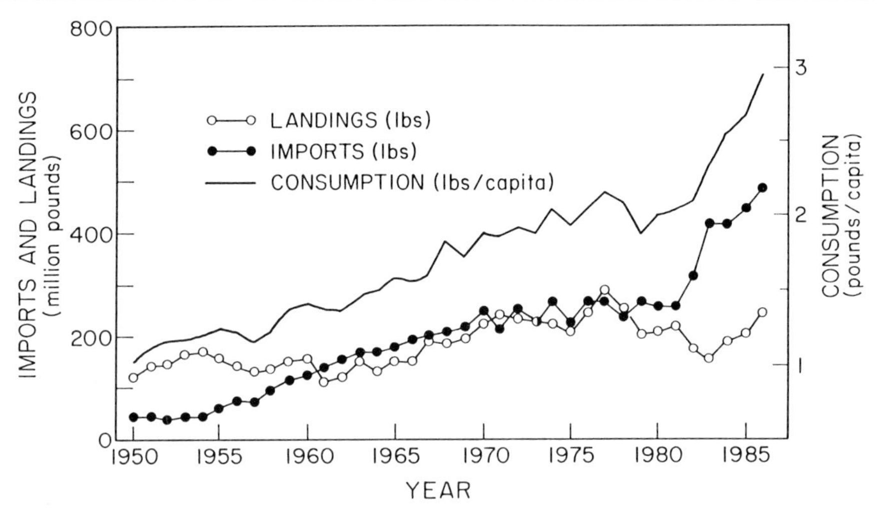

Figure 1. Total pounds of shrimp produced in the U.S. (Landings), imported, and the U.S. per capital consumption of shrimp between 1950 and 1986. Data from Vondruska (1987).

from shrimp cultured in captivity. As we have seen above, we cannot expect any further increase in wild-caught shrimp. Any increase in U.S. consumption from domestically produced shrimp must therefore come from cultured shrimp. Shrimp are now cultured commercially in several states including Hawaii, Texas, South Carolina, Louisiana and Alabama. During 1986, shrimp cultured in Hawaii exceeded 550,000 lbs (250 MT) heads-on, while estimated 1987 production decreased to 400,000 lb (181.8 MT) due to reduced production by two farms (Fassler 1987). During 1987, shrimp culture production for Texas is estimated at 1,142,500 lb (519.3 MT; Chamberlain 1988), and for S. Carolina at 355,700 lb (161.7 MT; Hopkins 1988). Total U.S. live weight production of cultured shrimp during 1987 was probably not greater than 2.3 million lb (954.5 MT), or roughly 0.7% of the total U.S. production, and 0.2% of the total U.S. consumption (Hopkins 1987).

These production figures for shrimp cultured in the U.S., compared with U.S. wild-caught shrimp, or shrimp imports, are obviously quite trivial. Since there is strong interest in the domestic culture of shrimp in the U.S., these findings beg the question; what conditions are constraining shrimp culture in the U.S.?

Worldwide, shrimp culture is not constrained by technology. Technology exists for seed production (Lawrence *et al.* ,1981; Liao 1985a; McVey 1983; Primavera 1985; Chamberlain 1985b; Aquacop 1979) and growout in ponds, tanks and raceways (Chiang and Liao 1985; Colvin 1985; Liao 1985b; Apud *et al.*, 1983; Hirono 1983; Fast *et al.*, in press, a). U.S. scientists and entrepreneurs are thoroughly familiar with existent technology, and can grow shrimp as well as anyone, technically. The problem is to transfer existent technology to the U.S., modify it, and otherwise develop technology which is appropriate to environmental, social and economic conditions in the U.S.

If the U.S. is to develop a shrimp culture industry of substance and magnitude, it will need to target production costs such that a reasonable profit can be realized with minimum risks, using world market prices for shrimp as a guide. World prices for shrimp are reflected in the wholesale prices per pound at New York (Table 1). The 1987 prices for 5 lb blocks of frozen, raw shrimp tails range from $3.70/lb for small shrimp (41-50/lb) to $7.65/lb for large shrimp (16-20/lb). In terms of pond-side prices for live weight shrimp (heads-on), prices range from $1.62/lb for 14 g shrimp to $3.41/lb for 44 g shrimp. It is unlikely that many U.S. shrimp farms can now grow shrimp at these prices, let alone at a lower cost where reasonable profits can be realized.

Before we further consider problem areas with shrimp culture systems in the U.S. and possible solutions, it is helpful to first; (a) review some common types of growout systems which have developed worldwide, (b) compare typical growout systems in Ecuador and Taiwan and (c) analyze a simulated transfer of intensive Taiwan style shrimp culture from Taiwan to Hawaii.

Shrimp Growout Systems

Species

Of more than 100 species of commercially important marine shrimp, 32 species have been cultured to some extent (Liao 1985b). Of these, only a few are suitable for large scale commercial production. Three species, *Penaeus vannamei*, *P. stylirostris* and *P. monodon* perhaps account for more than 80% of the world's cultured shrimp production (Table 2). These 3 species, together with the other 9 shown in

Table 1. Average wholesale prices per pound for raw headless (tails) Gulf Grown shrimp at New York (Vondruska 1987). The equivalent wholesale prices per pound for whole shrimp are shown in () parentheses under the headless prices. The estimated equivalent prices per pound, paid to the farmer (pond-side) for whole shrimp are shown in <> parentheses under the NY wholesale prices. The pond-side prices are taken as 68.6% of NY wholesale equivalent prices.

	Shrimp Sizes					
Tail Counts (#/lb)	16-20	21-25	26-30	31-35	36-40	41-50
Whole Body Weight (g)[1]	43.6-34.9	33.3-27.9	26.7-23.3	22.5-20.0	19.4-17.5	17.0-14.0
Year						
1984	$7.11	$5.98	$5.24	$4.65	$4.02	$3.64
	($4.62)	($3.89)	($3.41)	$3.02)	($2.61)	($2.37)
	<3.17>	<2.73>	<2.34>	<2.07>	<1.79>	<1.62>
1985	6.06	5.22	4.76	3.98	3.79	3.36
	(3.94)	(3.39)	(3.09)	(2.59)	(2.46)	(2.18)
	<2.70>	<2.32>	<2.12>	<1.78>	<1.69>	<1.50>
1986	6.71	6.37	5.84	5.24	4.79	4.33
	(4.36)	(4.14)	(3.80)	(3.41)	(3.11)	(2.81)
	<2.99>	<2.84>	<2.61>	<2.34>	<2.13>	<1.93>
1987[2]	7.65	6.26	5.21	4.38	3.95	3.70
	(4.97)	(4.07)	(3.39)	(2.85)	(2.57)	(2.40)
	<3.41>	<2.79>	<2.32>	<1.96>	<1.76>	<1.65>

[1] Assume that tail weight, shell on, is 65% of whole body weight

[2] Eleven month average

Table 2 account for perhaps more than 95% of all shrimp cultured in captivity. *Penaeus vannamei* and *P. stylirostris* are the principle shrimp species cultured in the Americas, while *P. monodon* is the species of choice for Asian pond culture systems.

Pond Systems

Most cultured shrimp are still produced in relatively primitive growout systems. These systems, known as extensive growout systems, are characterized by large ponds with very little materials, energy or cost inputs (Table 3). Extensive culture systems have appeared in many parts of the world as a first step in the evolution of more sophisticated and intensive culture methods.

The most primitive extensive systems consist of a dike with one gate. To stock and fill the pond the gate is opened and water enters at high tide carrying shrimp seed and other species with it. The gate is then closed and the water held for some time, perhaps 4 to 6 months without any special attention (Liao 1985a). The water is then drained by gravity flow and the crop harvested. As expected, this most primitive form of extensive shrimp culture gives variable yields since most management aspects are left to chance. Improved extensive culture can include such things as controlling stocking rate of shrimp seed, excluding predators, adding fertilizers, and increasing water exchange rate during growout to maintain good water quality. Extensive growout is also characterized by low yields of shrimp per unit area. In places where land, labor and shrimp seed are abundant and low-cost, this type of shrimp aquaculture can be quite profitable (Hirasawa, 1985). Most of the shrimp growout systems in Ecuador and Southeast Asia are extensive.

As farms intensify from the most primitive forms of extensive culture, one or more of the following practices are incorporated into the pond management practices:

(a) There are two gates on each pond with separate water supply and water discharge canals. Water is typically exchanged twice a month during the bi-monthly high tides by first draining perhaps half of the water through the discharge gate by gravity flow during low tide, closing this gate, and then allowing the pond to re-fill by gravity flow during the following high tide;

(b) Stocking a known number of PL shrimp into the ponds, usually at low density

Table 2. Some most commonly cultured species of marine shrimp in earthen ponds.

Scientific Name	Common Name	Original Range[1]	Specifical culture Characteristics	Typical Harvest size (g)	References
Penaeus monodon	Tiger prawn, grass shrimp, sugpo	Asia, Taiwan to India & Australia	low salinity (100 20°/oo; temp (25-30 C)	25-50 (x 30 g)	Liao (1985a, 1985b, 1987) Apud et al (1983), ASEAN (1978), Chiang & Liao (1985) Fast et al (in press), Motoh (1981)
Penaeus Vannamei	Mexican white shrimp	Pacific from S. America to Mecico	wide salinity range (10-50°/oo temp (25-30 C)	15 to 30 (x 21 g)	Hirono (1983), Pardy et al (1983) Sandifer et al (1986), Menz & Blake (1980)
Penaeus stylirostris	Pacific blue shrimp	Same as *P. vannamei*	same as P. *vannamei*, but grows faster & larger	15 to 35	Malca (1983), Pardy et al (1983)
Penaeus merguiensis	Banana shrimp	Indo-Pacific	tolerate high salinity & high temp (>40°/oo); >30°C). Harvest at 10-12g size.	8 to 12	New & Rabanal (1984)
Penaeus Indicus	Indian white shrimp	Indo-Pacific	similar to P. marguiensis	8 to 12	New & Rabanal (1984) Foster & Beard (1974) APUD et al (1983)
Penaeus japonicus	Kuruma-ebi Bamboo	Indo-Pacific	cool water species but can grow at higher temp also. Need high protein feed & sand to burrow.	30	Shigueno (1975), Hudinaga (1942), Liao & Chao (1983), Korringa (1976)

"continued"

"Table 2 continued"

Penaeus chinensis (P. orientalis)	White shrimp	China, Korea, Japan	grows well at cold temp in range of 15 to 25C.	25	Foster & Beard (1974)
Penaeus schmitti		Gulf of Mexico	- -	- -	Foster & Beard (1974)
Penaeus setiferus	White shrimp	S. Atlantic & Gulf of Mexico	- -	- -	Tatum & Trimble (1978)
Metapenaeus ensis	Greasy-backed shrimp Sand shrimp	Asia	- -	8 to 10	Liao & Chao (1983)
Penaeus penicillatus	Red-tailed prawn	Asis	grows well at full strength seawater and cooler temp.	- -	Liao & Chao (1983)
Penaeus semisulcatus	Red-legged prawn, Bear Shrimp	Asis	grows well at full strength seawater and cooler temp.	- -	Browdy & Samocha (1985)

[1] Holthuis (1980), Dall (1957)

Table 3. Characteristics of marine shrimp culture in ponds at three levels of intensity. Modified from Apud et al. (1983)

Characteristic	Level of Intensity		
	Extensive	Semi-intensive	Intensive
ProductionLevel (kg/ha/yr)	100 to 500	500 to 4,000	5,000 to 15,000
Stocking Rate (#/m2/crop)	0.1 to 1.0	1 to 10	20 or more
Feed	Natural (No supplement)	Natural + Supplement	Formulated
Water Exchange (%/day)	For evaporation and seepage replacement	1% to 5%	10% or more
Pond Size (ha)	>5	1 to 2	1 or less
Aeration	Natural	Water Exchange or mechanical	Continual mechanical and flushing

and often in polyculture with fish such as milkfish (*Chanos chanos*), mullet (*Mugil* spp) or tilapia. Shrimp PL's are usually collected from the wild but may, in more recent years, come from hatchery produced seed;

(c) The pond bottom is cared for between crops by a combination of lime (typically $CaCO_3$) applications, manure applications just before or after the pond is refilled, and applications of inorganic fertilizers;

(d) Plankton blooms, or benthic algal production is promoted during the culture period by the application of manures and/or inorganic fertilizers;

(e) Supplemental feeds are added either as fresh materials such as chopped trash fish, minced clams or organic wastes, or in the form of prepared and pelleted feeds.

Although it is convenient to classify the growout practices into categories such as extensive, semi-intensive, intensive, etc., these practices and the respective associated production levels are more of a continuum from the lowest to the highest production levels.

In places where land and labor are expensive, more intensive culture systems have evolved. The most intensive systems are characterized by small pond size, high feed and energy inputs, continuous management attention, and high yield. These intensive growout systems have primarily evolved in Taiwan, Japan and the U.S. (Hirasawa 1985; Salser *et al.* ,1978).

In Japan, intensive shrimp culture using the Shigueno system has concentrated on the production of *Penaeus japonicus* (Fig 2; Shigueno 1975; 1985). This is a cool water penaeid which commands a very high price in Japan of up to $80/kg. It also requires special culture techniques, such as a clean sand bottom, high energy

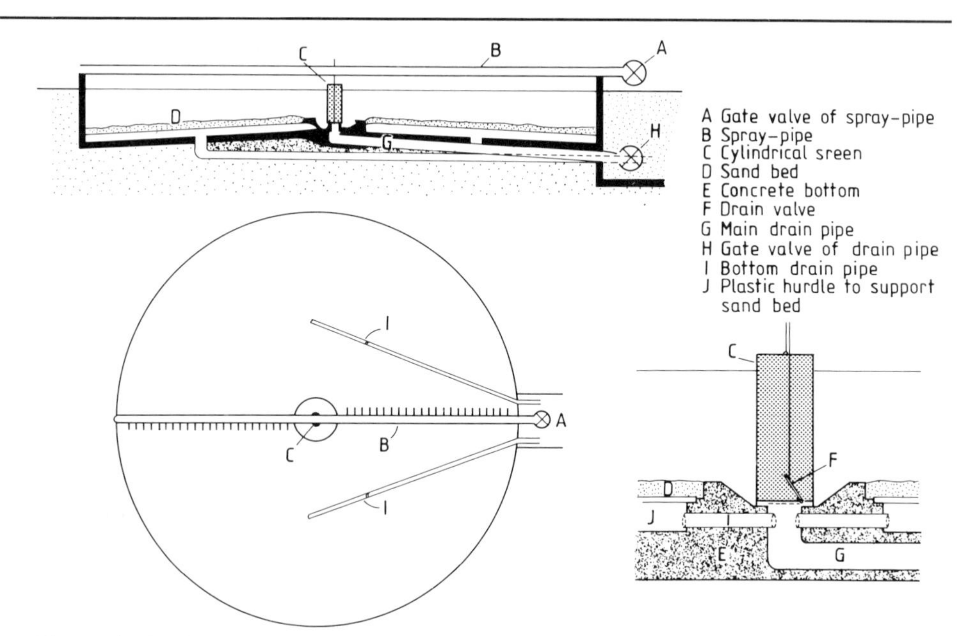

Figure 2. Tank structure and water supply system used for the culture of Kuruma-eabi prawns (*Panaeus japonicus* in Japan. (after Shigueno 1975; Kurata *et al.* 1980)

inputs for water circulation and aeration, and a costly high protein feed. Using this system, standing crops of 3.5 kg/m^2 and yields of 35,000 kg/ha/crop are possible (Kurata *et al.*, 1980). These high energy inputs, plus the relatively slow growth of this species make its culture marginally profitable at times even in Japan. The profitability of the Shigueno system is closely linked to the cost of energy.

More recently, an ultra-intensive shrimp culture system was developed in the U.S. through research efforts by the Coca-Cola Co. and the University of Arizona. This system, presently operating as Marine Culture Enterprise (MCE) in Hawaii, relies on covered raceways, very high water exchange rates, and pure oxygen injection (Salser *et al.*, 1978, Figure 3). Shrimp densities of 4 kg/m^2 or more are common, with occasional live weight standing crops of 70,000 kg/ha (Liao, 1985b; Colvin, 1985). The economic feasibility of this system is still under evaluation.

In Taiwan, intensive shrimp culture in ponds has evolved which yields production levels less than either the Shigueno or MCE systems. Taiwan has also concentrated almost entirely on the production of *Penaeus monodon*. In Taiwan, *P. monodon* is known as the grass shrimp while elsewhere it is commonly known as the giant tiger prawn, sugpo or black tiger prawn.

Tiger prawns in Taiwan are mostly cultured in earthen ponds with either concrete or earthen dikes. Production levels in these ponds can be very high. The country-

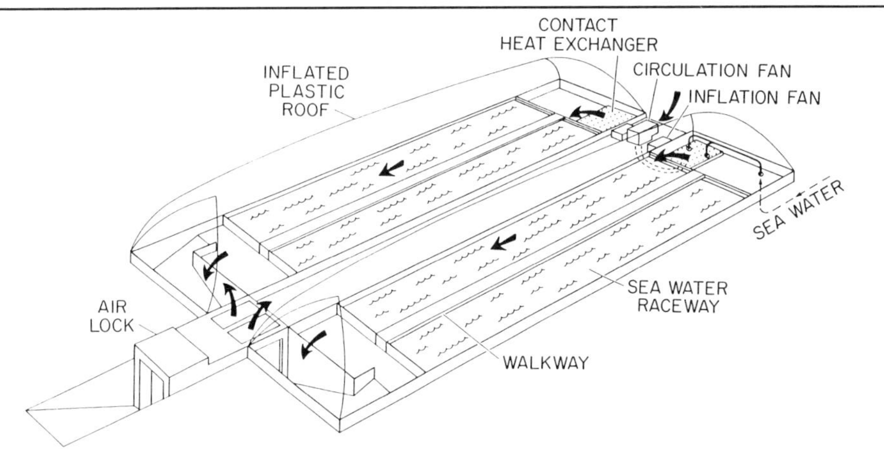

Figure 3. Aquacell used for ultra-intensive culture of marine shrimp by Marine Culture Enterprises (MCE) of Kahuku, Hawaii. (after Mahler *et al.* 1974).

wide average for Taiwan intensive culture systems during 1985 was about 5,000 kg/ha with some farms exceeding 20,000 kg/ha (Chiang and Liao 1985; Fast *et al.*, in press, a). From all accounts, profits from these tiger prawn farms in Taiwan are very good. Characteristics of these Taiwan style farms and their profitability are described more fully in following sections of this paper.

Although the Taiwan intensive culture system seems to work well in Taiwan, there have been few documented cases of this technology being used elsewhere at as high a yield and with as good a profit. The most notable exception to the lack of information on out-of-country yield is the farm operated by the San Miguel Corp. on Negros Island, The Philippines. This model demonstration farm represents a direct transfer of the best available Taiwanese intensive culture system technology for tiger prawns at the time it was built. Production from this farm was projected at about 10,000 to 13,000 kg/ha/yr (9,200 to 12,000 lb/A/yr), with best case estimates of 24,250 kg/ha/yr (22,250 lb/A/yr; Veloso, 1984; Liu and Mancebo, 1983). The average production yielded a net profit of $15,802/acre/yr with production costs and selling prices of $1.65/lb and $3.27/lb for whole shrimp respectively.

Although not documented, Taiwanese intensive culture systems are now widely used elsewhere in the Philippines as well as in Thailand and Indonesia. These recently developed farms are apparently quite profitable and in many cases are owned and/or operated by Taiwanese expatriates.

Increased maximum standing crop is associated with intensification of shrimp growout. With extensive culture systems maximum standing crop seldom exceeds 150 g/m^2 (Figure 4). With increased water exchange, aeration, quality feed applications, fertilization, controlled stocking densities, controlled predation and other appropriate management practices, maximum shrimp standing crop increases. The ultra-high intensity system used by Marine Culture Enterprises (MCE) yields the greatest standing crop of 4,000 g/m^2 or more. With today's technology and economic considerations, maximum standing crops of perhaps 1,000 g/m^2 (10,000 kg/ha) are probably a more realistic value for earthen ponds. In practice, 700 to 800 g/m^2 is a more likely target for commercial production using Taiwan type pond grow-out techniques.

A given maximum standing crop can be achieved by producing many small shrimp cultured over a short-time or fewer large shrimp cultured over a longer time.

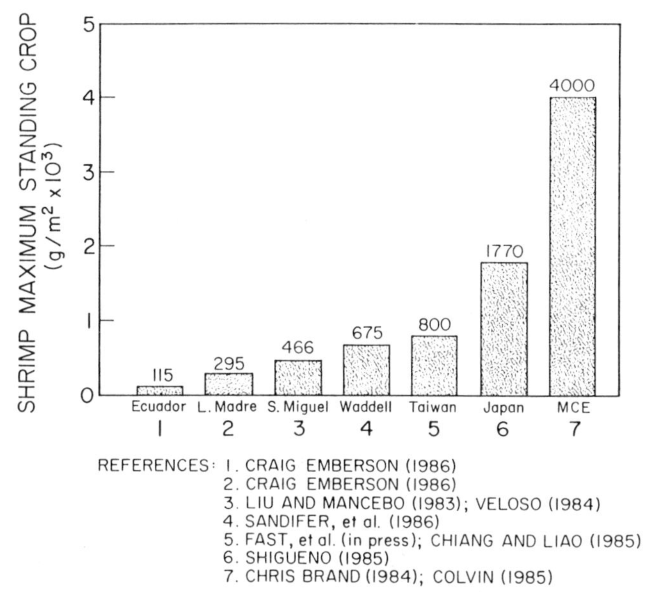

Figure 4. Maximum standing crops at harvest (g/M^2) for marine shrimp production systems ranging from extensive (Ecuador) to ultra-intensive (MCE).

While the total biomass is the same in both cases, the most profitable combination will depend on such things as price per weight for the different sized shrimp and seed costs. For a given shrimp species, shrimp size price structure, seed costs, and culture system, farms will tend to optimize stocking density, growout period, and shrimp size at harvest to produce the greatest profit. One of the real advantages of shrimp culture is that the culturist has considerable control over many of these factors. This flexibility, if appropriately used can increase profits.

Shrimp Culture Systems Comparison: Ecuador and Taiwan

During 1985, two countries, Ecuador and Taiwan accounted for perhaps half of the world's total cultured shrimp production, with 30,205 MT and 30,000 MT production figures respectively (Table 4). Total production continues to climb in both countries as new farms come into production, production intensifies from existing farms and other improvements occur in the industries.

To further understand the range of growout conditions and configurations of

Table 4. Pond production of marine shrimp in Taiwan and in Ecuador between 1968 and 1985. Production figures are in metric tons. Data from Chiang and Liao (1985), Liao (1987) and Anonymous (1987).

Year	Taiwan Production	Ecuador Production
1968	61	N/A
1969	69	N/A
1970	73	N/A
1971	76	N/A
1972	112	N/A
1973	119	N/A
1974	140	N/A
1975	150	N/A
1976	270	1,170
1977	1,100	1,350
1978	1,556	4,215
1979	4,123	4,698
1980	5,000	9,180
1981	6,000	12,100
1982	8,000	21,500
1983	15,000	35,600
1984	18,000	33,600
1985	30,000	30,205
1986	45,000	N/A

parameters for shrimp culture, it will be useful to compare the respective culture systems which have evolved in these two quite different environments and societies. In both countries the creation of a critical mass of infrastructure and support "services" has greatly facilitated the development of shrimp culture.

Although shrimp culture in both Ecuador and Taiwan runs the range from extensive to intensive, Ecuador culture systems can be thought of as "typically" semi-intensive, while Taiwan is "typically" intensive. We will compare these two types here.

Shrimp farms in Taiwan are typically small (4 ha), family run operations with many small ponds (Table 5). In Ecuador, farms are much larger (50 to 500 ha or more), corporate enterprises with vertical integration of farm, hatchery, nursery, feed mill and processing plant. In Taiwan these are all separate businesses, with

Table 5. Comparison between pond culture systems for marine shrimp in Ecuador (Hirono 1983) and Taiwan (Fast *et al.*, in press, a).

	Ecuador	Taiwan
1. Growout Intensity	Semi-intensive	Intensive
2. Business Form	Corporation	Family business
3. Farm Size (ha)	Large; 50-500	Small; 4
4. Pond Size (ha)	Large: 10-40	Small; 0.5
5. Shrimp Species (Penaeus	*P. Vannamei* *P. stylirostris*	*P. Mondon*
6. Water Source	Surface waters	wells
7. Salinity Control	Very little	Good, blend fresh and salt water
8. Principal Power Source	Diesel engines	Electric service
9. Power Cost (diesel fuel/elect)	$0.08/liter	$0.048/kw-hr
10. Land Costs	Low	Very high
11. Pond construction	Earthen	Concrete Wall
12. Water Depth (m)	1	2
13. Aeration	None	Paddlewheel
14. Seed Source	Wild & hatchery	Hatchery/nursery
15. Nursery	Some	Most
16. Seed Cost	$3-$6/1,000 PL5-10	$3-$5/1,000 Pl-5 $40/1,000 PL20-40
17. Stocking Densities (PL/m^2)	5 to 12	30-40
18. Crops/yr	2.4	2+
19. Survival	75%	75%
20. Feed Conversion	2.6	1.7
21. Feed Cost/kg	$0.28 to $0.40	$0.88
22. Harvest size (g) (whole shrimp)	16 to 24	30-35
23. Yield (kg/ha/crop) (kg/ha/yr)	800 to 1,500	5,000
24. Labor Costs	Low	Low
25. Water Exchange	High	High

horizontal rather than vertical integration.

The Ecuador shrimp industry is built on two species of shrimp, *Penaeus vannamei* (Mexican white shrimp) and *P. stylirostris* (Pacific blue shrimp) (Table 5). The former is generally preferred due to its hardier nature and better survival, although its growth rate is less than the latter. Both species grow slower than *P. monodon*, the mainstay of Taiwan intensive shrimp culture farms. Although *P. mondon* grows much faster than the Ecuadorian species at salinities in the range of 15 to 20°/oo, its growth rate at higher salinities may not be much greater than *P. vannamei* and *P. stylirostris*. Growth of *P. vannamei* is reported to be relatively unaffected by salinity over a range of perhaps 10 to 50°/oo (Rieban 1986).

Another major difference between Ecuadorian and Taiwanese farms is water quality management. Ecuadorian farms typically have a single, surface water source on an estuary or river. Very large volumes of water are pumped at the proper tidal stage to take advantage of water salinity. The water is pumped into a reservoir canal which is elevated above pond level. Water is flushed through the earthen ponds into a drainage canal on the opposite side of the ponds, where it drains back into the water source (Figure 5). The usual practice is to flush 10% of the water volume per day with capacity of up to 55% for some ponds. For a 1,000 ha farm, 1,000,000 m^3 of water must be pumped daily (183,500 gpm equivalency on a 24 hr basis). Diesel powered water pumps with 250 hp engines and 60 to 90 cm pipes, capable of pumping 100,000 l/min. or more are common. Few Ecuadorian farms have electricity at pond-side and almost none have emergency aeration capabilities other than by water exchange. This system evolved in Ecuador due in large part to low cost, low lying land near saline water and very low fuel costs (Ecuador is an oil exporting country). Diesel fuel costs about $0.08/l there.

In Taiwan, by contrast, the small ponds typically receive water from two wells; one freshwater well and one saltwater well. Due to expensive land prices, concrete walls are typically used to maximize production area on small land parcels (Figure 6). Water is blended in each pond to achieve the desired salinity and water exchange rate is adjusted during the growout cycle to give an overall water exchange of about 10%/day during growout. Complimenting water exchange in Taiwan ponds is mechanical aeration usually with paddlewheel aerators. The power source for both

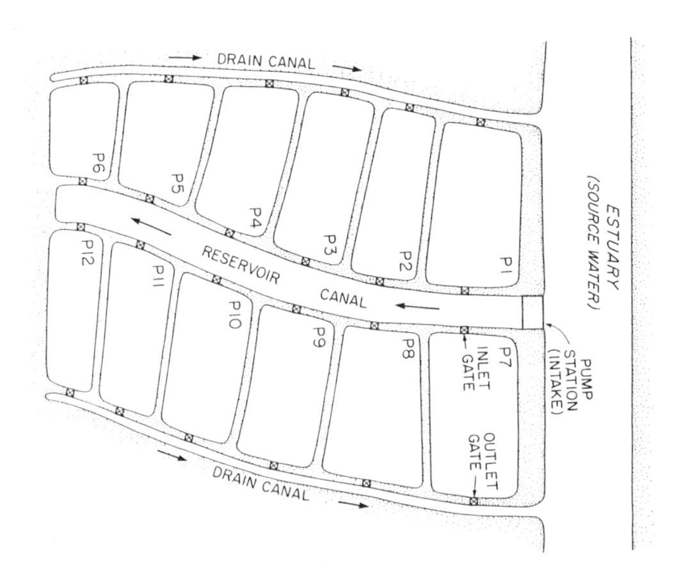

Figure 5. A typical marine shrimp semi-intensive arm in Ecuador. The farm typically has one water intake structure, a reservoir canal feeding the ponds, and drainage canals that return waste water to the source water.

pumps and aerators is electricity which is both reliable and inexpensive in Taiwan ($0.048/kw-hr). Both Ecuador and Taiwan rely on low cost energy and low cost labor.

In terms of water quality management, the biggest difference between the Ecuadorian and Taiwanese shrimp growout systems is the use of mechanical aeration in Taiwan and the lack of mechanical aeration in Ecuador. Although there are many other differences between these two growout systems, perhaps aeration is the single most important factor contributing to the 3 to 5 fold greater production per unit area in Taiwan.

From our work at the University of Hawaii, we have concluded that aeration is not a substitute for water exchange. Aeration without water exchange can prevent shrimp mortalities but both aeration and water exchange are needed in order to increase shrimp growth and yield (A.W. Fast, unpublished data). Water exchange at

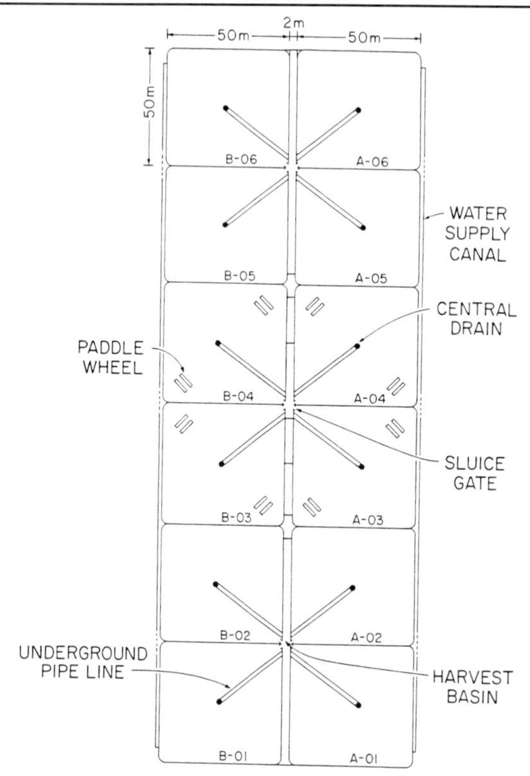

Figure 6. A typical Taïwanese style concrete walled pond design with center drains (after Liu and Mancebo 1983).

a very high rate, without aeration might accomplish the same result but at a much greater cost. A combination of aeration/water exchange should minimize operating costs while at the same time removing the water quality constraint to shrimp growth and yield.

While objective criteria for aerator operation exist, we do not have objective rules for water exchange in combination with aeration. Water exchange rates are usually determined either by some subjective means, or on a pre-determined schedule that is not necessarily related to water quality at a given time. This area of pond management is one that needs considerable attention before we consider pond management a science rather than an art form.

Pond depths of the large Ecuadorian shrimp ponds (18 to 20 ha) range from 0.6

m at one end to 1.5 m at the drain-harvest end, with an average depth of 1.0 m. In Taiwan, pond depths increased from 1 m to 2 m as the farms intensified (Fast *et al.* in press, a). The reasons for this increase are not fully known.

Fast *et al.* (in press, b) and Carpenter *et al.*, (1986) conducted growout experiments in the Philippines with *P. monodon* in replicated 1,000 m^2 ponds at 0.5, 1.0 and 1.7 m water depths in an effort to determine how pond depth affects pond dynamic processes and shrimp yield. They found that temperature, oxygen and vertical stratification were all significantly affected by pond depth, but that there was not a significant difference in shrimp yield between treatments. They concluded that the lack of significance in shrimp yield was due to low shrimp stocking densities (4 PL/m^2), and they further speculated that significant differences in yield would occur as stocking density increased and production intensified.

Seed source in Ecuador is traditionally from the wild although a tremendous expansion of hatchery capacity has taken place during the past 5 years. In Taiwan virtually all shrimp are from hatcheries and almost all farmers purchase nursed shrimp (PL20 or older) from separate nursery operations. Post larvae from hatcheries (PL5 to 10) are about the same in both countries but the cost of nursed *P. monodon* in Taiwan increased drastically to $40/1,000, apparently due to the sometimes high mortality of *P. monodon* between PL5 and PL20. Most semi-intensive farms in Ecuador nurse their own shrimp in nursery ponds for about 40 to 60 days. Ecuadorian nursery ponds account for about 10 to 12% of pond area on a given farm. Survival during the nursery phase averages about 60% (Boyd 1987).

Stocking rates in Taiwan are 3 to 8 times higher than in Ecuador but survival rates during final growout are about the same in both places. In Ecuador the shrimp are harvested at 16 to 24 g as opposed to 30 to 35 g in Taiwan, even though the overall growout period (about 140 days) and crops per year are about the same in both places. The respective stocking densities, species growth rates, and size at harvest account for most of the yield differences between the two growout systems.

Feed costs per kg are much greater in Taiwan. This is due in part to the higher quality of the Taiwan feed and to the importation of most of the feed ingredients . In Ecuador, anchovy fish meal is locally available at low cost. Although Taiwan feed costs more than feed in Ecuador, it gives better feed conversion ratios (FCR) and allows greater standing crops due to its nutritional superiority.

The foregoing discussion demonstrates major differences in growout strategies in Ecuador and Taiwan, yet shrimp farming is very profitable in both places due to country specific conditions. The lesson for us in the U.S. then, is that we need to develop shrimp culture technology which is specific to conditions in the U.S. As we will discuss further in the next section, transfer of either of these technologies to the U.S. may be inappropriate. Furthermore, it is quite likely that appropriate shrimp culture technology within the U.S. will be region specific. The most appropriate system for Hawaii most likely will not be the most appropriate for the mainland U.S.

Economics of Technology Transfer

Before trying to create an appropriate shrimp growout technology from scratch for the U.S., it would be prudent to first simulate the transfer of existing technology from elsewhere. An analysis of the simulation will thereby provide clues for what aspects of the existing technology are appropriate and what aspects must be replaced with something more appropriate.

With these thoughts in mind, we simulated the transfer of Taiwanese intensive culture systems to Hawaii. The reason that the Taiwanese shrimp aquaculture technology caught our interest is that it seemed relevant to certain of the more important economic constraints that we have in Hawaii. Land costs are high in Taiwan, as they are in Hawaii. Production systems which maximize production per unit land area should be appropriate to both areas. Seed, feed and certain other aspects of shrimp production in Taiwan are relatively sophisticated; something that we in the U.S. should also be able to master. Although labor costs and other aspects of shrimp culture in Taiwan differ substantially from conditions in the U.S., there are enough aspects of the Taiwanese shrimp culture system that seem relevant to conditions in the U.S. that we decided to evaluate the possible transfer of this technology to Hawaii.

The simulated transfer of intensive shrimp culture systems from Taiwan to Hawaii and the economic comparisons were based on a 4 ha size farm. This was the average size farm reported by Fast *et al.* (in press, a). The simulated transfer to Hawaii attempted to duplicate as much as possible farm construction and operations in Taiwan although some changes were necessary due to building code, environmental law and land ownership differences. The economic comparison assumed that a shrimp industry infrastructure had developed in Hawaii and that shrimp were sold at world market prices. Other conditions and assumptions are detailed by Fast *et al.* (in press, a).

Construction costs

The cost of concrete walled pond construction in Taiwan is based on a recent study by Chiang and Liao (1985) and is estimated at $44,100/ha. In Hawaii construction costs were estimated separately for earthen and concrete ponds, both with 4 ha of water surface area, and in Taiwanese style. Construction cost estimates for earthen and concrete walled ponds in Hawaii are $383,105 and $867,268 respectively for a 4 ha farm (Table 6). On a per ha basis, this is $95,776 and $216,817

Table 6. Estimated initial capital investment for construction of 4 ha (water surface area) intensive shrimp culture concrete walled and earthen dike ponds in Hawaii, built in Taiwanese style. Data from Fast *et al.* (in press, a).

	CONCRETE WALLED PONDS	EARTHEN DIKE PONDS
Construction Costs		
Permit	$ 30,000	$ 30,000
Excavation	6,300	121,470
Concrete work	558,600	N/A
Electricity	25,200	18,000
Pipes/plumbing	73,500	54,800
Design Costs (5%)	36,268	11,435
House	60,000	60,000
Storage/office	8,000	8,000
Well	30,000	30,000
Misc.	10,000	10,000
Sub-total	$837,868	$353,705
Equipment Costs		
Pumps	15,000	15,000
Aerators	14,400	14,400
Sub-total	$ 29,400	$ 29,400
Total Cost	$867,268	$383,105

respectively. We did not estimate construction costs for earthen ponds in Taiwan, but the costs for this style farm is far less than for concrete walled ponds. The greater construction costs in Hawaii are due largely to higher labor costs, high cost of professional services (e.g., electrician, plumber, design engineer) and to the permit processes.

Operating costs

Annual operating costs in Taiwan on a 4 ha, concrete-walled intensive shrimp farm were $77,239/ha (Table 7). Seed (Post larvae) and feed costs were the largest expenses and accounted for 64% of operating costs. In Hawaii annual operating costs for the same type farm were $149,088/ha. With earthen diked farms in Hawaii annual operating expenses were $129,079/ha. While seed and feed costs were slightly higher in Hawaii due to more crops/pond/year, the major items contributing to the much greater operating expenses in Hawaii were labor, electricity, and capitalization costs (interest and depreciation).

Table 7. Estimated annual operating costs per ha on 4 ha (water surface area) intensive shrimp culture farms in Taiwan and Hawaii. Only concrete walled ponds were considered for Taiwan, while both concrete walled and earthen dike ponds were considered for Hawaii. Data from Fast *et al.* (in press, a).

	Taiwan (w crops/year)		Hawaii (2.5 crops/year)			
	Concrete Pond		Earthen Pond		Concrete Pond	
Seed	$27,200	35%	$34,000	26%	$34,000	23%
Feed	22,400	29	28,050	22	28,050	19
Labor	5,000	7	18,213	14	18,213	12
Harvest	630					
Electricity	5,194	7	15,538	12	15,538	10
Chemicals	1,600	2	2,000	2	2,000	1
Land Lease	4,000	5	7,178	6	5,127	3
Interest	5,292	7	11,493	9	26,468	18
Depreciation	2,205	3	5,946	5	11,998	8
Sales Tax			586		586	
Misc.	3,678	5	6,075	5	7,108	5
TOTAL	$77,239		$129,079		$149,088	

Gross revenues and profits

An average yield of 7,500 kg/ha/crop and a farm price of $6.25/kg for 31 g live weight shrimp were assumed in both places. The gross revenues for concrete walled ponds in Taiwan were $93,750/ha/yr, with product costs of $77,239/ha/yr (Table 8). This resulted in a net profit of $16,511/ha/yr for concrete walled farms in Taiwan. In Hawaii gross revenues were greater due to more crops per year, but very high production costs resulted in profit losses of $-11,891/ha/yr and $-31,900/ha/yr for earthen and concrete walled ponds respectively. Earthen ponds in Hawaii would lose less money since the capitalization costs were less.

Conclusions

The transfer of Taiwanese intensive shrimp culture technology to Hawaii would not be profitable unless production level can be increased and/or production costs can be reduced. Only slight increases in production are likely without the development of improved technology. A more promising approach to improving profitability in Hawaii would be to reduce production costs.

Comparing the costs of production between Taiwan and Hawaii, it is apparent that costs of labor and electricity in Hawaii are about 3 times higher than that in Taiwan. Interest and depreciation also cost 2 to 5 times more in Hawaii (depending on type of ponds) mainly because of higher construction costs. Efforts in reducing these costs are necessary.

In Hawaii the costs of construction of concrete ponds are more than double that of

Table 8. Estimated annual gross revenue, production costs, and net profits for concrete walled intensive shrimp ponds in Taiwan, and concrete walled and earthe diked ponds in Hawaii. Data from Fast *et al.* (in press, a).

		Hawaii	
	Taiwan	Earthen Ponds	Concrete Ponds
Production/ha/year	15,000 kg	18,750 kg	18,750 kg
Farm price/kg	$6.25	$6.25	$6.25
Gross revenue	$93,750	$117,188	$117,188
Cost of Production	$77,239	$129,079	$149,088
Profit/ha/year	$ 16,511	($-11,891)	($-31,900)

earthen ponds. The concrete pond operation is unlikely to be justified economically in Hawaii unless the high construction cost can be off-set by other costs reductions and by higher yield.

It is also likely that seed and feed costs can be substantially reduced. These are large expenses in both Taiwan and Hawaii.

In the next section I will discuss more fully some approaches to the development of improved shrimp culture technology for the U.S. which were based in part on the results of this simulated transfer of Taiwanese shrimp culture technology to the U.S.

Development of Appropriate Shrimp Culture Technology for the U.S.

Shrimp culture systems in the U.S. are for the most part modeled after growout systems developed for channel catfish (*Ictaluras punctatus*) in the southeastern U.S. (Huner and Dupree 1984; Boyd 1985; Tucker 1985). These growout systems include earthen ponds ranging in size from 0.25 to 10 ha. They lack a reservoir canal and usually have wells of a single salinity without the capability to control salinity. Mechanical aeration, if it exists at all, is generally by paddlewheel aerator. Water exchange capabilities vary but perhaps average about 5%/day for the whole farm. On the U.S. mainland, one crop a year is typical while in Hawaii two or three crops/ year are common. *Penaeus vannamei* accounts for perhaps 90% of all farm production in the U.S. Shrimp are harvested at small sizes, in the range of 12 to 18 g heads-on. The industry, if it can truly be called that, is very small with not more than 25 primary farms employing perhaps 250 to 300 people. There is much diversity in farm design, farm management, management strategies and production success. Shrimp farming in the U.S. is obviously still seeking a profitable technology.

From the foregoing, and from what is common knowledge about the profitability of marine shrimp culture in the U.S. it is apparent that we need to develop improved shrimp culture technology if we hope to make U.S. cultured shrimp prices competitive with foreign produced shrimp. Unless we can achieve that goal we have little hope of reducing the influx of foreign shrimp into the U.S., and thereby reducing the large negative trade imbalance in the U.S. for this commodity.

The question now is; what must the U.S. do to make shrimp cultured in the U.S. price competitive with foreign grown shrimp? The answer is; we must reduce production costs per pound of production. There are many aspects of production which can be worked with to achieve production cost reductions, some of which we will discuss below but perhaps labor and energy costs lend themselves most to reductions in the U.S. More specifically, we need to reduce the costs of labor and energy in the growout phase of shrimp production. With the simulated transfer of Taiwanese intensive culture technology to Hawaii, cost reductions of 72.5% for labor and 66.6% for energy in Hawaii would be necessary to bring these costs in line with Taiwan. These cost reductions alone would make the technology transfer profitable in Hawaii.

In many ways research on marine pond production in the U.S. has lagged behind other aspects of shrimp culture such as reproductive biology and larviculture. In part this lag was due to emphasis on the critical limiting factor and to the scarcity of marine pond research facilities in the U.S. Even now there are only 3 or 4 pond research facilities in the U.S. capable of doing appropriate R & D on marine shrimp growout technology.

The initial limiting factor for shrimp culture was seed production and early research efforts focused on this problem. Although more work is needed on seed production, it is no longer the constraint to shrimp culture that it once was. There has now been a shift of the principal constraint from seed to pond growout. The main constraint now to economic viability of U.S. cultured shrimp is in the costs to rear shrimp to market size after they leave the hatchery. For the most part this means rearing in earthen ponds.

The following are specific areas of shrimp culture that need development of technology appropriate to the U.S.:

Pond Dynamics R & D and Microcomputer Applications.

Earthen ponds represent a complex of interactions that we are only starting to understand. What we do know though, is that for a given set of conditions in a pond, shrimp growth will not exceed a given rate nor will shrimp biomass exceed some maximum. Shrimp growth and biomass can often be increased above this limit by improved feed applications, increased water flow-through, aeration as needed, and attentive monitoring of conditions. The problem is that these management practices can be expensive, consume much energy, and are labor intensive.

Furthermore, we have not as yet developed a quantifiable understanding between shrimp growth, standing crop, and these pond dynamic parameters.

Specifically, we need to better understand and be able to predict what is needed to maintain high water quality. The term "water quality" is a catchy phrase for a poorly understood condition. We know some things that affect water quality such as dissolved oxygen (DO), temperature and carbon dioxide. We can quantify these and largely predict their effects. There are other important aspects of water quality that we know little about, such as the metabolites or other substances which accumulate in the water during intensive culture practices. We know that these factors can be removed through water flushing with higher quality water but we presently cannot identify a specific need for flushing or flushing rate. Whether or not we achieve a thorough understanding of these pond dynamic processes we need to better quantify the need for water exchange. Water exchange with intensive Taiwan systems accounts for more than 60% of the energy costs, while aeration accounts for most of the remaining 40% energy costs (Fast *et al.*, in press, a). At present pumping schedules are highly subjective. Low water exchange rates of up to 5% per day in channel catfish ponds had no apparent effect on water quality or fish production (McGee and Boyd 1983) while high exchange rates of more than 50% per day in tanks cause too much phytoplankton loss (Ryther 1971; Ryther *et al.*, 1972). The appropriate water exchange rate at a given time in intensive marine shrimp culture systems will undoubtedly depend on a number of factors such as the health of the algae, available nutrients, and metabolite concentrations. The amount of water which should be exchanged on a given day could vary from 0% to 50% or more but at present we do not know how to determine this need.

While we do not have a good understanding of all the parameters involved, we do know that water exchange, especially during time of high shrimp standing crop and high feed applications is essential for the maintenance of suitable water quality and high shrimp yield. It may be possible to develop criteria for water exchange which achieve the desired results while at the same time greatly reducing water exchange and energy use. This may also reduce the need for aeration as well and thus result in additional energy savings.

We at the University of Hawaii have concentrated our research efforts on the application of microcomputer technology to ponds. We feel that this application is essential for the development of new and better pond management technology both through the use of microcomputers to gain a better understanding of pond dynamic processes and secondly through microcomputer use to operate farms on a real-time basis.

A better understanding of in-pond processes should lead to improved management strategies. As a specific example we are using data loggers to continuously monitor such parameters as DO, temperature, pH, and fluorescence. From these and other data we can estimate components of pond respiration, photosynthesis, and perhaps algal densities (Madenjian *et al.* ,1987a; Madenjian *et al.*, 1987b; Rogers and Fast 1988; Rogers *et al.*, 1988). This could lead to real-time assessment of algal health and the prediction of algal die-off well before it occurs. Algal die-off is one of the main water management problems for pond operators. It can not only lead to mass mortalities of the culture species but it negatively impacts growth even if mass mortalities do not occur. If we can develop predictive techniques we can apply corrective measures to prevent algal die-off. Corrective measures could include increased water exchange, pumping water from another pond with healthy algae, fertilization or some other corrective measure.

Maintaining a healthy, rapidly growing algal population is perhaps one of the best ways to maintain good water quality. A healthy, rapidly growing algal population will produce more oxygen than it consumes, absorb certain undesirable metabolites such as ammonia, produce food for the cultured shrimp and otherwise maintain good culture conditions. The algal growth models (or oxygen budget models) of Romaire and Boyd (1979), Boyd *et al.*, (1978), Baumert and Uhlmann (1983) and Laws and Malecha (1981) taken together indicate that algal densities for healthy algae are optimal for culture purposes in the range of 100 to 300 mg/m^3 chlorophyll a with 250 mg/m^3 producing the highest oxygen concentrations at dawn. Below this range, insufficient oxygen is produced to meet pond respiration needs, while above this range algal self-shading may reduce photosynthesis and oxygen production while at the same time algal respiration is quite high. Comparison of the above mentioned phytoplankton oxygen models further indicates that under very dense algal standing crop conditions nutrient limitations rather than algal self-shading may be even more critical (Smith and Piedrahita 1988). Our understanding of how to maintain healthy algal crops in outdoor tanks or ponds is not well developed (Goldman 1979 a,b) but we do know that algal "cropping" perhaps in combination with selective fertilization should help maintain healthy algae (Smith 1987; Smith and Piedrahita 1988). Algal cropping may be achieved through the use of polyculture with phyto-phagenous fish and bivalves (clams and oysters) and/or through the exchange of new source water and the flushing out of algae from the pond (Ryther *et al.*, 1972). Research on these approaches has only just begun and could take many years before practical techniques are developed. The development of practical techniques will almost certainly require a combination of the following: old fashioned trial and error, but well reasoned experiments with appropriate species combinations; development of appropriate pond process models; and the use of microcomputer technology to assess treatment effects. The ultimate objectives of this line of research are to develop technologies for maintaining high water quality and maximal shrimp growth while at the same time reducing production costs and reducing the risk of crop loss.

The first step in developing technologies for algal management is understanding the critical relationships discussed above. Data loggers and computer analysis of the data generated is pivotal to this development process. Once we understand the interactions to the point where we can manipulate them by process control it is almost inconceivable that control of algal health could be effectively accomplished other than by the real-time use of microcomputers to constantly assess pond conditions and to take appropriate actions. Non-computer techniques may evolve but these almost certainly would not be as effective as computer managed systems.

Microcomputers can be used to continuously measure water quality and to perform routine pond management operations such as pH and oxygen measurements, lime addition decisions, feeding, aeration/circulation operation decisions, and water exchange decisions. All but the last item have been quantified or sufficiently understood that microcomputers can now be used for these purposes. Such applications, perhaps more than any other thing, should result in the necessary labor and energy savings needed to make U.S. cultured shrimp competitive on the world market.

One microcomputer monitoring configuration is shown in Figure 7. This includes pumping water from a number of ponds sequentially to a single point where DO and other parameters are measured. Temperature will normally be measured in-pond with thermisters but could also be measured at the central location. An alternative of pumping to one location is to have probes in each pond but probe cost and technology is perhaps not cost effective for this application at this time. As data is collected from each pond it can be analyzed as noted above, and management decisions made. A simple example of this for aerator/circulator operation is illustrated in Figure 8. Aeration is normally accomplished with paddlewheel or propeller-aspirator aerators (Tucker and Boyd 1985; Ahmad and Boyd 1987; Boyd and Ahmad 1987) by air injection or some other means (Boyd 1983a,b) while circulation is normally accomplished by axial-flow or slow-moving paddlewheel devices (Fast *et al.* 1983; Busch and Goodman 1981). This decision model is based on absolute DO as well as rate of change in DO. Other decision logic could be developed for

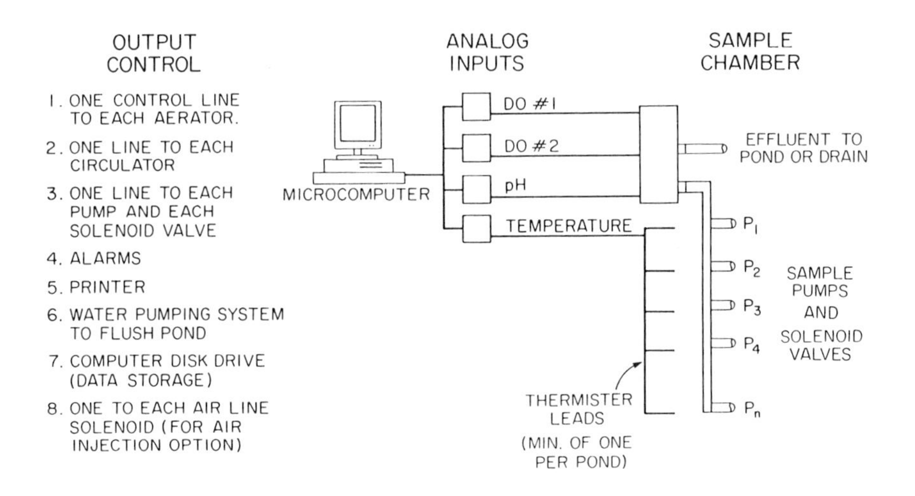

Figure 7. Schematic of microcomputer pond monitoring and management system. The microcomputer and electronic probes are located at one point, and pond water is pumped from a series of ponds sequentially to the probes

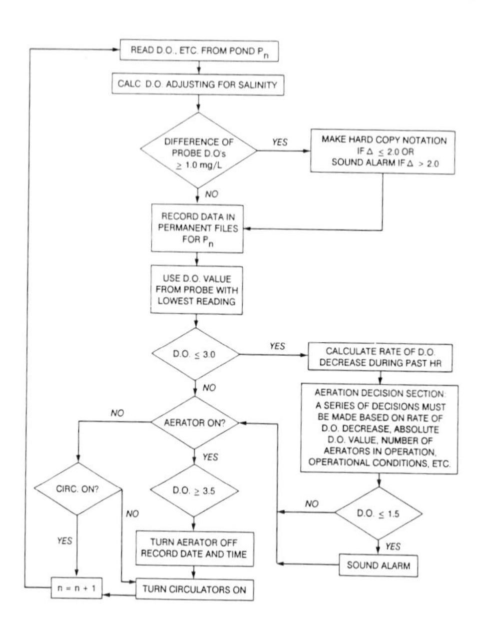

Figure 8. A simple conceptual model for decision making logic in a microcomputer pond monitoring/management system. This model is for aeration/circulation decisions, while other models would service decisions on water exchange, feeding, fertilization and/or other pond management functions.

water exchange, fertilization and/or other management options.

The application of the above microcomputer arrangement for pond monitoring and management could lead to new pond designs. One such new design is where eight triangular shaped ponds radiate from a central hub (Figure 9). The hub area would contain the microcomputer monitor, central harvest basin and utilities (water and electricity) controls. These modular units of eight ponds could be replicated and data could be sent to a separate location for interpretation and decision making.

Again, the overall objectives of these microcomputer applications are to reduce labor costs, reduce energy costs and to reduce risk of crop failure, while at the same time optimizing shrimp growth and yield. The object is to improve profitability.

Farm design, equipment applications, and management procedures

As seen above innovative farm designs may be better than existing designs if they reduce labor, energy consumption and risks while maintaining high yield (Chien and Liao 1988). We need improved farm equipment and farm layouts such that labor to feed, harvest and maintain the facilities are minimized. Automated or semi-automated feeding systems could greatly reduce the feed labor component.

Our pond research group at the University of Hawaii has also pioneered work on low-energy pond circulation equipment. Fast *et al.*, (in press, b) have shown that even 1.0m deep shrimp ponds will strongly stratify and that the degree of stratification will increase with pond depth. There is also a trend towards deeper ponds as shrimp culture intensified (Fast *et al.* , in press, a). The present practice to overcome this stratification is to operate pond aeration devices even during times when surface dissolved oxygen is supersaturated. This practice not only uses more energy than

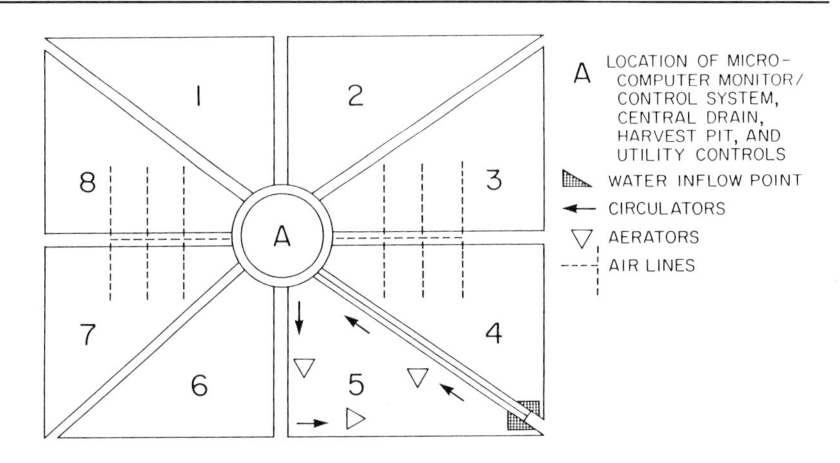

Figure 9. Plan view of eight pond system which incorporates the PAMS pond monitoring, decision making and control. The microcomputer and proble chamber is centrally positioned at location A. Water from each of the eight ponds is pumped to the probel chamber for analysis. The central drain and harvest pit is also positioned at location A. For illustration purposed, three aerators and four circulators are shown in pond 5 only, while aeration lines are only shown in ponds 3, 4, 7, and 8. Water inflow points are shown in ponds 4 and 5. This design centralizes most of the pond monitoring and utility functions. Pond size is independent of design, but might range from 1 to 5 acres per pond for intensive shrimp culture. Most aquaculture ponds now are either square or rectangular.

necessary but it also drives oxygen from the pond, thus creating a greater need for artificial aeration during the night. The solution to the problem was the development of low-energy pond circulation devices that borrowed technology originally developed for lake destratification (Garton 1981). This included the design and use of several models of axial flow water pumping devices that use a fan blade and speed reduction motor to pump large volumes of water within the pond (Fast *et al.*, 1983; Rogers and Fast 1988). These devices have been used to artificially mix both freshwater prawn ponds and brackish-water shrimp ponds. Used appropriately in combination with aeration devices this aeration/mixing system should reduce total energy consumption while at the same time creating suitable oxygen and temperature water quality conditions.

Harvest labor requirements could be reduced through the development of drain harvest techniques where the shrimp "run" with the water and are removed from the drain water through the use of live fish pumps and sorting tanks. Some shrimp, such as tiger prawns, tend under certain conditions to burrow in the mud rather than run with the drain. A better understanding of this behavior could perhaps, through timing the drain, drain outlet design, draining rate and/or by pre-conditioning the water somehow, lead to all of the shrimp running with the drain water. This alone will greatly reduce harvest labor needs but coupled with improved sorting and cleaning, such as with live fish pumps and live tanks, would further reduce harvest labor costs while maintaining a premium product.

Another aspect of farm design which affects labor and perhaps energy consumption, is farm size. Should the farm be designed for a "single family", or for a larger "corporate" operation? If the former, should they be clustered into "cooperatives" in order to realize economics of scale, while at the same time realizing economics of efficiencies associated with smaller farms?

Maintenance of farm equipment and premises (e.g. grass cutting) can consume large amounts of labor. Farms, equipment and maintenance approaches need to be developed which minimize these labor consumers.

Multiphasic Growout

The greater the number of crops of a given yield and shrimp size per year per unit area, the greater the potential profit. One of the principal constraints to number of crops is the number of times that the pond reaches carrying capacity during the year or, put another way, the average percent of carrying capacity during the year. If the pond is stocked with PL5's and the growout period averages from 120 to 150 days, then only two crops maximum are generally possible. If, however, a 2-phase growout system is used (nursery + growout), then perhaps 3 or more crops could be produced. Stamp (1978) predicts that two transfers of prawns during a 150 day growout cycle will increase production by 30%. A 3-phase growout could further increase the number of crops per year. The solution then, requires the development of management strategies and facilities designed for easily managing a multiple step growout process (ASEAN 1978; Chien and Liao 1988).

We need improved nursery technology in the U.S. in order to maximize production from a given land area without incurring greatly increased labor or energy costs. Good progress has been made in this area by the Shrimp Mariculture Program at Texas A&M University where a primary objective is to head-start shrimp in an effort to maximize production during a 5 or 6 month annual production cycle in ponds (Sturmer and Lawrence 1986).

Basic Data on Shrimp Biology and Behavior

Very little is known about the basic biology of shrimp and how this relates to growout in earthen ponds. The specific effects of temperature and salinity on the growth potential through market size is not known for even one species of cultured shrimp. This basic information is critically important when making decisions about farm siting, water supply, and the best species to culture at a given farm during different times of the year. We need to develop this basic information soon if we expect to increase production and profits from shrimp farms in the U.S. or elsewhere.

We also know very little about shrimp behavior in ponds. Factors affecting their distribution, response to organic sediment accumulations, tendency to burrow under different conditions, and factors affecting their tendencies to “run” during pond draining are all virtually unknown. We need to know more about these shrimp behaviors for each species if we expect to improve shrimp growout technology.

Although the controlled temperature and oxygen growth responses are either totally lacking or ill defined for all shrimp species, there are some data available on a few species (Zein-Elden and Aldrich 1965). Griffin *et al.*, (1981) used these data in a bioeconomic model for *P. setiferus* to predict profits from various management strategies and water quality conditions. Greatest profits were realized at minimum oxygen concentrations of 3 to 4 mg/l with less profit above or below these minimum values. Maintaining oxygen concentrations greater than 4 mg/l required substantially greater energy to artificially aerate the water (by water exchange) while minimum oxygen concentrations below 3 mg/l not only suppressed shrimp growth but also increased the risk of crop loss. With *P. setiferus*, salinity also has an important impact on growth and profit. A salinity decrease of 10% from 30°/oo reduced profits by 27%.

<u>Crop Rotation and the Winter Crop</u>

Another constraint to maximization of crops is the problem caused by seasonally low water temperatures and slow growth of the more commonly cultured shrimp species during the winter months. In Hawaii, pond temperatures typically range between 18 and 23 °C for 3 to 4 months each year. During this time growth of *P. vannamei, P. monodon* and most of the warmwater penaeids is very slow. A solution to this problem is to identify cold or cool water species which will grow well in this temperature range. *Penaeus chinensis* and *P. penicillatus* from Asia are two possible candidates. The overall solution to this problem is to develop a growth rate matrix based on temperature and salinity for each commonly cultured species. This matrix can then be used to predict the best combination of species to culture at a given location based on past temperature conditions and likely salinities. Such optimization could improve profitability.

<u>Seed Costs</u>

Although seed production is no longer considered a technical constraint for shrimp culture, it is a major operating cost. Seed costs of $40/1,000 for PL20 or older account for 35% and 27% of operating costs in Taiwan and Hawaii respectively (Fast *et al.*, in press, a). These costs are high compared with other countries of SE Asia where prices for comparable PL's cost less than 1/2 this amount. In the U.S., as in Taiwan, we may also experience high costs for this sized PL due to the anticipated need for closed cycle seed production. Seed production costs need to be reduced in the U.S.

Feed Costs

Shrimp feed costs of $0.88/kg ($0.40/lb) account for 29% and 22% of projected operating costs of tiger prawns in Taiwan and the U.S. respectively (Fast *et al.*, in press, a). Compared with U.S. feed costs for other species such as rainbow trout (*Salmo gairdneri*) and channel catfish, where nutritional needs are better defined, shrimp feed costs are excessively high. Good quality trout and catfish feeds are now produced in the U.S. at retail prices of about $0.23 and $0.18/lb respectively with feed conversions in the range of 1.1 to 1.5. Perhaps a better understanding of shrimp nutritional needs and alternative feed substances will reduce shrimp feed costs as well.

Overall

There are many other items which could affect shrimp culture profitability. A thorough discussion of all of these is beyond the scope of this paper. I believe, however, that I have identified the major items now causing U.S. cultured shrimp to be economically uncompetitive with shrimp cultured elsewhere. Research and development efforts on these priority items will hopefully resolve these impediments and thus place U.S. grown shrimp on a competitive basis in the world market.

Projected World Shrimp Supply and its Impact on Shrimp Culture

The world's shrimp catch, like that of the United States, is thought to be at or near maximum sustainable yield with a total catch of about 1.75 million MT (live weight) (Asian Development Bank 1983). Although individual country catches vary from year to year, the total is expected to remain more constant. Any increase in the total is expected to come from shrimp culture in ponds.

Substantial increases in world shrimp culture are projected. During 1982 the world production of cultured shrimp was 171.1 million lbs (77,773 MT) or 4.4% of total world production (Table 9). Projected world production of cultured shrimp by 1990 is 531.9 million lbs (241,773 MT), or 13.8% of total world production. There is already evidence that the total production of pond cultured shrimp has reached or exceeded the 1990 prediction (Hopkins 1987, Pecham 1988). Most of this production increase is expected from Asia (Figure 10). Further shrimp culture production increases are expected well past the year 1990 as production technology improves,

Table 9. World's production of cultured shrimp (live-wt) during 1982, and projected production in 1990. Data from National Marine Fisheries Service (1984).

Continent/Country	Year 1982	1990	Primary Species
Asia			
India	33	110	*P. indicus & P. mondon*
Indonesia	23	88	*P. merguiensis & P. mondon*
Taiwan	21	66	*P. mondon*
Thailand	22	55	*Macrobrachium r. & P. mondon*
Philippines	8.6	44	*P. mondon*
China	3.1	6.2	*P. orientalis*
Japan	4.4	6.3	*P. japonicus*
Malaysia	.3	4.4	N/A
Sri Lanka	----	.9	*Macrobrachium r.*
Pakistan	----	.8	N/A
Korea (ROK)	.2	.2	N/A
Australia	----	.01	N/A
Others	----	.6	N/A
Total	117.8	381.8	
Latin America			
Ecuador	47.3	88	*P. vannamei*
Brazil	.4	8.8	*P. japonicus*
Peru	1.2	7.7	*P. vannamei*
Panama	3.3	6.6	*P. vannamei*
Columbia	----	6.6	*P. vannamei*
Hondurus	.5	5.5	*P. vannamei*
Mexico	----	4.4	*P. vannamei*
Venezuela	----	3.3	N/A
Belize	----	3.3	*P. vannamei*
Bahamas	----	3.1	N/A
Guatemal	.2	2.2	*P. vannamei*
French Guiana	----	1.7	*Macrobracium r.*
Dominican Republic	----	1.1	N.A
Puerto Rico	----	1.1	*Macrobrachium r.*
Costa Rico	----	1.1	*P. vannamei*
Haiti	----	.9	N/A
Martinique	.1	.7	*Macrobrachium r.*
Guadeloupe	----	.6	*Macrobrachium r.*
El Salvador	----	.4	*Macrobrachium r.*

" continued"

"Table 9 continued"

Nicaragua	----	.4	N/A
Dominica	----	.4	N/A
Suriname	----	.2	N/A
Jamaica	.05	.2	N/A
Cuba	----	.1	*P. schmitti*
Guyana	----	.1	N/A
Uruguay	----	----	N/A
Argentina	----	----	N/A
Chile	----	----	N/A
Others	----	.4	N/A
Total	53.4	148.6	
Africa and the Middle East			
Ivory Coast	----	.2	N/A
Saudi Arabia	----	.2	N/A
Senegal	----	.02	N/A
Kenya	----	.02	N/A
Others	----	.6	N/A
Total	----	1.0	
Europe	----	.2	N/A
World Total[1]	171.1	531.9	

N/A - Not Available

[1]million pounds, live weight

---

and as more land is brought into production.

The anticipated impact of these increases in shrimp supply is to cause shrimp prices to remain static or even decrease. Even static real prices during rising real costs will mean less net profit to the producer.

The effect of this production increase on fishermen (trawler fleet) is that many of the less efficient or less lucky operators will go out of business, with the result that the total catch per boat, or catch per unit effort of those remaining will increase (National Marine Fisheries Service 1984). Up to some point, total catch from the wild should remain about the same.

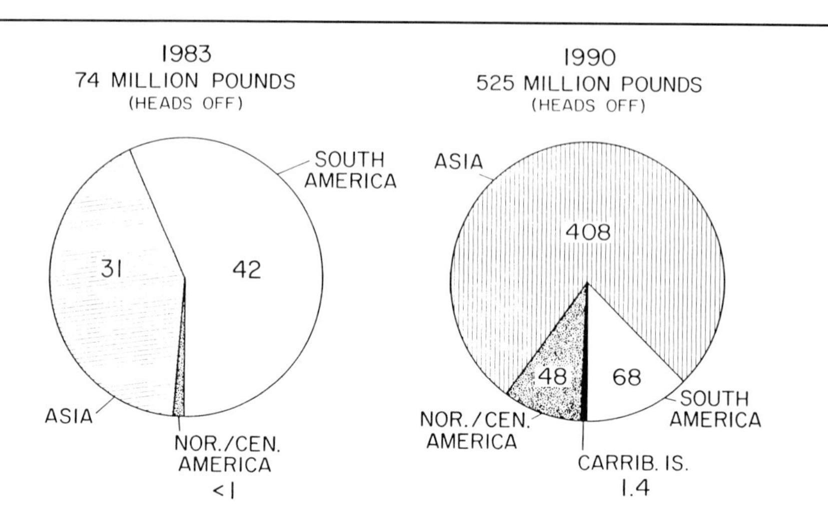

Figure 10. World shrimp farm production estimated for 1983 and 1990 by region. (after Vondruska 1987).

The effect of large production increases on shrimp culture operations will be that the most efficient operations will survive, while the less efficient farms will be either unprofitable or less profitable. Unlike the wildcatch, "catch per unit effort", or pond yield will not increase simply because there are less farms. The relevance of this situation for the production of cultured shrimp in the U.S. is that we must reduce production costs below even those which would make us competitive on the world market today (Table 1). The development of appropriate shrimp culture technology for the U.S. must be even more economically viable than most of those existing elsewhere today.

Acknowledgements

Pond dynamics research at the University of Hawaii has been supported by the University of Hawaii Sea Grant College Program; the Aquaculture Development Program; Hawaii State Department of Land and Natural Resources; and by the U.S. Agency for International Development through a Title XII Aquaculture Cooperative Research Support (CRSP) contract with the University of Hawaii. Indirect support was provided by Chulalongkorn University while preparing this paper.

LITERATURE CITED

Ahmad, T. and C. E. Boyd. 1987. Design and performance of paddle wheel aerators. Aquacult. Engineer. 7(1987):1024.

Anonymous. 1987. Shrimp culture in Ecuador. In. Acuacultura del Ecuador. Publicion de la Camera de Productores de Camaron. Numero 3, Enero de 1987. Guayaquil, Ecuador.

Apud, F., J. H. Primavera, and P.T. Torres, Jr. 1983. *Farming of prawns and shrimps*. Extension Manual No. 5 (3rd edition). Aquaculture Department, Southeast Asian Fisheries Development Center, Tagbauan, Iloilo, Philippines.

Aquacop. 1979. Penaeid reared broodstock: closing the cycle. Proc. World Maricul. Soc. 10:445-452.

ASEAN. 1978. Manual on Pond Culture of Penaeid Shrimp. ASEAN 77/SHR/CUL3. Published by ASEAN National Coordinating Agency of the Philippines, Ministry of Foreign Affairs. 122 pp.

Asian Development Bank. 1983. The international shrimp market for shrimp. Vol. 3, ADB/FAO Second Fish Market Study. Manila, Philippines. 79 pp.

Baumert, H. and D. Uhlmann. 1983. Theory of the upper limit to phytoplankton production per unit area in natural waters. Int. Rev. Gesamten Hydrobiol., 68:753-783.

Boyd, C. E. 1985. Hydrology and pond construction. In. C.S. Tucker (eds.) Channel catfish culture. Elsevier, New York. pg. 107-134.

Boyd, C. E. 1983a. Theory of aeration. In. Mississippi State Univ. Southern Series Bull. 290; pgs. 13-20.

Boyd, C. E. 1983b. Emergency aeration. *Ibid*, pgs. 21-23.

Boyd, C. E. and T. Ahmad. 1987. Evaluation of aerators for channel catfish farming. Alabama Agricult. Exper. Station. Bull. No. 584, Auburn Univ., Auburn, AL, 52 pgs.

Boyd, C. E., R. P. Romaire and E. Johnston. 1978. Predicting early morning dissolved oxygen concentrations in channel catfish ponds. Trans. Am. Fish. Soc., 107:484-492.

Boyd, C. E. 1987. Personal communications. Auburn University, Auburn, AL.

Brand, C. 1984. Personal Communications. Marine Culture Enterprises. Kahuku, HI.

Browdy, C. L. and T. M. Samocha. 1985. Maturation and spawning of ablated and nonablated *Penaeus semisulcatus* de Haan. J. World. Maricul. Soc. 16:236-249

Busch, C. D. and R. K. Goodman. 1981. Water circulation, an alternative to emergency aeration. Proc. World Maricult. Soc. pg. 8-10.

Carpenter, K. E., A. W. Fast, V. L. Corre, Jr., J. Woessner, and R. L. Janeo. 1986. The effects of water depth and circulation on the growth of *Penaeus monodon* in earthen ponds. In. J. L. McLean, L.B. Dizon and L.V. Hosillos (eds.). The First Asian Fisheries Forum. Asian Fisheries Soc., Manila, Philippines. pp. 21-24.

Chamberlain, G. 1988. Personal Communications. Texas Cooperative Extension Service. Corpus Christi, Texas

Chamberlain, G. W. 1985a. Shrimp farming in the Caribbean and Latin America. Coastal Aquaculture 2(1):7.

Chamberlain, G.W . 1985b. In. G.W. Chamberlain, M.G. Haby and R.J. Miget (eds.); Texas Shrimp Farming Manual; Texas Agri. Ext. Service, Corpus Christi, TX, 20 pgs.

Chiang, P. and I-Chiu Liao. 1985. The practice of grass prawn (*Penaeus mondon*) culture in Taiwan from 1968 to 1984. Jour. World Maricult. Soc. (16):297-315.

Chien, Y. H. and I. C. Liao. 1988. The evolution of prawn growout systems and their management in Taiwan. Presented at Aquaculture Engineering Technologies for the Future. Univ. Sterling, Scotland. June 20-23. 39 pgs.

Colvin, L. B. 1985. Intensive growout systems for shrimp. In. G. W. Chamberlain, M. G. Haby and R. J. Miget (eds.); Texas Shrimp Farming Manual; Texas Agri. Ext. Service, Corpus Christi, TX; pgs. IV-1/14.

Dall, W. 1957. A revision of the Australian species of penaeids (crustacea: decapoda: penaeidae). Aust. J. Mar. Freshwater Res. 8:136-230.

Emberson, C. 1986. Personal Communications. ORCA Sea Farms, Molokai, HI.

Fassler, R. 1987. Personal Communications, Hawaii State Aquaculture Development Program, Honolulu, HI.

Fast, A. W., D. K. Barclay and G. Akiyama. 1983. Artificial circulation of Hawaiian prawn ponds. Sea Grant Coop. Rept. UHIHI-SEAGRANT-CR-84-01, 83 pgs.

Fast, A. W., Y. C. Shang, G. L. Roger and I. C. Liao (in press, a). Description of Taiwan intensive shrimp culture farms, and simulated transfer to Hawaii. Univ. Hawaii Sea Grant College Program.

Fast, A. W., K. E. Carpenter, V. L. Estilo and H. J. Gonzales. (in press, b). Effects of pond depth and artificial mixing on dynamics of Philippine brackishwater shrimp ponds. J. Aquacult. Engineer.

Foster, J. R. M. and T. W. Beard. 1974. Experiment to assess the suitability of nine species of prawns for intensive cultivation. Aquaculture 3:355-368.

Garton, J. E. 1981. Destratification experiments to determine design factors for the Garton pump. In. Destratification of lakes and reservoirs to improve water quality. F.L. Burns and I.J. Powling (eds.), Australian Govern. Publ. Service, Camberra. pg 354-375.

Goldman, J. C. 1979a, Outdoor algal mass cultures. I. Applications. Water Res.,

13:1-19.
Goldman, J. C. 1979b. Outdoor algal mass cultures. II. Photosynthetic yield limitations. Water Res. 13:119-136.
Griffin, W. L., J. S. Hanson, R. W. Brick and M. A. Johns. 1981. Bioeconomic modeling with stochastic elements in shrimp culture. J. World Maricul. Soc. 12(1):94-103.
Hirasawa, Y. 1985. Economics of Shrimp culture in Asia. In. Y. Taki, J.H. Primavera and J.A. Llobrera (eds.): Proceed First Int. Conf. Cult. Penaeid Prawns/Shrimp. SEAFDEC, Iloilo City, the Philippines. Pgs. 131-150.
Hirono, Y. 1983. Preliminary report on shrimp culture activities in Ecuador. J. World Maricul. Soc. 14:451-457.
Holthuis, L.B. 1980. F.A.O. Species catalogue, Vol 1. Shrimps and prawns of the world. 261 pp.
Hopkins, J. S. 1987. Summary of shrimp farming research and development efforts in South Carolina during 1987. Waddell Maricult. Center, Bluffton, S.C., 29 pp.
Hopkins, Steven. 1988. Personal Communications. Waddell Aquaculture Center, Bluffton, SC.
Hudinaga, M. 1942. Reproduction, development and raring of *Penaeus japonicus* Bate. Jap. J. Zool., 10:305-393.
Huner, J. V. and H. K. Dupree. 1984. Pond culture systems. In. H.K. Dupree and J. V. Huner (eds.) The third report to the fish farmer. U.S. Fish and Wildlife Service, Washington, D.C. pgs. 6-16.
Kurata, H., K. Yatsuyanagi, and K. Shigueno, 1980. Kuruma shrimp culture in Japan. UJNR, Aquaculture Panel 9th Joint Meeting, Mimeographed, 7 pp.
Korringa, P. 1976. Farming marine fishes and shrimps. Elsevier Scientific Publ., New York. 208 pp.
Lawrence, A. L., M. A. Johns and W. L. Griffin. 1984. Shrimp mariculture, state of the art. TAMU-SG-84-502. Texas A&M Univ. Sea Grant.
Lawrence, A. L., G. W. Chamberlain, and D. L. Hutchins. 1981. Shrimp mariculture general information and commercial status. Proc. Trop. Subtrop. Fish. Tech. Conf. of the Americas, 6:20-24.
Laws, E. A. and S. Malecha. 1981. Application of a nutrient-saturated growth model to phytoplankton management in freshwater prawn (*Macrobrachium rosenbergii*) ponds in Hawaii. Aquaculture 24:91-101.
Liao, I-Chiu. 1987. Future technology in prawn production. Contribution B NO. 51 from the Tungkang Marine Laboratory. Presented at the Special Session, the 18th Annual Meeting of the World Aquaculture Society. Guayaquil, Ecuador, 18-23.
Liao, I. C. 1985a. A brief review of larval rearing techniques for penaeid prawns. Pages 65-78. In. Y. Taki, J.H. Primavera, and J.A. Llobrera (eds.), Proceedings at the First International Conference on the Culture of Penaeid Prawns/Shrimp, Aquaculture Department, SEAFDEC, Iloilo, Philippines.
Liao, I-Chiu. 1985b. General introduction to the prawn system in Taiwan. Contribution B No. 34 from the Tungkang Marine Laboratory. Presented at

Joint US-ROC Workshop on Aquaculture Engineering and Simulation, Honolulu, HI. January 27-February 1, 1985.

Liao, I. C. and N. H. Chao. 1983. Development of prawn culture and its related studies in Taiwan. Pages 127-142. In. G. L. Rogers, R.Day and A. Lim (eds.), Proceedings of the First International Conference on Warm Water Aquaculture-Crustacea, Hawaii, U.S.A.

Liu, M. S., and V.J . Mancebo. 1983. Pond culture of *Penaeus mondon* in the Philippines: Survival, growth and yield using commercially formulated feed. J. World Maricul. Soc. 14:75-85.

Madenjian, C. P., G. L. Rogers, and A. W. Fast. 1987a. Predicting nightime dissolved oxygen loss in prawn ponds of Hawaii: Part I. Evaluation of traditional methods. J. Aquacult. Engineering 6 (1987): 191-208.

Madenjian, C. P., G. L. Rogers, and A. W. Fast. 1987b. Predicting nightime dissolved oxygen loss in prawn ponds of Hawaii: Part II. A new method. J. Aquacult. Engineering 6 (1987):209-225.

Mahler, L. E., J. E. Groh and C. N. Hodges. 1974. Controlled environment aquaculture. Proc. World Maricul. Soc. 5:379-384.

Malca, R. P. 1983. Penaeus shrimp pond grow-out in Panama. In. J.P. McVey (ed.). Handbook of mariculture. I. Crustacean aquaculture. CRC Press, Boca Raton, FL. pg 169-183.

McGee, M. V. and C. E. Boyd. 1983. Evaluation of the influence of water exchange in channel catfish ponds. Trans. Amer. Fish. Soc., 112:557-560.

McVey, J. P. 1983. CRC handbook of mariculture. Volume I. Crustacean aquaculture. CRC Press, Boca Raton, FL 442 p.

Menz, A. and B. F. Blake. 1980. Experiments on the growth of *Penaeus vannamei* Boone. J. Exper. Marine Biol. Ecol. 48:99-111.

Motoh, H. 1981. Studies on the fisheries biology of the giant tiger prawn, *Penaeus monodon* in the Philippines. Technical Report No. 7, Aquaculture Department, SEAFDEC, Tigbauan, Iloilo, Philippines.

National Marine Fisheries Service. 1984. Shrimp situation and outlook. June. St. Petersburg, FL, 21 pgs. mimeo + append.

New, M. B., and H. R. Rabanal, 1984. A review of the status of penaeid aquaculture in South East Asia. p. 307-326. In. P. C. Roghrisberg, B.J . Hill, and D. J. Staples (eds.) Second Australian National Prawn Seminar. 468 pp. Published in Brisbane by NPS2 P. O. Box 120, Cleveland 4163, Australia.

Pardy, C. R., W. L. Griffin, M. A. Johns and A. L. Lawrence. 1983. A preliminary eocnomic analysis of stocking strategies for penaeid shrimp culture. J. World Maricult. Soc. 14:49-63.

Peckham, C. 1988. The jumbo market for shrimp; Farming catches on. In. R. Rosenberry (ed.), Aquacutlture Digest, San Diego, California. (July 1988), pg. 14.

Primavera, J. H. 1985. A review of maturation and reproduction in closed thelycum penaeids. In. Y. Taki, J.H. Primavera and J.A. Llobrera (eds.): Proceed. First Int. Conf. Cult. Penaeid Prawns/Shrimp. SEAFDED, Iloilo City, the Philippines. pgs. 47-64.

Rieban, B. 1986. Personal Communications. Laguna Madre Shrimp Farms, Harlington, TX.

Rogers, G. L. and A. W. Fast. 1988. Potential benefits of low energy water circulation in Hawaiian prawn ponds. Aquaculture Eng. 7(1988).

Rogers, G. L., C. P. Madenjian and A. W. Fast. 1988. A computer data acquisition system for aquaculture pond management. Presented at World Aquacult. Soc. Meeting in Honolulu, HI, Jan 4-9, 1988.

Romaire, R. P. and C. E. Boyd. 1979. The effects of solar radiation on the dynamics of dissolved oxygen in channel catfish ponds. Trans. Amer. Fish. Soc. 107:473-478.

Ryther, J. H., W. M. Dunstan, K. R. Tenore and J. E. Huguenin. 1972. Controlled eutrophication - Increasing food production from the sea by recycling human wastes. Bil-Science 22(3):144-152.

Ryther, J. H. 1971. Recycling human wastes to enhance food production from the sea. Environmental Letters 192:79-87.

Salser, B., L. Mahler, D. Lightner, J. Ure, D. Danald, C. Brand, N. Stamp, D. Moore and B. Colvin. 1978. Controlled environment aquaculture of penaeids. In. P.N. Kaul and C.J. Sindermann (eds.), Drugs and food from the sea. Univ. Oklahoma, Norman, OK. pg. 345-355.

Sandifer, P. A., J. Hopkins and D. Stokes. 1986. The potential for intensive culture of *Penaeus vannamei* in South Carolina. Presented at World Maricul. Soc. Annual Meeting. Jan 19-24, 1986. Reno, NV.

Shigueno, K. 1985. Intensive culture and feed development in *Penaeus japonicus*. In. Y. Taki, J.H. Primavera and J.A. Llobrera (eds.): Proceed. First Int. Conf. Cult. Penaeid Prawns/Shrimp. SEAFDEC, Iloilo City, the Philippines, pgs. 115-122.

Shigueno, K. 1975. Shrimp culture in Japan. Association for International Technical Promotion, Tokyo, Japan, 153 pp.

Smith, D. W. 1985 Biological control of excessive phytoplankton growth and enhancement of aquacultural production. Can. J. Fish. Aquat. Sci. 42:1940-1945.

Smith, D. W. 1987. Biological control of excessive phytoplankton growth and enhancement of aquacultural production, Ph.D. Dissertation. Univ. Calif. Santa Barbara, CA 196 pp.

Smith, D. W. and R. H. Piedrahita. 1988. The relationship between phytoplankton and dissolved oxygen in fish ponds. Aquaculture 68:249-265.

Stamp, H. H. E. 1978. Computer technology and farm management economics. Proc. 9th Annual Meeting World Maricul. Soc. 383-392.

Sturmer, K. N. and A. L. Lawrence. 1986. Evaluation of raceway intensive nursery rearing systems for penaeid shrimp. Annual meeting, World Aquacult. Soc., Abst. 49.

Tatum, W. M. and W. C. Trimble. 1978. Monoculture and polyculture studies with pompano (*Trachinotus carolinus*) and penaeid shrimp (*Penaeus aztecus*, *P. duorarum* and *P. setiferus*) in Alabama, 1975-1977. Proc. World Maricult. Soc. 9:433-446.

Tucker, C. S. 1985. Channel catfish cutlure. Elsevier, New York. 657 pps.
Tucker, C. S. and C. E. Boyd. 1985. Water Quality, In. C.S. Tucker (ed). Channel catfish culture. Elsevier, New York. pg. 135-228.
Vondruska, J. 1987. Shrimp situation and outlook. U.S. National Marine Fisheries Service, St. Petersburg, FL. Mimeo.
Veloso, A. P. 1984. San Miguel Corporations' initial venture into aquaculture in the Philippines. Paper presented to AIESEC Symposium on Agri-business Development strategies for the Philippines, January 19, 1984. Manila.
Zein-Eldin, Z. P. and D. V. Aldrich. 1965. Growth and survival of postlarval *Penaeus aztecus* under controlled conditions of temperature and salinity. Biol. Bull. 129:199-216.

Technological Advances in Intensive Pond Culture of Shrimp in the United States

Paul A. Sandifer[1], J. Stephen Hopkins[1], Alvin D. Stokes[1], and Gary D. Pruder[2]
[1]Marine Resources Division, South Carolina Wildlife and Marine Resources Department, P. O. Box 12559, Charleston, SC 29412
[2]Oceanic Institute, Makapuu Point, P. O. Box 25280, Honolulu, HI 96825

Introduction

While intensive pond culture of the black tiger prawn, *Penaeus monodon*, has become highly developed in Taiwan (Liao, 1984), no similar aquaculture system exists for shrimp in the Western Hemisphere. Nevertheless, it is clear that if shrimp farmers in North America are to compete with those in areas such as Taiwan, China, Ecuador, Indonesia, and the Philippines, they must produce higher yields (Sandifer *et al.*, 1988). High yields require development of efficient intensive production systems. The necessity of developing an intensive shrimp culture system for the U.S. was recognized over a decade ago by Parker and his associates in Texas (Parker *et al.*, 1974), but little work followed their pioneering efforts until the Waddell Mariculture Center team initiated intensification experiments during summer 1985 (Sandifer *et at.*, 1987). Shortly afterwards, the Oceanic Institute undertook studies on the intensification of pond shrimp production in Hawaii (Wyban and Sweeney, 1989).

The present paper reviews significant recent advances in intensification of pond production of shrimp in the U. S.

Approach

Both the Waddell Mariculture Center and the Oceanic Institute adopted the basic elements of the Taiwanese system for intensive cultivation of *P. monodon* (Liu and Mancebo, 1983; Chiang and Liao, 1985; Liao, 1985; Liao and Murai, 1985), but directed their efforts toward the species of choice in the Americas, *P. vannamei*. These key elements include:

1. **Use of relatively small, easily managed ponds.** Ponds utilized at the Waddell Mariculture Center are rectangular, with a 2:1 length:width ratio, and lined on the bottoms and sides with 1 mm thick high density polyethylene to prevent seepage. The lined pond bottoms are covered with a 26 cm deep layer of sand to provide a natural substrate for the shrimp. Experimental ponds are 0.10, 0.25 and 0.50 ha in size. In addition, earthen ponds to 1.5 ha in size on private farms in South Carolina were used in cooperative studies.

The Oceanic Institute has concentrated its work on round ponds. The Oceanic Institute round pond approach was developed by combining and modifying the Japanese (Shigueno) intensive round pond system for *P. japonicus* and the Taiwanese grow-out strategies for *P. monodon*. Its experimental round pond is a 337 m^2 unit with vertical concrete walls and an earthen (clay) bottom. A larger (0.2 ha) commercial prototype pond built by the Institute on a private farm has gunnite coated, sloped earthen walls. Details of pond management procedures for the two systems are given in Sandifer *et al.*, (1987, 1988) and Wyban and Sweeney (1989).

2. **Use of high stocking densities.** Prior to the work reviewed here, stocking densities with *P. vannamei* rarely exceeded 12-15 postlarvae/m^2. However, based on the Taiwanese experience, densities of approximately 40 shrimp/m^2 were targeted initially. To date, densities of 10, 12, 20, 40, 60 and 100 postlarvae/m^2 have been evaluated at the Waddell Mariculture Center, and the Oceanic Institute has utilized densities to 112 shrimp/m^2.

3. **Use of paddlewheel aerators.** Paddlewheels not only increase the dissolved oxygen concentration, they also can establish a circulation pattern which mixes the pond water and concentrates settled organic matter and debris near the pond center for ease of removal via a drain. One horsepower electric paddle-wheel units were used in all the shrimp pond culture experiments reviewed here, with rates of 4-10 hp/ha at the Waddell Mariculture Center (Sandifer *et al.*, 1988) and 30 hp/ha at the Oceanic Institute (Wyban *et al.,* 1988).

4. **Use of regular water exchange.** The water exchange rate is increased over the growing season as standing crop biomass in the ponds increases. At the Waddell Mariculture Center, attempts have been made to keep average exchange rates over the entire season to <20%/day because of cost considerations, and most have been

maintained within the range 8-16%/day (Sandifer *et al.*, 1988). The Oceanic Institute uses a considerably higher exchange rate, averaging 61%/day (Wyban *et al.*, 1988).

5. **Use of high quality prepared feed**. The Waddell Mariculture Center used a 40% protein shrimp diet manufactured by Zeigler Brothers, Inc., (Gardeners, PA) while the Oceanic Institute used a 38% protein commercial shrimp grower feed from Taiwan (Hanaqua). Both are reasonable quality, commercially available feeds. Despite the recommendation that high quality feeds be provided, there is some difference of opinion as to the importance of prepared feed as the primary source of shrimp nutrition in intensive pond culture. In preliminary analyses of changes in benthic populations in shrimp ponds over a growing season, Hopkins *et al.*, (1988) found that, at high densities, the shrimp apparently graze down the populations of macro- and meiobenthic organisms within the first two months after stocking, and that populations of these organisms are not reestablished during the growing season. These observations suggested that the prepared feed becomes an important, perhaps the primary, source of shrimp nutrition following exhaustion of benthic prey. However, Leber and Pruder (1988) found that unspecified factors in pond water provided an important growth-enhancing effect on shrimp reared in laboratory microcosms, suggesting that prepared feeds do not provide all or most of the nutrition required for shrimp growth. Similarly, Aquacop (1984) reported that *P. vannamei* adapted especially well to intensive culture conditions involving "bacterial flock" mediums. It is likely that the prepared feed provides both direct (via ingestion and digestion) and indirect (via fertilizing effect on the pond) nutrition for the shrimp. This area requires considerable further research.

6. **Careful pond management**. Along with high stocking densities, proper pond management is essential for successful intensive pond culture of shrimp. Because of the high stocking densities, the subsequent high standing crop biomass, and the rate at which dissolved oxygen can decrease to lethal levels, regular monitoring of each pond is essential.

Approach and Results

A preliminary experiment demonstrated that yields of *P. vannamei* from pond culture could be increased significantly using the approach outlined above (Sandifer *et al.*, 1987). This study utilized two 0.1 ha ponds at the Waddell Mariculture Center during summer 1985. The ponds were stocked at 40 and 45 postlarvae/m^2, and yielded an average of 6,757 kg/ha of whole shrimp.

Sandifer *et al.*, (1988) also demonstrated that pond size in the range 0.1 to 0.5 ha had no effect on shrimp growth or production at stocking densities of 12 and 40 shrimp/m^2. Intensification efforts have been moderately successful in considerably larger ponds on private farms, producing > 4,000 kg/ha (Hopkins, 1988). Thus, there is no de facto reason to restrict intensification efforts to very small ponds. However, it is necessary that the ponds be manageable in terms of mixing, water exchange, feeding and harvesting.

The Waddell Mariculture Center has conducted several studies to determine the effects of stocking density on shrimp growth and survival rates. An initial tank study conducted in 1985 in 6-m diameter outdoor tanks compared densities of 10, 20 and 40 nursery-reared juvenile shrimp/m^2 (Sandifer *et al.*, 1987). The effect of density on growth rate was obvious, but not pronounced; final mean shrimp sizes were 33.9 ± 3.6 g, 32.5 ± 2.8 g, and 26.7 ± 3.0 g, at the three densities, respectively. Survival was unaffected by stocking density, but standing crop biomass increased two-fold as stocking density doubled from 10 to 20 shrimp/m^2 (226 to 442 g/m^2) and three-fold (to 685 g/m^2) as density increased to 40/m^2 (Sandifer *et al.*, 1987).

A second tank experiment involved stocking densities of 60, 80 and 100 postlarvae/m^2. The shrimp were 0.02 g in mean weight at initiation of the study. Survival was higher at the lowest density (81% vs. 68.5 and 64.5%, respectively), although not significantly so, and overall there was no effect of density on shrimp size at termination (Sandifer *et al.*, 1988). In all cases, mean shrimp size was approximately 12 g at harvest. Again, standing crop biomass increased directly with stocking density, reaching a mean (± one std. dev.) of 779 ± 198 g/m^2 at 100 shrimp/m^2. One replicate at this density produced 1,006 g/m^2, suggesting the possibility of pond harvests of 10 MT/ha/crop.

In 1986 the Waddell Mariculture Center undertook a proof-of-concept demonstration of intensive pond culture of *P. vannamei* (Stokes *et al.*, 1987). A secondary objective of the study was to attempt to decrease levels of aeration and water exchange to reduce operating costs. Aeration was reduced from 10 hp/ha in 1985 to 4 hp/ha in 1986, and average daily water exchange was reduced from 16-17% to 8%. Eight ponds totaling 2.5 ha were used in the study. Four 0.25 ha and two 0.5 ha units were stocked at 40 postlarvae/m^2 and two 0.25 ha ponds were stocked at 60/m^2.

During the 1986 growing season, South Carolina experienced a major drought and heat wave which, along with the reduced inputs for aeration and exchange, caused some partial kills due to low dissolved oxygen levels. Despite the problems, standing crop biomass at harvest averaged 5,185 kg/ha at the 40/m^2 density and 6,026 kg/ha at 60/m^2. Also, there was no difference in mean yields from the 0.25 and 0.5 ha ponds stocked at 40 postlarvae/m^2 (5,143 kg/ha vs. 5,263 kg/ha). Overall, this pilot-scale study produced a total of 13,600 kg of live shrimp in one crop from 2.5 ha of water.

The 1986 pond data also indicated little difference in growth rate for shrimp stocked at densities of 40 and 60 shrimp/m^2. Growth rates for shrimp at similar sizes (<0.1 to 12g) were then compared in all the tank and pond experiments conducted in 1985 and 1986 in which postlarvae were used (Table 1). These data indicated a substantial decrease in average daily growth rate as density increased from 12 shrimp/m^2 to 40/m^2, but relatively little further decline as density increased from 40 to at least 100 shrimp/m^2.

The 1987 Waddell Mariculture Center pond studies provided further evidence of only minor impacts of stocking density on growth and survival rates, while standing

Table 1. Average daily growth rates for *Penaeus vannamei* as a function of stocking density. (Growth rates are compared for shrimp over approximately the same size range, [≤ 0.1 to 12 - 13 g] in all cases since maximum size attained in the 1986 tank study was 12 g. Data from the 1987 pond study are not included because of the unusually slow growth experienced at all densities that year).

Pond studies (1985 and 1986)						
stocking density (#/m^2)	12	42	40	60		
growth rate (mg/day)	168	113	120	110		
Tank Study (1986)						
stocking density (#/m^2)				60	80	100
growth rate (mg/day)				105	105	106

crop biomass again increased directly with stocking density. Densities of 20, 40 and 60 postlarvae/m^2 were stocked into triplicate 0.25 ha ponds, and one 0.1 ha pond was stocked at 100/m^2 (Sandifer *et al.*, 1988). There was a slight decline in mean size and survival at harvest as density increased from 20 to 60 postlarvae/m^2 (Table 2), although the differences were not significant. In addition, mean harvest size and survival in the 100/m^2 pond were similar to those in the lowest density.

A summary of three years of pond production data from the Waddell Mariculture

Table 2. Summary of 1987 high density pond culture of *Penaeus vannamei* at the Waddell Mariculture Center. (Data are presented as means ± one std. dev. Duration = 152-179 days. Initial mean shrimp weight was 0.001 g in densities of 20, 40 and 60/m^2 and 0.03 g in the 100/m^2 treatment). The 100/m^2 treatment was not replicated. (After Sandifer *et al.*, 1988).

Stocking Density (#/m^2)	Mean Weight (g)	Survival (%)	Harvest Biomass (kg/ha)	Feed Conversion
20	13.4 ± 0.7	92.4 ± 2.9	2,487 ± 182	2.0 ± 0.2
40	12.4 ± 0.5	89.6 ± 3.3	4,433 ±132	2.3 ± 0.1
60	12.1 ± 1.3	87.6 ± 6.8	6,320 ± 449	2.2 ± 0.2
100	13.5	94.0	12,680	2.1
means	12.7	90.3		2.2

Center (Table 3) indicates that:

(1) while mean size at harvest was much greater at 12/m^2 than at any higher density, there were essentially no differences in harvested mean size as density increased

from 20 to 100 shrimp/m^2;

(2) survival tended to decline slightly, but not significantly, with increasing stocking density (with the exception of the high survival in the one pond at 100 postlarvae/m^2); and

(3) standing crop biomass at harvest increased directly with stocking density, with a maximum yield of 12.7 MT/ha/crop achieved to date in a pond stocked at 100 shrimp/m^2.

Table 3. Summary of production results for *Penaeus vannamei* stocked as postlarvae (0.001 - 0.1 g) at different densities in earthen ponds at the Waddell Mariculture Center, 1985-1987.

Stocking Density (#/m^2)	Number of Observations (N)	Mean Weight (g)	Survival (%)	Harvest Biomass (kg/ha)
12	3	19.7 ±0.9	100*	2,477 ±98
20	3	13.4 +0.7	92.4 ±2.9	2,487 ±182
40	11	15.3 ±2.2	85.1 ±10.8	5,266 ±970
60	5	13.7 ±2.8	78.7 ±18.3	6,202 ±702
100	1	13.5	94.0	12,680

* Survival rates were 99 - 114%, indicating overstocking.

Data from the Oceanic Institute round pond experiments show similar successes in intensification. In contrast to the Waddell Mariculture Center, the Oceanic Institute used larger juveniles (approximately 1 g in mean weight) to stock grow-out experiments. Data from three recent trials in the 337 m^2 round pond are summarized in Table 4. Survival rates were high and similar to those obtained in the larger rectangular ponds stocked with postlarvae at the Waddell Mariculture Center. However, the use of nursed juveniles, coupled with much higher growth rates obtained at the Oceanic Institute, produced 3-5 g larger shrimp at harvest in a shorter growing period (10-15 weeks vs. 22-25 weeks in SC), and yields ranging from about 13,000 to 19,000 kg/ha/crop (Wyban *et al.*, 1988; G. Pruder, personal communication). Results from two production trials in a commercial scale (0.2 ha) round pond showed slight reductions in mean shrimp weight and biomass at harvest (Table 5) compared to results from the experimental round pond, but the data were comparable to those from similar size ponds at the Waddell Mariculture Center.

In 1987, approximately 89% of the 171,864 kg of live shrimp produced from

Table 4. Production results from three intensive grow-out experiments with *Penaeus vannamei* in the Oceanic Institute's experimental round pond (337 m^2) during 1987.

Parameter	Trial 1	Trial 2	Trial 3
stocking wt. (g)	1.23	1.21	0.75
stocking density (#/m^2)	100	100	100
harvest wt. (g)	16.5	19.8	16.4
duration (days)	85	74	106
survival (%)	77.8	98.0	92.3
food conversion	2.3	2.0	2.5
harvest biomass (kg/ha)	12,868	19,404	15,803

South Carolina farms was harvested from three private semi-intensive farms, plus the Waddell Mariculture Center, although these operations comprised only about 6% of the total area cultivated (Sandifer *et al.*, 1988). Higher stocking densities, coupled with aeration and careful management, have led to substantial increases in harvest biomass on these private farms. Cooperative efforts with farmers have resulted in one private grower producing 5,050 kg/ha from a 2480.3 ha pond and another harvesting 4,625 kg/ha from a 1.5 ha unit (Sandifer *et al.*, 1988). The Oceanic Institute has experienced similar success in initial intensification efforts with a private farm.

Table 5. Production results from the first two grow-out trials in the 0.2 ha commercial prototype round pond at Amorient Aquafarm in Hawaii, 1987 and 1988. (Data from Wyban *et al.*, 1988 and G. Pruder, personal communication).

Parameter	Trial 1	Trial 2
stocking wt. (g)	0.9	1.1
stocking density (#/m^2)	75	112
duration (days)	88	102
survival (%)	67	94
harvest wt. (g)	18	14.6
production (kg/ha)	9,120	15,404

Conclusions

1. A technology for intensive pond culture of shrimp in North America has been developed and proven. This technology needs further refinement, but results to date demonstrate that it works repeatedly and that it can be transferred successfully to the private sector.

2. Pond size did not affect growth or survival of *P. vannamei.*

3. Shrimp production has been substantially intensified in earthen ponds of up to 1.5 ha in size, and it appears likely that ponds several ha in size can be managed for intensive shrimp production. The intensive culture technology works well in rectangular and round ponds.

4. Stocking density affected growth rate, but this effect was not pronounced at densities above 20 shrimp/m^2. Further, survival was essentially unaffected by stocking density in well managed ponds.

5. Assuming proper management, pond yields are directly related to stocking density. Generally, the greater the stocking density, the greater the standing crop biomass at harvest.

6. Nursery systems will be necessary if larger (>18 g) shrimp are to be produced in intensive pond culture. This is clear from tank studies at the Waddell Mariculture Center (see Sandifer *et al.*, 1987) and from the round pond experiments at the Oceanic Institute.

7. Yields of 12-15 MT/ha/crop have been achieved in experimental units that are of minimum commercial size. Prior to 1985, the state-of-the-art of shrimp production in the U. S. was generally < 2 MT/ha/crop. Already, private farms are producing twice this level in some ponds (Emberson, 1986; Hopkins, 1988; Sandifer *et al.*, 1988), and private farm yields approaching 20 MT/ha/crop are anticipated within the next decade.

8. The potential economic impact of a commercial intensive production technology in the U. S. is enormous. For South Carolina alone, RPI International, Inc. (1988) estimated that if only 10% of the apparently suitable land were put into semi-intensive shrimp culture, 5,600 ha would be producing 3,368 kg/ha with a farm gate value of nearly $95 million and a state-wide economic impact of over $470 million (Table 6). Obviously, if the production/ha increased to the levels reported here, the

Table 6. Projected economic benefits of a shrimp mariculture industry to South Carolina at different levels of land development.[1]

% Development of Available Land	Shrimp Farm Area (ha)	Annual Yield of whole shrimp[2] (kg x 10^3)	Annual Farm Sales[3] ($ x 10^3)	Total Economic Impact[4] ($ x 10^3)
5	2,813	9,474	47,275	236,375
10	5,626	18,948	94,552	472,760
25	14,064	47,368	236,364	1,181,820
50	28,128	94,735	472,728	2,363,638

[1] Adapted from RPI International, Inc. (1988)
[2] Based on an average harvest of 3,368 kg shrimp/ha (3,000 lb/acre)
[3] Based on an average price of $4.99/kg whole shrimp ($2.27/lb)
[4] Using an economic multiplier of 5x

magnitude of the industry would increase several-fold. For comparison, Ecuador has roughly 90,000 ha in shrimp culture, producing about 800 kg whole shrimp/ha/year for a total harvest in 1987 of approximately 67,000 MT live weight (U. S. Department of Commerce, 1988). A similar annual production level could be achieved in South Carolina with 5,600 ha operated intensively producing 12 MT/ha/yr, or if approximately 35% of the potentially available land were put into semi-intensive shrimp culture yielding 3.4 MT/ha/yr (Table 6). The economic potential for shrimp farming in Texas, other southern states, Hawaii, and the U. S. Caribbean is probably similar to that of South Carolina. Potential outputs from a hypothetical 25 ha intensive shrimp farm in the U. S. are indicated in Table 7.

Frontiers in Intensive Culture Technology

The frontiers of intensive shrimp culture technology are essentially limitless. One area of immediate opportunity is further increasing stocking density above 100 shrimp/m^2 to produce even higher harvest levels. Stocking densities of 200 and 400 postlarvae/m^2 are currently under evaluation at the Waddell Mariculture Center.

Table 7. Potential outputs from a hypothetical 25 ha intensive shrimp farm.

* Harvest 5 -15 MT/ha/crop
(1 crop/year in continental U.S.;
2+ crops/year in tropics)
*Annual farm yield: 125 - 750 MT
*Pond-side value (@$5/kg whole
shrimp): $625,000 - $3,750,000

A second major need is for increases in mean size (and therefore value) of the shrimp at harvest, without sacrificing the high yields possible through intensive culture. Increased growth rates should eventually be attained through improved diets and development of faster growing shrimp via genetic and hormonal manipulations. However, the most immediate need is to determine how to consistently obtain the growth rate of which the animal is capable. Menz and Blake (1980) reported a growth rate of 0.28 g/day (approximately 2 g/wk) for *P. vannamei* in the wild. Researchers at the Oceanic Institute frequently obtain similar, and occasionally higher, rates for short periods of time in the laboratory (Leber and Pruder, 1988). Nevertheless, the best rate obtained in the experimental round pond has been 1.75 g/wk, and the average for the three 1987 trials was 1.34 + 0.36 g/wk. Growth rates obtained in the prototype commercial round pond have been a little lower to date, averaging 1.36 and 0.92 g/wk for the first two production trials (Table 5). Considering animals of similar size (that is, looking only at that part of the growth curve after the shrimp reach > 1 g), the Waddell Mariculture Center growth rates are generally on the order of 1 g/wk. Thus, at the present level of intensive production technology, only about 50-75% of the demonstrated growth potential of the animal is being realized in intensive culture systems. There appears to be considerable opportunity for increasing growth rates through improving prepared diets, increasing the contribution of natural nutriments from the pond biota, and improving pond management procedures to provide better growing conditions.

A complementary approach to producing larger shrimp is through refinement and implementation of cost-effective nursery systems which allow farmers to get a head-start on the growing season. Lawrence (1988) described open nursery ponds and greenhouse-covered raceway nursery systems for penaeid shrimp. However, open nursery ponds are not suitable for broad use in the continental U. S. because of temperature considerations, and in-ground nursery raceways covered by agricultural greenhouses may not be cost effective (Juan *et al.*, 1988). Because of these and other considerations, Waddell Mariculture Center staff are experimenting with a covered nursery pond. In this approach, a small pond (0.1 ha) is enclosed with a plastic "bubble" covering that is trenched into the earth around the pond perimeter and then inflated with a small air blower. An alternative approach under consideration would utilize a floating plastic cover. Preliminary work with an inflated cover indicates that solar heating is sufficient to maintain minimum temperatures for postlarval shrimp survival and growth even in late winter in South Carolina. In the event of major storms, the "bubble" can be deflated, which allows it to settle onto the pond surface where it is believed to be less likely to be damaged by high winds, hail, tree limbs, or wind-borne objects. As soon as the storm passes, the cover can be reinflated. This approach has the advantage of low cost and simple implementation. Considerable further testing and refinement are needed for enclosed nursery systems to become cost-effective components of intensive pond growout systems.

Another technological frontier in need of further exploration is improvement of aeration and circulation equipment. Most studies on aerator efficiency to date have been conducted in tanks where the hp/ha levels may be orders of magnitude greater than utilized in the "real world" of ponds. The Waddell Mariculture Center approach has been to test aerators in ponds, following the methodology used by Boyd and Ahmad (1987) for tank evaluations. In addition, mixing rates are determined using rhodamine dye and a fluorometer. Results to date indicate that, at the hp/ha rates required for the intensive pond production of shrimp (i. e., > 8 hp/ha), complete mixing of the pond occurs in four to six hours using paddlewheels. Initial data also indicate that direct mixing by the paddlewheels may account for only about 5% of the total oxygen transfer, with the balance apparently resulting from photo-synthetic activity and oxygen diffusion across the pond surface as the water is continuously mixed. Thus, it appears likely that aeration equipment designed to circulate rather than splash water should be more effective in maintaining desired dissolved oxygen levels. Further work is urgently needed to identify and refine the most cost-effective means of mechanical aeration for shrimp ponds.

An additional area of investigation that may lead to increased profitability for the U. S. shrimp farmer is production of another crop either in conjunction with or alternating with the shrimp (*i.e.*, biculture). Possibilities being explored include overwintering clam (*Mercenaria mercenaria*) seed from New England hatcheries (Battey and Manzi, 1988) and the cultivation of oysters (*Crassostrea virginica*) in South Carolina shrimp ponds (Manzi *et al.*, 1988).

New England clam hatcheries find that growth and survival of seed clams in the North are poor in winter due to low water temperatures. Cooperative studies with several private hatcheries have resulted in promising techniques to rear clam seed in South Carolina shrimp ponds over winter and then ship the larger seed North for growout the following spring prior to the beginning of the shrimp growing season (Battey and Manzi, 1988).

A second form of biculture being evaluated produces market size oysters from seed. Seed oysters are set naturally on plastic stakes which are then transferred to shrimp ponds. The seed oysters are nourished by the abundant phytoplankton which is produced by the fertilizing effects of feeds and shrimp waste products. After the shrimp harvest, the pond is refilled and oyster growout continues for the 16-30 months required to reach market size. This growout period can be interrupted periodically as needed to stock or harvest shrimp without harming the oysters (Manzi *et al.*, 1988).

There are several advantages of using mollusks as biculture crops. They can easily withstand the desiccation associated with draining the pond to harvest shrimp. Removal of phytoplankton by the filter-feeding oysters may also help to dampen diel oscillations in dissolved oxygen concentrations. In addition, shrimp will actively feed on the oysters' pseudofeces, which are bound packets of material oysters have filtered from the water but are unable to digest.

Acknowledgements

We thank the entire staff of the Waddell Mariculture Center and the shrimp team at the Oceanic Institute for their dedication and outstanding efforts. Work reported here was supported in part by the U. S. Department of Agriculture through a subcontract from CSRS grant #85-CSRS-2-2537 to the Oceanic Institute, for the U. S. Marine Shrimp Farming Program of the Gulf Coast Research Laboratory Consortium. Contribution No. 271 from the South Carolina Marine Resources Center.

Literature Cited

Aquacop. 1984. Review of ten years of experimental penaeid shrimp culture in Tahiti and New Caledonia (South Pacific). Journal of the World Mariculture Society 15: 73-91.

Battey, C. R. and J. J. Manzi. 1988. Overwintering hard clams, *Mercenaria mercenaria*, in South Carolina shrimp ponds. Journal of Shellfish Research 7(1): 214 (Abstract).

Boyd, C. E. and T. Ahmad. 1987. Evaluation of aerators for channel catfish farming. Bulletin 584, Alabama Agricultural Experiment Station, Auburn University.

Chiang, P. and I. C. Liao. 1985. The practice of grass prawn (*Penaeus monodon*) culture in Taiwan from 1968 to 1984. Journal of the World Mariculture Society 16: 297-315.

Emberson, C. R. 1986. Intensive pond culture of *Penaeus vannamei* in Hawaii. Presented at Second Latin American Conference on Aquaculture. Sept. 1986, Salvador, Brazil.

Hopkins, J. S. 1988. Shrimp farming in the United States. Presented at Shellfish Institute of North America and National Blue Crab Industry Association convention, 28 February-2 March 1988, Charleston, SC (unpublished MS).

Hopkins, J. S., M. L. Baird, O. G. Grados, P. P. Maier, P. A. Sandifer, and A. D.

Stokes. 1988. Impacts of intensive shrimp production on the culture pond ecosystem. Journal of the World Aquaculture Society 19: 37A (Abstract 122).
Juan, Y. S., W. L. Griffin, and A. L. Lawrence. 1988. The production costs of juveniles in an intensive greenhouse raceway nursery system. Journal of the World Aquaculture Society 19: 149-160.
Lawrence, A. L. 1988. Intensive shrimp nursery pond production systems. Presented at Frontiers of Shrimp Research, 23-24 June 1988, Washington, DC.
Leber, K. M. and G. D. Pruder. 1988. Using experimental microcosms in shrimp research: the growth- enhancing effect of shrimp pond water. Journal of the World Aquaculture Society. 19: 197-203.
Liao, I.-C. 1984. Status and problems of grass prawn culture in Taiwan. Tungkang Marine Laboratory Conference Proceedings 1: 81-98.
Liao, I.-C. 1985. A brief review of the larval rearing techniques of penaeid prawns. Pages 65-78 In. Y. Taki, J. H. Primavera, and J. A. Llobrera, editors. *Proceedings of the First International Conference on the Culture of Penaeid Prawns/Shrimp*, Iloilo City, Philippines, 1984.
Liao, I.-C. and T. Murai. 1985. Some essential factors in prawn culture development with special reference to achievements in Taiwan. Presented at II Simposio Brasileiro Sobre Cultivo de Camarao, 9-13 Sept. 1985, Parnaiba, Piaui, Brazil.
Liu, M. S. and V. J. Mancebo. 1983. Pond culture of *Penaeus monodon* in the Philippines: survival, growth and yield using commercially formulated feeds. Journal of the World Mariculture Society 14: 75-85.
Manzi, J. J., C. B. O'Rourke, M. Y. Bobo, G. H. Steele, and R. A. Smiley. 1988. Results of an oyster-shrimp biculture study. Journal of Shellfish Research 7 (1): 205 (Abstract).
Menz, A. and B. F. Blake. 1980. Experiments on the growth of *Penaeus vannamei* Boone. Journal of Experimental Marine Biology and Ecology 48: 99-111.
Parker, J. C., F. S. Conte, W. S. MacGrath, and B. W. Miller. 1974. An intensive culture system for penaeid shrimp. Proceedings of the World Mariculture Society 5: 65-79.
RPI International, Inc. 1988. Shrimp mariculture: a concept for economic development in South Carolina. Unpublished report, 68 pp. appendices.
Sandifer, P. A., J. S. Hopkins, and A. D. Stokes. 1987. Intensive culture potential of *Penaeus vannamei*. Journal of the World Aquaculture Society 18(2): 94-100.
Sandifer, P. A., J. S. Hopkins, and A. D. Stokes. 1988. Intensification of shrimp culture in earthen ponds in South Carolina: progress and prospects. Journal of the World Aquaculture Society 19: 218-226.
Stokes, A. D., P. A. Sandifer, and J. S. Hopkins. 1987. Effects of pond size and management practices on intensive culture of *Penaeus vannamei* in South Carolina. Journal of the World Aquaculture Society 18(1): 5A (Abstract 2).
U. S. Department of Commerce. 1988. Aquaculture and capture fisheries: impacts in U. S. seafood markets. Report prepared pursuant to The National Aquaculture

Improvement Act of 1985 (P.L. 99-198), 230 pp.
Wyban, J. A. and J. N. Sweeney. 1989. Intensive shrimp growout trials in a round pond. Aquaculture 76: 215-225.

Wyban, J. A., J. N. Sweeney, and R. A. Kana. 1988. Shrimp yields and economic potential of intensive round pond systems. Journal of the World Aquaculture Society 19: 210-217.

Information Services in Aquaculture

James P. McVey, Ph.D.
National Sea Grant College Program
1335 East-West Highway
Silver Spring, MD 20910
Deborah T. Hanfman
National Agricultural Library
Beltsville, MD 20705

Information Services in Aquaculture

For more than a decade, information services have expanded to meet the demands of the rapidly growing aquaculture industry. Today, individuals seek aquaculture information to satisfy their interests and research needs. They include the farmer, county agent, extension service specialist, research community, private investor, educator, information specialist, and librarian.

Aquaculture has been identified as an alternative type of agriculture. Farmers now have the opportunity to increase production efficiency and profitability by the addition of aquaculture products.

Information services in aquaculture offer the basic knowledge, expertise, and resources to help potential and practicing aquaculturists make wise decisions regarding aquaculture endeavors. Information services vary according to their specialities as well as to their intended audience. This paper will provide an overview of the wide variety of services currently available in aquaculture. The Farm and Security Act of 1985, under Title XVII, Subtitle D, emphasizes the importance of the Federal Government in information collection and dissemination. Since 1985, Federal agencies have joined together to enhance and expand information services in aquaculture. For this reason, the role of the Federal Government will be stressed as a major component in information transfer.

Types of Information Products

Information products are the key sources for obtaining knowledge on aquaculture-related subjects. They are the backbone to effective information services. There are many types of products currently available to the public that offer a wealth of information on both plant and animal aquaculture. Individuals can obtain informative materials ranging from historical background to current perspectives on the aquaculture industry. Specialized subjects in aquaculture can be easily addressed through information products developed by experts in the field.

Although the majority of information products include newsletters, journal articles, reports, proceedings and books, there are additional sources of information that need to be identified. Microfilm offers an alternative method for storing literature citations, articles, and specialized collections, while requiring less storage space than its printed counterparts. With the advent of the computer age, the acceptance of microcomputers in the workplace has placed new emphasis on electronic transfer of information. Products no longer need to be confined to paper.

Below are various types of aquaculture information products. Selected samples from each product type are listed in Appendix A.

Newsletters

Newsletters are becoming major resources for individuals seeking quick and current information about the aquaculture industry. Many provide overviews of new culturing methods and research activities. They usually focus on news or information of interest chiefly to a special group of people. They offer aquaculture news in brief, are sometimes controversial, and provide an affordable way to obtain information.

Magazines and Journals

These types of information products provide articles of both practical and technical aquaculture. The titles of magazines and journals are usually broad in scope allowing for a wide variety of topics and professional expertise to be addressed by authors in the form of articles and regularly appearing columns.

Books

Books are nonperiodical literary publications on particular subjects. There are many books available on aquaculture today. Some focus on broad subjects such as marine aquaculture or shrimp mariculture, while others deal with more specialized subjects such as cultivation of Pacific molluscs or toxicity testing of aquatic species.

Technical Reports

A technical report gives details and results of a specific investigation of a scientific or technical problem. It is usually compiled by specialists having expertise in that particular subject area. Many times technical reports do not undergo intensive review and editing. A technical report may be published as a single publication or in a series.

Proceedings

A proceedings is the published record of a meeting of a society or other organization, frequently accompanied by abstracts or reports of papers presented. The theme of the conference determines the subject matter included in the proceedings. Proceedings are very informative in that they provide some of the most current aquaculture-related information. Papers represent the work of students, aquaculture specialists, information specialists, and renowned scientists.

Bibliographies

Bibliographies are the basic tools needed to provide access to the vast amount of literature available on aquaculture. They can serve the entire aquaculture community by providing resources that can suit their particular needs. Subject areas can be very broad, such as a general bibliography on aquaculture, or very specific, such as a bibliography on "strawberry disease" of rainbow trout. Many are compiled from computer searches of various aquaculture-related databases, while others are manually created.

Bibliographies contain the basic components needed to retrieve a document. They include the article or book title, author or editor, date of publication, publisher, journal title, issue or series number, pages, and many times location information (i.e., call number, ISSN or ISBN numbers, NTIS or GPO order numbers, etc.). Some contain abstracts or short annotations.

Directories

Directories offer the information seeker contact sources in the field of aquaculture. They are one of the most often called-for resources in libraries and information facilities since they direct individuals to appropriate companies, associations, Federal or State agencies, universities, etc. Directories offer access to specific programs or activities in aquaculture. They are comprised of a listing of persons or organizations, systematically arranged, usually in alphabetic or classed order, giving address, affiliations, etc. for individuals, and addresses, officers, functions, services, and similar data for organizations.

Library Microfiche Collections

Microfiche collections store reproductions of a printed page, a document, or other type of information on film that is easily stored, accessed, capable of enlargement and relatively inexpensive. Microfiche collections provide a space saving alternative to printed documents. With proper equipment, microfiche can be duplicated or produced as printed copy.

Slides

Slide collections provide an easy way to identify with particular species and subject areas in aquaculture. Some focus on species identification, while others address such areas as pond management in aquaculture. Slide collections offer the public an informative way to learn about aquaculture. Some of them have accompanying sound and are available for purchase or loan. Others are developed in-house by researchers for use in presentations. Although many collections exist today, the majority are used internally or for regional outreach activities. Future hopes are for improving the nation's collection of slides through gift, exchange, and purchase activities.

Audiovisuals

Audiovisuals are another source of information for the aquaculture community. They are materials in audio and visual formats that convey information primarily by sound and image rather than text. They include movies and videotapes, photographs, and recordings, and require the use of special equipment in order to be seen or heard. Audiovisuals can serve as excellent teaching devices in aquaculture. The need for improved access to these collections should be stressed.

Computer Databases and Software

Databases are currently available for quick access to information. Many of the commercial databases are bibliographic in nature, but can provide the user with a wealth of literature citations on his/her subject of interest. With a microcomputer, modem, communications software, and a subscription to a database service such as DIALOG or BRS (Bibliographic Retrieval Service), an individual can easily locate resources to specialized subjects. Databases can also be accessed in-house using other storage media, which will be discussed in more detail below under "Optical Disk Technology".

Computer software is becoming more available for aquaculture. Management software has been developed for the aquaculturist to better organize his/her data. Expert systems are being developed utilizing various expert system software packages or programming languages in artificial intelligence. Expert systems are computer systems that mimic advisory work done by human experts in a particular field of knowledge. An expert system on "tilapia culture" is being developed and explored at the Tennessee Valley Authority in Muscle Shoals, Alabama. An expert advisory system called "AquaRef" was developed by staff of the Aquaculture Information Center of the National Agricultural Library, to guide users to aquaculture reference tools. More developments in areas of management software and expert systems could substantially improve aquaculture techniques and farmer profitability.

Optical Disk Technology

Optical disks offer new and exciting opportunities for storage, access, and dissemination of aquaculture information. There are many types of optical disks currently available or in stages of development. They are briefly outlined below:

Laser Discs. Laser discs are the large 12-inch discs originally designed for motion picture storage. Currently, laser discs are being produced for storing rare photographs or specialized image collections. A laser disk can store 55,000 photograph images on one side.

From 1985 to the present, the National Agricultural Library in Beltsville, MD, has been exploring the use of laser discs in storing printed publications for both text and image access. The disc holds up to 800 megabytes (800 million characters) of data.

CD-ROMs. A CD-ROM (Compact-Disk, Read-Only Memory) is a small 4 3/4-

inch disk that holds digital information. CD-ROMs have a storage capacity of 550 megabytes, or 1,500 floppy disks, or 275,000 pages. Many CD-ROMs are commercially available that provide access to current and retrospective bibliographic databases. The advantage for the user is unlimited searching capability without the telecommunication costs that are normally charged for remote database searching. CD-ROMs are becoming quite popular in the information industry, and offer persons in remote locations the opportunity of having a wealth of bibliographic information at their fingertips.

The microcomputer hardware that is required to enable the user to access the data on CD-ROMs is becoming more standardized, so that multiple disks produced by different companies can be accessed using the same hardware. Scientists and extension specialists are purchasing CD-ROM players for their microcomputers, and many university and federal libraries are providing CD-ROMs for patron usage in their reference areas.

Existing Information Services

National

Information services in the United States are extensive in number and specialty. The Federal Government plays a large role in administering and overseeing information services in aquaculture. The three departments within the Federal Government that were mandated aquaculture responsibilities in the National Aquaculture Improvement Act of 1985 are the U.S. Department of Agriculture, U.S. Department of Commerce, and U.S. Department of the Interior. These Departments will be described below in terms of their involvement in aquaculture information services. Appendix B provides the address and phone number of each described agency.

U.S. Department of Agriculture

The U.S. Department of Agriculture (USDA) was designated in the National Aquaculture Improvement Act of 1985 as the lead agency for promoting aquaculture programs in the Nation. The Department recognizes the need for expanded growth of aquaculture for food production. As a result, programs involving aquaculture research, development, and education have increased to support the needs of the industry.

Approximately 12 agencies and organizations within the USDA are involved in aquaculture activities. Four of these offer strong programs in aquaculture information services and are described below:

National Agricultural Library. The National Agricultural Library (NAL) has established an Aquaculture Information Center to provide information and reference-related support to a wide variety of patrons. The Aquaculture Information Center was Congressionally mandated by the Food Security Act of 1985 (Title XVII, Subtitle D) to serve as a repository for national aquaculture information. The center also supports the international needs of aquaculturists through its information services.

Some of the aquaculture information services provided by the center include the following: manual and computer searches of the aquaculture literature; production of bibliographies, directories, and other information products; collection of development activities, networking activities; and specialized projects for improving information transfer in aquaculture.

Aquaculture publications, which represent a portion of the total 2 million volumes of agricultural-related documents, are housed within the general library collection. Lending services are conducted through interlibrary loan on a cost-recovery basis.

Cooperative State Research Service (CSRS). CSRS supports formula-funded research efforts in aquaculture in the Land-Grant College system and at State agricultural experiment stations. The service administers an Aquaculture Special Research Grant Program on a competitive basis; it also awards special grants in aquaculture on special research problems that Congress identifies as important to the Nation.

CSRS administers Regional Aquaculture Centers in cooperation with the Extension Service for improving research, extension, and demonstration projects in aquaculture. It houses the Office of Aquaculture responsible for coordinating all aquaculture programs in the U.S. Department of Agriculture.

Office of Aquaculture. The Office of Aquaculture, housed within CSRS, is the coordinating unit for all Department-wide aquaculture activities. It provides leadership for the Federal-wide Joint Subcommittee on Aquaculture and coordinates the operations and activities of the Regional Aquaculture Centers.

Cooperative Extension Service. The Cooperative Extension Service is the primary educational arm of the U.S. Department of Agriculture. The system functions as a nationwide educational network and includes professional staff in almost all of the nation's 3,150 counties. The system is tied directly to land-grant colleges and universities, and provides for the transfer of aquaculture information through programs at the state and county levels. USDA extension agents in aquaculture are often jointly funded by state Sea Grant programs for marine species. The Cooperative Extension Service relies heavily on information generated by aquaculture research and helps to interpret the results to speed application of this information.

Many states have developed extension educational programs in aquaculture. County agents and extension specialists support information needs in each state on aquaculture-related subjects. Programs offered through Extension personnel include workshops for new fish farmers, training on specialized topics, demonstrations, farm visits, and field days. Educational material is also published and distributed as a means for improving information transfer.

U.S. Department of Commerce

The U.S. Department of Commerce has been involved in aquaculture research for over 20 years and cooperates with the two other departments to strengthen aquaculture programs. Several agencies have played major roles in the planning, development, and implementation of aquaculture activities and services. These agencies are described below.

National Sea Grant College Program. The National Sea Grant College Program conducts research, extension and educational programs with universities in all coastal and Great Lakes states. Each year, Sea Grant funds between 110-120 aquaculture projects valued at $7-8 million when matching funds are included. Sea Grant chooses projects on the basis of scientific merit, peer review and the present needs of the aquaculture industry. Information products obtained from research are used by the Sea Grant Marine Advisory Service and the National Sea Grant Depository, which in turn serve the public and other state and Federal agencies.

The National Sea Grant Depository is a unique collection housing all publications generated by the National Sea Grant College Program. The collection was established at the University of Rhode Island's Pell Marine Science Library in 1970. It covers a wide variety of subjects, including aquaculture, fisheries, and marine education. The Depository collection includes technical reports, journal reprints, newsletters, marine advisory reports, manuals, guides, directories, bibliographies, proceedings, and atlases, charts and maps produced by Sea Grant-funded researchers. Telephone, mail, and inter-library loan requests are welcome to anyone in need of information.

Each Sea Grant College Program and Institution has an extension or outreach component called the Marine Advisory/Extension Service (MAS). Some of these programs are administratively linked to the state Cooperative Extension Service; others are administratively and programmatically the sole responsibility of Sea Grant. Regardless of their administrative organization, the MAS has a presence in each coastal and Great Lakes state with the exception of Pennsylvania. Sea Grant MAS agents and specialists, about 350 individuals in total, conduct extensive outreach programming in aquaculture as well as many other areas. Their major objective and responsibility is to educate their aquaculture clientele groups in the total array of disciplines that comprise an aquaculture venture, from the biological and technical to the economic and financial. Their primary areas of interest are in marine and estuarine species, with the exception of the Great Lakes where cool- and cold-water, freshwater finfish receive attention. The chief source of information to support their activities is Sea Grant research which is driven by state-of-the-art needs, but all other available information is drawn in to supplement the programming which is conducted using all appropriate extension education techniques. Emphasis is placed in developing new species and improved, efficient technologies to maximize economic benefit and to develop new markets, including international. Levels of programming activity range from intimate one-on-one interaction with individual aquaculturists to large-scale integrated production demonstrations for extensive education.

National Environmental Satellite, Data, and Information Service/National Oceanic and Atmospheric Administration (NOAA). The National Environmental Data Referral Service (NEDRES) is a database provided by the National

Environmental Satellite, Data, and Information Service. It offers a rapid and easy way to identify and locate environmental data held by people and organizations in the United States and Canada. This database can be directly accessed by users at their own terminal or computer. NEDRES covers the full spectrum of environmental sciences, including materials that support marine sciences and aquaculture research and development.

The National Environmental Satellite, Data, and Information Service (NESDIS) has actively served as a liaison to various aquaculture activities. A Memorandum of Understanding between NOAA and the National Agricultural Library established a project that strengthens worldwide collection and dissemination of aquaculture information through its support of an international database.

NESDIS also facilitates networking of their affiliated NOAA Library with the Aquaculture Information Center at the National Agricultural Library as a means to improve communication in both the national and international sectors. NESDIS represents the United States' interests in aquaculture through the Aquatic Sciences and Fisheries Information System (ASFIS), an international computer-oriented information system.

National Marine Fisheries Service/NOAA. The National Marine Fisheries Service (NMFS) directs its aquaculture efforts toward managing common property resources and contributing to the restoration and protection of endangered species and stocks. It accomplishes this primarily through research activities at several NMFS laboratories in the United States. Many NMFS labs maintain their own libraries that offer informational services to the public and to university students and staff, as well as other Federal agencies.

National Technical Information Service (NTIS). NTIS is the central source for public sale of U.S. Government-sponsored research and development reports in all subjects, including those related to aquaculture. It also serves as a source for sale of foreign technical reports prepared by national and local government agencies and their contractors or grantees.

U.S. Department of the Interior

U.S. Fish and Wildlife Service. The U.S. Fish and Wildlife Service of the U.S.

Department of the Interior focuses information services on freshwater, anadromous, estuarine, and exotic species of ecological, recreational, and commercial importance. It administers a fishery resource base comprising more than 100 organizational activities throughout the 50 States. These include national fish hatcheries, research laboratories and centers, biological stations, development centers, diagnostic units, vessel bases, and a training facility. The Service is comprised of seven geographical regions and an eighth national region for research and development.

The Fish and Wildlife Service provides technical and scientific information and assistance upon request. The National Fisheries Center-Leetown, West Virginia, administers various geographically scattered aquaculture activities to provide timely research and development information on the production of quality fish for management and commercial needs. Research involves fish health, genetics, nutrition and disease diagnostics, and culture of freshwater and anadromous species of fish.

The Leetown facility also includes a library which services Fish and Wildlife researchers in various geographic regions as well as the scientific community. The library houses periodicals, books, and other documents. An abundance of materials on fish health can be found in the collection.

The Extension component of the Center is committed to a continuing program of cooperation with the extension community in furthering mutual needs by informing and educating the public on fish and wildlife resources; it addresses the field of aquaculture both from the standpoint of extension education and 4-H. The Center in Leetown is open to the public.

Several Fish and Wildlife research centers are located in the U.S. and focus on specialized topics such as fish contaminants, culture and diseases of Pacific salmon, and drug and chemical registration for fish culture and management.

International

Food and Agricultural Organization (FAO). The United Nations sponsors the Food and Agricultural Organization that has an information service that includes fisheries and aquaculture. FAO supports meetings and workshops dealing with aquaculture and the information service provides reprints and other hard copy on specific aquaculture topics. The United Nations, through FAO, also supports a full-time aquaculture specialist who is responsible for tracking aquaculture developments in member countries. FAO maintains directories of ongoing projects in aquaculture and is beginning to make yearly assessments of aquacultural production throughout the world.

International Association of Marine Science Libraries and Information Centers. IAMSLIC is an international organization of marine-related libraries and information centers which was established in 1975 in Woods Hole, Massachusetts. It is a nonprofit organization that promotes cooperation and sharing of resources among libraries and information centers specializing in marine science. A Union List of Serials in marine science libraries, called MUSSEL, was created in 1984 in microfiche and serves as the only source list of serial titles available in marine science today.

Pacific Science Association. The Pacific Science Association's Scientific Committee on Marine Science is composed of members from several countries and has recently become active in promoting scientific cooperation in the marine sciences in the Pacific Basin. The Committee does this in part by holding symposia and by serving as an exchange center for scientific literature at the Pacific Oceanological Institute in Vladivostok. Scientists can send reports and publications to the center, and in turn be able to request copies of other materials deposited in the center. Materials available in the center are published in the committee's annual yearbook.

South Pacific Commission. The South Pacific Commission, which receives funding from the United Nations, publishes a bimonthly newsletter on fisheries and aquaculture topics. They also maintain a library that provides information services for participating Pacific nations. Projects in aquaculture and aquaculture specialists are funded periodically to demonstrate the potential of aquaculture in the region.

Svanoy Foundation International Aquaculture Centre. The Svanoy Foundation serves as a non-governmental nonprofit intermediary for the promotion of aquaculture development, especially through enhanced participation of the private sector. It was originally established on the island of Svanoy on the west coast of Norway, off Floro, and in 1985 moved its main operational centre to Rome, Italy.

Activities include the application of cold water aquaculture techniques to the improvement of culture technologies for warm water species. Its main focus is on developing countries, although services are equally available to industrially advanced countries. The Centre serves as an information source for its cooperating institutions, agencies, and industries.

Other Information Services

Many states maintain their own marine research and educational facilities. Several state facilities specialize in aquaculture research and extension. The James M. Waddell, Jr. Mariculture Research and Development Center in South Carolina, the Claude Peteet Mariculture Centre in Alabama, the Florida Department of Natural Resources Aquacultural Facility in St. Petersburg, Florida, and the University of Texas Marine Science Institute at Port Aransas, Texas, are examples of facilities that maintain libraries and provide information services on aquaculture. Other state marine research facilities have smaller aquaculture components as part of their overall programs.

North Carolina Biotechnology Center. Through its Marine Biotechnology Program, the North Carolina Biotechnology Center is playing a leading role in the development of marine biotechnology and aquaculture in North Carolina. Its primary goal is economic development through the establishment and growth of marine-related industries. It has also made a major commitment to help develop Western North Carolina's fledgling trout industry.

The Center funds research, industrial development, and conferences and workshops in marine biotechnology and aquaculture. Grants support research ranging from the pharmacological properties of North Carolina marine organisms to the development of advanced fish feeds.

Current Status and Future Trends in the Information Industry

Considering the vast amount of products and services currently available on aquaculture, it is vitally important that all information systems work together to enhance worldwide information exchange. Although information specialists are still far from establishing effective communication links, which can be attributed in part to geographical barriers and degrees of technological advancements, there is an increasing awareness for improved networking. Networking will help prevent duplication of efforts and promote new and inventive ways to get information out to the public and private sectors.

Some information linkages have already taken place on a small scale. The Joint Subcommittee on Aquaculture which joins the aquaculture interests of more than 16 Federal Government agencies has supported information needs of the community in various activities. It has established an Information Task Force for improving information exchange in aquaculture.

An Aquaculture Forum held in November 1987 gathered aquaculture specialists from around the nation to solicit input to the revision and update of the National Aquaculture Development Plan. The forum addressed goals, opportunities, constraints, and action strategies for the continued development of the U.S. aquaculture industry.

An Aquaculture Library Network was established in January 1987 as a vehicle for information exchange between librarians and information specialists in aquaculture. National programs that enhance information collection and dissemination in aquaculture have been supported by this networking system, and may offer future opportunities for improving information exchange worldwide. This system also functions to support regional linkages within the U.S., through interactions with representatives of five Regional Aquaculture Centers established for research, development, and demonstration programs.

Cooperative efforts in database development and access are visible in the Aquatic Sciences and Fisheries Information System (ASFIS), an international network of centers and referral offices for aquatic sciences and fisheries. The network links together bibliographic data and abstracts originating in various countries into a total database, called "Aquatic Sciences and Fisheries Abstracts" (ASFA) database. This is a comprehensive database on aquaculture and is accessible worldwide by remote access and on a CD-ROM.

Advancements in computer technologies are taking place on a large scale. Opportunities for mass storage of aquaculture documents are being explored through optical scanning techniques. Optical disks are becoming commonplace in the office as a new means to collect, store, and retrieve data. Electronic mail systems provide individuals with improved communication linkages to persons throughout the world, and along with bulletin boards are serving basic needs for information exchange.

There are a number of institutions and organizations that provide expertise in aquaculture. They include the sea grant colleges, land-grant colleges, and private nonprofit institutions. Associations are becoming more active in aquaculture as well, with new ones formed as the industry expands in size and needs. Sea Grant MAS and USDA Extension personnel and county agents are key components to information exchange and provide the necessary expertise for successful aquaculture ventures.

We can look forward to new and exciting challenges in the future to improve the way in which we collect, store, and exchange information. It is not inconceivable that microcomputers and optical disks will be common information tools in our workplace. Interactive systems for education and training such as interactive videodiscs and expert systems will be available.

The need for advanced research and development in aquaculture will continue to grow to support the demands of the rapidly growing industry. Specialists will be needed for their practical and technical expertise. Librarians will offer the essential resources for patrons to make knowledgeable decisions about aquaculture endeavors. All of these information services can be integrated into a networked system that we can use to meet the goals of the national and international aquaculture community.

Appendix A: Types of Information Products

The information products listed below represent only a small selection of aquaculture materials currently available. They are provided only for supplementary information and do not imply preference or endorsement by the authors of this paper.

Newsletters

AQUAFARM Letter. Published every other Friday, Feb.-Nov. Washington AQUAFARM Letter, Box 14260, Benjamin Franklin Station, Washington, D.C.

20044.

California Aquaculture. Published monthly. Aquaculture Extension, Univ. of California, Davis, CA 95616.

California Aquatic Farming. Published monthly. P. O. Box 1004, Niland, California 92257.

The Caribbean Aquaculturist. Published quarterly. Dept. of Marine Sciences., Univ. of Puerto Rico, P. O. Box 5000, Mayaguez, PR 00709.

Carolina Aquaculture News. Published bimonthly by AAS-Aquaculture Advisory Service. AAS, Dept. N, P.O. Box 1294, Garner, NC 27529.

Coastal Aquaculture. Published monthly. Texas A&M Research and Extension Center, Route 2, Box 589, Corpus Christi, TX 78410.

Extension Service Update. Published quarterly by the Extension Service, USDA, Washington, D.C.

Florida Aquaculture Association. Published quarterly. P. O. Box 3989, Tallahassee, FL 32315-3984.

National Shellfish Association Newsletter. Published quarterly. College of Marine Studies, Univ. of Delaware, Lewes, DE 19958.

State Sea Grant Newsletters. Many Sea Grant Programs publish periodic newsletters with information on aqua-culture. For information contact: National Sea Grant College Program, 1335 East West Highway, Silver Spring, MD 20910; (301)427-2451.

The World Aquaculture Society Newsletter. Published quarterly. For subscription information, write: World Aquaculture Society, 16E. Fraternity Lane, Louisiana State University, Baton Rouge, LA 70803.

Magazines and Journals

Many of the magazines and journals listed below can be located at a local public or university library. Addresses regarding subscription information are supplied after each title.

Aquaculture. Elsevier Science Publishers B.V., Journals Dept., P.O. Box 211, 1000 AE Amsterdam, The Netherlands. 32 issues per year.

Aquacultural Engineering. Elsevier Applied Science Publishers Ltd, Crown House, Linton Road, Barking, Essex IG11 8JU, England.

Aquacultural Magazine. Aquaculture Magazine, P. O. Box 2329, Ashville, NC 28802. Published bi-monthly. Annual Buyer's Guide published in January of each year.

Aqua-O2 News. Aeration Industries International, Inc., 4100 Peavey Road, Chaska, MN 55318; 1-800-328-8287.

Bulletin Francais de la Peche et de la Pisciculture. M. le Regisseur de Recettes du C.S.P., Centre du Paraclet, B.P. no 5, F 80440 BOVES, France.

Canadian Journal of Fisheries and Aquatic Sciences. Supply and Services Canada, Ottawa, Ontario, Canada K1A 059.

Canadian Aquaculture Magazine. Published bimonthly. Canadian Aquaculture, 4611 William Head Road, Victoria, B.C. V8x3W9.

Catfish Journal. Official Publication of Catfish Farmers of America. Aquacom, Inc., P.O. Box 1700, Clinton, MS 39056, (601) 924-4407.

Catfish News. Aquacom, Inc., P.O. Box 4566, Jackson, MS 39216.

Farm Pond Harvest. Farm Pond Harvest, P.O. Box AA, Momence, IL 60954. Published four times a year.

Fish Farming International. Arthur J. Heighway Publications, Ltd., Heighway House, 87 Blackfriars Road, London SE1 8HB. Published quarterly.

Journal of Aquatic Plant Management. Aquatic Plant Management Society, Inc., P.O. Box 16, Vicksburg, MS 39180.

Journal of Fish Biology. Academic Press Limited, Foots Cray, Sidcup, Kent DA14 5HP, U.K. Published monthly.

Journal of the World Aquaculture Society. Membership information is available from the World Aquaculture Society, Membership Department, 16 E. Fraternity Lane, Louisiana State University, Baton Rouge, LA 70803. (Membership includes a Journal and Newsletter subscription.)

Progressive Fish-Culturist. American Fisheries Society, 5410 Grosvenor Lane, Suite 110, Bethesda, MD 20814. Published quarterly.

Salmonid. Salmonid, 515 Rock St., Little Rock, AR 72202; (501) 372-3595. Published quarterly.

Books

Aquaculture: The Farming and Husbandry of Freshwater and Marine Organisms. Bardach, John E.; Ryther, John H.; and McLarney, William O. 1972. Published by John Wiley & Sons, Inc. 868 pp.

CRC Handbook of Mariculture. Volume I: Crustacean Aquaculture. 1983. McVey, James P.; Moore, J. Robert. 1986. Published by CRC Press, Inc. 442 pp.

Crustacean and Mollusk Aquaculture in the United States. 1985. Huner, J. V. and E. E. Brown. The AVI Publishing Co., Inc.

Disease Diagnosis and Control in North-American Marine Aquaculture. Sindermann, J. C.; and Lightner, D. V. 1988, 2nd ed. Published by Elsevier Science Publishers. 426 pp.

Fish and Shellfish Pathology. Ellis, Anthony E. 1985. Published by Academic Press. 412 pp.

Mussel Culture and Harvest: A North American Perspective. Developments in Aquaculture and Fisheries Science, Volume 7. Elsevier Science Publishers. 379 pp.

Seaweed Cultivation for Renewable Resources. K. T. Bird and P. H. Benson. 1987. Published by Elsevier Science Publishers. 379 pp.

Technical Reports

Crawfish Culture in Maryland. Harrell, R. M. 1987. UM-SG-MAP-87-02. Sea Grant College, Univ. of Maryland, 1224 H. J. Patterson Hall, College Park, MD 20742.

FAO Fisheries Technical Paper, Marketing the Products of Aquaculture. Shaw, Susan A. 1986. FAO, via delle Terme di Caracalla, 00100 Rome, Italy.

NOAA Technical Report NMFS SSPF-703. Vandruska, John. 1976. U.S. Dept. of Commerce, NMFS, Universal Bldg., 1825 Connecticut Ave., N.W., Washington, DC 20235.

Technical Papers of the Florida Sea Grant College Program. University of Florida, Gainesville, FL 32611. [Titles vary and are identified by their Technical Paper No.]

University of Hawaii Sea Grant Advisory Reports, Published periodically. Hawaii Sea Grant Program. University of Hawaii, 1000 Pope Road, Rm. 220, Honolulu, Hawaii 96822.

Proceedings

Recent Innovations in Cultivation of Pacific Molluscs. Morse, D. E., K. K.

Chew, and R. Mann. 1984. Elsevier Science Publishers. 404 pp.

The Aquaculture of Striped Bass. McCraren, Joseph P. 1984. Univ. of Maryland Sea Grant Press. 262 pp.

Proceedings of the Thirty-Seventh Annual Gulf and Caribbean Fisheries Institute (Cancun, Mexico, Nov. 1984). May 1986. 248 pp.

Proceedings of the National Shellfisheries Association. June 1978. Volume 68. Published for the National Shellfisheries Association, Inc., by The Memorial Press Group, Plymouth, Massachusetts. 94 pp.

Bibliographies/Abstracts

Bibliography of Publications of the National Fisheries Research Center-Leetown 1981-1985. McKenzie, Lora C.; Catrow, Violet J.; and Mann, Joyce, A., September 1987. Biological Report 87 (5). Published by the Fish and Wildlife Service. U.S. Department of the Interior. 33 pp. Request a copy from: National Fisheries Research Center-Leetown, Box 700, Kearneysville, WV 25430.

Practical Aquaculture Literature: A Bibliography. August 1985. BLA Number 35. Published by the National Agricultural Library, Beltsville, MD. 40 pp. Copies may be obtained from: National Technical Information Service, 5285 Port Royal Road, Springfield, VA 22161; (703) 487-4650; (NTIS Order Number = PB 86-103447).

Shrimp Mariculture 1979-1987. Hanfman, Deborah T. April 1988. Quick Bibliography Series: QB 88-47. Published by the National Agricultural Library, Beltsville, MD. 7 pp. For a current listing of free bibliographies, write: National Agricultural Library, Aquaculture Information Center, ATTN: Bib. List, Room 304, Beltsville, MD 20705.

Sea Grant Abstracts. Published quarterly by National Sea Grant Depository, Pell Library Building, Bay Campus, Univ. of Rhode Island, Narragansett, RI 02882.

Directories

Aquaculture: A Guide to Federal Government Programs. November 1987. Published by the National Agricultural Library, Beltsville, MD. Prepared by the Joint Subcommittee on Aquaculture. 34 pp. For a free copy, write: National Agricultural Library, Aquaculture Information Center, ATTN: Guide, Room 304, Beltsville, MD 20705.

Directory of Aquaculture Information Resources. December 1982. BLA Number 26. Published by the National Agricultural Library, Beltsville, MD. Compiled by Aspen Systems Corporation. 53 pp. To order, contact: National Technical Information Service, 5285 Port Royal Road, Springfield, VA 22161; (703) 487-4650. (NTIS Order Number= PB 83-169474.)

Directory of Marine Sciences Libraries and Information Centers. 1981. Winn, Carolyn P. Published by Woods Hole Oceanographic Institution. 137 entries; no pagination. For ordering information, write: Research Librarian, Woods Hole Oceanographic Institution, Woods Hole, Massachusetts 02543.

National Aquaculture Directory, 1986. December 1986. Ayers, James W. Published by National Marine Fisheries Service, NOAA, U.S. Dept. of Commerce, 11215 Hermitage Rd., Suite 200, Little Rock, Arkansas 72211-3809. 222 pp. To order, contact the publisher or contact: National Technical Information Service, 5285 Port Royal Road, Springfield, VA 22161; (703) 487-4650. (NTIS Order Number = PB 87-110268).

Library Microfiche Collections

Virginia Institute of Marine Science Microfiche (VIMS) Collection on Aquaculture. Collection contains more than 11,000 microfiche records of aquaculture documents cited in the AQUACULTURE database (available on line through DIALOG Information Service, Inc. as File 112) and is housed at the National Agricultural Library. For information regarding document delivery service, contact Lending Branch, National Agricultural Library, Room 300, Beltsville, MD 20705; (301) 344-3755.

Aquaculture Slide Collections

"American Fisheries Society Catalog of Fish Slides". This collection is composed of 453 slides of fish species which were donated to the society by many individuals. Copies are to be used only for projection for nonprofit educational purposes. For more information, contact: Don Flescher, Chairman, AFS Fish Photo Committee, National Marine fisheries Service, Woods Hole, MA 02543.

"Slide Collection on Fish Farming in Colorado". These slides along with the companion audio cassette show the potential fish farmer the very basics—from digging and stocking the fish pond to maintaining water quality. For more information, contact the Colorado State University Cooperative Extension Service at (303) 491-6411.

Audiovisuals

"Blue Revolution." A nine-part television series produced for PBS. Each part is one hour long and consists of the following segments: The Ocean Planet; Harvest from the Sea; Drugs from the Sea; Quest for Ocean Minerals; Tapping the Ocean's Energy Wealth; The Ocean as a Waste Dump; The Blue Highways; Struggle for the Seas; The Blue Revolution. For information on availability, contact: The Mare Nostrum Foundation, 93 Main Street, Suite 400, Annapolis, MD 21401; (301) 263-3324.

"Farmers of the Sea." An hour-long documentary on Aquaculture Worldwide, produced by the Oregon Sea Grant College Program and broadcast on the PBS NOVA program. For more information contact Oregon State University, Corvallis, Oregon 97331, or phone 503/754-4531.

Computer Databases and Software

AGRICOLA (AGRICultural OnLine Access) Database. On line version is updated monthly on DIALOG system; CD-ROM version is updated quarterly. Online version, called CAIN, is also available through the Bibliographic Retrieval Service (BRS). Provides bibliographic citations to agricultural literature, including aquaculture. Database represents indexed material housed at the National Agricultural Library and dating 1970 to present.

For additional information regarding database content, contact producer: National Agricultural Library, Information Systems Division, Beltsville, MD; (301) 344-3813. For information on how to access the database remotely, contact DIALOG at (800) 3-DIALOG, or BRS at (800) 345-4BRS.

AGRICOLA is currently available on CD-ROM through subscription with SilverPlatter. For information regarding subscriptions, contact: SilverPlatter Information, Inc., 37 Walnut Street, Wellesley, MA 02181; (617) 239-0306.

AQUACULTURE Database. Online version is currently available on the DIALOG system. Although this database is no longer being updated, the National Oceanic and Atmospheric Administration has maintained the database on line since 1980. AQUACULTURE provides approximately 11,000 citations to the literature dating 1970-1984. For information on how to access the database remotely, contact DIALOG at (800) 3-DIALOG.

AQUAREF: An Expert Advisory System on Aquaculture Information. AquaRef is an expert advisory system on aquaculture. It serves as a reference tool to direct users to appropriate resources in the field of aquaculture. The system was created in 1986 by staff of the Aquaculture Information Center of the National Agricultural Library and is available on a diskette at no charge to interested individuals. For more information, contact the Aquaculture Information Center at (301) 344-3704.

ASFA (Aquatic Sciences and Fisheries Abstracts) Database. Online version is updated monthly on DIALOG system; CD-ROM version updated semi-annually.Provides comprehensive citations to worldwide aquaculture literature, including abstracts. Database represents indexed material dating 1970 to present. For database information contact producer: Cambridge Scientific Abstracts, 7200 Wisconsin Ave., Bethesda, MD 20814; (301) 951-6750.

"A Records System for Catfish Production Management Decision-Making. This system is compatible with the following microcomputers: Radio Shack Model III; Radio Shack Model II; Radio Shack Model 16; IBM Personal Computer". 1983.

The program is designed to build and report catfish production records. Copies of documentation and diskettes containing the program are available through: Leader, Computer Applications & Service Dept., Box 5405, Mississippi State, MS 39762; (601) 325-3225.

Optical Disk Technology

National Agricultural Text Digitizing project produced by National Agricultural Library in cooperation with Land Grant Universities and agencies of JSA. Contact: Debbie Hanfman: NAL, Rm. 304, Beltsville, MD 20705; 301/344-3704.

Other Information Products

Aquaculture Digest. A published monthly report on marine fish and shellfish farming. It includes newsworthy items such as new products, business developments, employ-ment opportunities, meetings, and innovative ideas. For subscription information, contact: Aquaculture Digest, 9434 Kearny Mesa Road, San Diego, CA 92126; (619) 271-0133.

Books-In-Print. An information resource that provides listings of books in a wide variety of disciplines, including aquaculture. The title, cost, and publisher information is provided for each book. Subject and author listings provide the user with quick and easy access to published materials.

Books-In-Print is a standard resource of libraries and available in the reference area for use by librarians and the public.

National Agricultural Text Digitizing Project. This digitizing project is currently underway at the National Agricultural Library to explore in-house scanning and text conversion for creating CD-ROMS for aquaculture topics.

Appendix B:

Agencies and Organizations that Provide Information Services

Claude Peteet Mariculture Center, P.O. Drawer 458, Gulf Shores, Alabama 36542.

Cooperative State Research Service (CSRS). U.s. Depart-ment Of Agriculture. J.S. Morrill Bldg., 15th and Independence Ave., S.W., Washington, DC 20251; (202) 447-5468.

Cooperative Extension Service. U.S. Department of Agriculture. For further information contact: Your County Agricultural Extension Office; or your State Cooperative Extension Service; or The National Program Leader, Aquaculture, Extension Service, Room 3871, South Building, Washington, DC 20250; (202) 447-5468.

Fish and Wildlife Service. U.S. Department of the Interior. For information on the regional activities of the Fish and Wildlife Service, contact: Office of Extension Education, U.S. Fish and Wildlife Service, 18th and C Streets, N.W., Matomic Bldg., Room 543, Washington, DC 20240; (202) 653-8783; for information on the reference and technical services provided by the National Fisheries Center-Leetown, contact: Technical Information Services, National Fisheries Center-Leetown, U.S. Department of the Interior, Route 3, Box 700, Kearneysville, WV 25430; (304) 725-8461.

Food and Agriculture Organization of the United Nations. For information, write: Department of Fisheries (FIDI), Food and Agriculture Organization of the United Nations, Via delle Terme di Caracalla, 00100 Rome, Italy.

Gulf Coast Research Lab, Ocean Springs, MS 39564.

Hawaii Institute ff Marine Biology, P. O. Box 1346, Kaneohe. HI 96744.

International Association of Marine Science Libraries and Information Centers (IAMSLIC). For information, write: Marilyn Guin, ATTN: IAMSLIC, Oregon State University, Marine Science Center, Newport, OR 97365.

James M. Waddell, Jr. Mariculture Research & Development Center. For information contact: The James M. Waddell, Jr., Mariculture Research and Development Center, Post Office Box 809, Bluffton, SC 29910; (803) 757-3795.

National Agricultural Library. Aquaculture Information Center. U.S. Department of Agriculture. 10301 Baltimore Blvd., Room 304, Beltsville, MD 20705. For general aquaculture information, contact center at (301) 344-3704; for document delivery information, contact the Lending Branch at (301) 344-3755.

National Environmental Data Referral Service (NEDRES). National Oceanic And Atmospheric Admini-stration. U.S. Department of Commerce. For information, contact: NEDRES Program Office, Assessment and Information Services Center, NOAA/NESDIS (E/AIx3), 3300 Whitehaven St., N.W., Washington, DC 20235; (202) 634-7722.

National Oceanic and Atmospheric Administration (NOAA) Library. U.S. Department of Commerce. For information, contact: Carol Watts, NOAA Library, NOAA/NESDIS, 6011 Executive Blvd., Suite 306, Rockville MD 20852; (301) 443-8287.

National Sea Grant Program. U.S. Department of Commerce. For information on the Sea Grant Depository, contact: National Sea Grant Depository, Pell Library Bldg., The University of Rhode Island, Narragansett Bay Campus, Narragansett, RI 02882-1197; (401) 792-6114; for general information on the National Sea Grant

College Program and State Sea Grant College Programs, contact: Associate Program Director (Aquaculture), National Sea Grant College Program, U.S. Department of Commerce, 1335 East West Highway, Silver Spring, MD 20910; (301)427-2451.

North Carolina Biotechnology Center. For information, contact: North Carolina Biotechnology Center, P.O. Box 13547, 79 Alexander Drive, 4501 Bldg., Research Triangle Park, NC 27709-3547; (919) 541-9366.

Office of Aquaculture. U.S. Department of Agriculture. 14th and Independence Ave., S.W., c/o Hamilton Bldg., Room 635, Washington, DC 20250; (202) 535-0960.

Pacific Science Association. Scientific Committee on Marine Sciences. For inform-ation, write: Scientific Committee on Marine Sciences, Pacific Oceanological Institute, Far Eastern Science Center, USSR Academy of Sciences, 7 Radio Street, 690032 Vladivostok, U.S.S.R.

Svanoy Foundation. International Aquaculture Centre. For information, write: Svanoy Foundation, International Aquaculture Centre, Piazza Luigi Sturzo, 9, 00144 Roma, Italy.

University of Texas at Austin Marine Science Institute. For information, contact: Marine Science Institute, The University of Texas at Austin, Port Aransas, Texas 78373-1267; (512) 749-6711.

University of Puerto Rico Marine Laboratory, Mayaguez, Puerto Rico 00708.

AND WHAT MIGHT SHRIMP RESEARCH HAVE ACCOMPLISHED BY THE YEAR 2000?

L. H. Kleinholz
Department of Biology
Reed College
Portland, OR 97202

As is evident from the topics chosen for this program, important considerations in the culture of crustaceans are nutrition, diseases, reproduction and growth. My own exposure to the applied aspects of these subjects is so very limited it would be presumptuous of me to venture critical comments about them. Let me instead offer two kinds of suggestions about these sessions: (1) some general comments about what I gather are concerns of those engaged in aquaculture (2) research approaches for the near future in the subjects with which I am familiar, endocrine factors involved in the physiology of reproduction and growth (Figure 1).

In the first category, better lines of communication than now exist might be established between those engaged in the commercial aspect of aquaculture, those doing research in improving technological methods for aquaculture, and biologists whose research activities lie in physiological and biochemical studies with the animals being cultured. I am certain each can learn much from the other. How this can best be undertaken may require some experimental study itself and probably over a longer range of time than the limit set by the title of my talk. One such approach is a center for study of biomedical marine resources to be established at the Marine Biological Laboratory of Woods Hole, Massachusetts. The operations and studies at such a center may be instructive.

Also in this first category I would put the answer to a question posed by one of the participants in this Workshop. The question, approximately, was, "When can you endocrinologists show us how to get shrimp that are 95% "tails" (abdomens) and 5% "heads" (cephatothoraces)?"; the goal is obviously stated as an extreme, but the intent is clear: what can be done about obtaining animals whose commercially

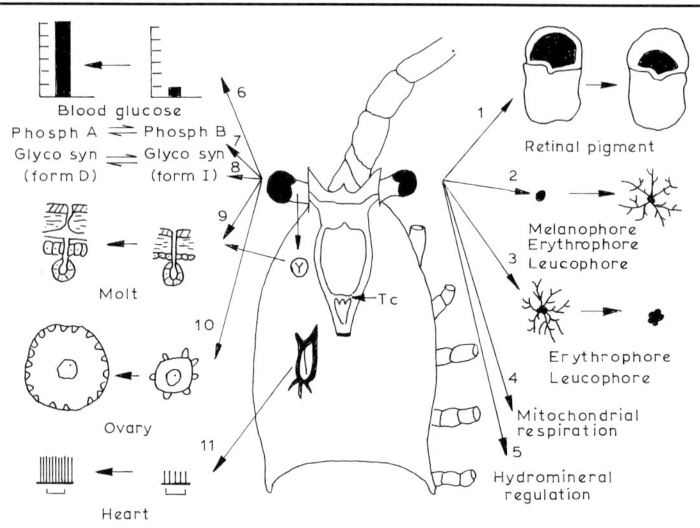

Fig. 1. Summary of physiological effects in crustaceans, produced by neurosecretory hormones and factors. (1) Light-adaptation of distal and reflecting retinal pigments; (2) Dispersion of pigment granules in integumentary melanophores, erythrophores and leucophores; (3) Concentration of erythrophore and of leucophore pigments; (4) Two inhibitors and one stimualtor of succinate oxidation in mitochondria from rat liver; (5) Sodium, chloride and water fluxes affected; best known factors are from cerebral and thoracic ganglia; (6) hyperglycemic hormone; (7) Activates muscle phosphorylase; (8) hyperglycemic hormone; (9) two molt inhibitors, one is a peptide and the second is xanthurenic acid; (10) inhibitor of gonad; (10b) stimulator of gonad known from cerebral and thoracic ganglia; (11) cardioexcitors form the pericardial organ. Tc, tritocerebral commissure and postcommissural organs; Y, Y-organ or molt gland. (Modified from Kleinholz and Keller, 1979)

valuable anatomical portions will be larger than at present? A practical way to this target is probably by selective breeding, rather than by a direct endocrinological approach. The technology is already available in the cryopreservation of gametes and with artifical insemination. We have only to examine what has been accomplished in establishing various sizes and shapes in some of our present breeds of domestic animals, dogs or cattle for example, to see the ranges possible. Selective breeding may also be an approach to raising animals that are resistant to various diseases, as has been achieved with numbers of plant species. Selective breeding is however a long-range program, the pros and cons for which will undoubtedly be considered by cost-efficiency advisers and enterepreneurs who might be interested in such a venture.

In moving to my second category, endocrine factors involved in the physiology of reproduction and growth, I shall give brief background accounts of the present state of research, summarize some of the presentations already made here, and offer suggestions for near-term research that might resolve some of the questions still facing us in these areas.

An ovary-inhibiting hormone from the eyestalk, often called gonad-inhibiting hormone (GIH) and more recently referred to as vitellogenesis-inhibiting hormone (VIH) (Soyez *et al.*, 1987), was first reported by Panouse (1946). It has been studied extensively in malacostracan crustaceans and reviewed recently (Kleinholz and Keller, 1979; Quackenbush, 1986; Charniaux-Cotton and Payen, 1988). Briefly, primary oocytes, after GIH removal by eyestalk ablation, grow considerably and proceed into secondary vitellogenesis, a stage where vitellogenin, a yolk precursor present in the hemolymph, may be converted to the major protein of yolk, vitellin.

A gonad-stimulating factor (GSH) accelerated yolk deposition and ovarian growth when thoracic ganglia of the crab, *Potamon*, were implanted into immature females of the same species. Feeding such animals thoracic ganglia from male *Sesarma* also increased ovarian growth, but to a lesser degree (Otsu, 1960; 1963). Increased ovarian growth has been obtained with implants of thoracic ganglia or injection of ganglionic extracts in a stomatopod as well as several other brachyuran species (Descaraman and Subramoniam, 1983; Eastman-Reks and Fingerman, 1984; Hinsch and Bennett, 1979). In the prawns *Paratya* and *Parapenaeopsis* extracts of cerebral as well as thoracic ganglia stimulated ovarian growth but brain extracts were more effective than thoracic ganglia in *Paratya* (Kulkarni *et al.*, 1981; Takayanagi *et al.*, 1986). Joshi and Khanna (1984) found thoracic ganglia extracts, injected into male *Potamon* during spermatogonial quiescence, produced hypertrophy of the androgenic glands along with onset of spermatogenesis, an increase in gonad index, as well as enlarged testicular tubules and vas deferens. Similar stimulation of the male reproductive system in *Metapenaeus* followed injection of extracts of cerebral or thoracic ganglia (Rao *et al.*, 1986). Evidence for GIH seems well established through classical endocrine procedures, but deficiency studies with the gonad-stimulating factor from ganglia of the central nervous system are lacking, probably because of difficulty with such surgery.

The mandibular organs, located in the anterior portion of the cephalothorax in decapod crustaceans, secrete methyl farnesoate (MF), structurally related to a group of juvenile hormones which occur in the corpora allata of insects. Since juvenile hormone in some insects stimulates vitellogenesis, a similar role has been proposed for the mandibular glands of crustaceans (Laufer *et al.*, 1987) (Figure 2).

GIH is characterized in several studies as apparently a peptide, neither species- nor sex-specific in action, but of several different molecular sizes (Table 1, Figure 3), estimated from gel filtration and electrophoresis. Little is known of the biochemical characteristics of GSH, but retention of gonadotropic activity after thoracic ganglia were fed to *Sesarma* (Otsu, 1963) indicates it might be non-proteinaceous. Clearly, more adequate characterization of GSH is desirable.

The physiological pathway of GIH in inhibiting secondary vitellogenesis requires closer examination. Among the possibilities open to exploration are the following. (1) GIH may act directly on oocytes, inhibiting uptake of vitellogenin or synthesis of the yolk protein, vitellin; this might be studied by *in vitro* culture of oocytes (Jugan and Soyez, 1985) with vitellogenin or vitellin (Derelle *et al.*, 1986) and a

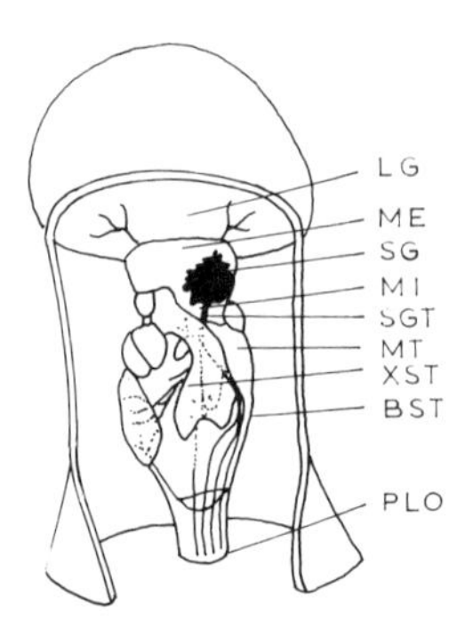

Fig. 2. Dorsal view of the right eyestalk of Cambarus virilis to show the neurosecretory system. Fiber tracts leading to the sinus gland from neurosecretory cells in the stippled areas are shown. BST, brain-sinus gland tract; LG, lamina ganglionaris, the first optic ganglion of the eyestalk; ME, medulla externa; MI, medulla interna; MT, medulla terminalis; PLO, peduncle of optic lobe; SG, sinus gland; SGT, sinus gland tract; XST, tract from the X-organ in the medulla terminalis to the sinus gland. (Modified from Bliss *et al.*, 1954)

Table 1

Characterization of gonad-inhibiting hormone

Source	Test animal	Mol. wt.	Author
Cancer ES[1]	*Crangon*	ca 2,000	Bomirski *et al.* (1981)
Panulirus ES[1]	*Uca*	ca 5,000	Quackenbush & Herrnkind (1983)
Homarus SC[2]	*Palaemonetes*	ca 7,500	Soyez *et al.* (1987)

[1]eyestalks

[2]sinus glands

Fig. 3. Ventral view of the anterior end of *Sesarma reticulatum*. The carapace and integument have been removed on the left side to expose the Y-organ or molt gland. Photograph supplied by Professor L. M. Passano.

range of concentrations of pure GIH. (2) In a relation like that between the molt-inhibiting hormone from the eyestalk and the molt gland, GIH may inhibit release of a gonad-stimulating hormone (GSH) from ganglia of the central nervous system (and/or the mandibular glands?); whether GSH then directly stimulates vitellogenesis is similarly testable *in vitro* as in (1) above. Perhaps of practical application in the husbandry of penaeids would be blocking the physiological action of GIH by injecting into appropriate females antisera raised against GIH (Meusy *et al.*, 1987), followed by injection of GSH. (3) GIH might also concomitantly inhibit synthesis of vitellogenin in the midgut gland (hepatopancreas) and its release into the hemolymph, whereas GSH might stimulate such synthesis and release.

Methyl farnesoate (MF) is present in the hemolymph of both male and female *Libinia*. It is secreted by mandibular glands *in vitro,* the highest rates of secretion being correlated with those glands removed from females in advanced stages of ovarian growth. Secretion *in vitro* of MF has also been reported for mandibular organs of *Homarus*, *Uca* and *Carcinus* (Laufer *et al.*, 1987; Borst *et al.*, 1987). MF level in hemolymph of male *Libinia* is greatly increased after bilateral eyestalk removal, whereas *in vitro* incubation of mandibular glands with eyestalk extract reduces MF synthesis. Of further interest is the report that culture of mandibular organs with 10^{-7} M synthetic RPH/PDH (retinal pigment light-adapting and chromatophoral pigment dispersing hormone) an octadecapeptide with a molecular weight of about 2,000 causes an 80% inhibition of MF secretion. Similar culture with ECH (the octapeptide, M.W.= about 1,000, that concentrates pigment of crustacean erythrophores) doubles the secretion of MF (Laufer *et al.,* 1987).

The demonstration of MF secretion by the mandibular organs is thorough and convincing, and underlines the similarity in this respect to the corpora allata of insects. However, evidence that MF might regulate vitellogenesis is less direct, coming chiefly from the described correlation of high titers of the terpenoid in hemolymph with high synthetic activity *in vitro* of mandibular glands from females in advanced stages of vitellogenesis. Thus far, Hinsch's (1980) observation that implants of mandibular organs into juvenile *Libinia* stimulate ovarian development is the only direct association between the two organs so that studies with additional species are desirable. This result is presumptive but not conclusive evidence that MF may be the responsible agent; would injection of MF into animals shortly before secondary vitellogenesis stimulate yolk deposition in the oocytes, or might culture of oocytes with vitellogenin and MF bring similar results?

The various factors, endocrine and environmental, that may participate in regulating vitellogenesis make critical experimental design a primacy in further experimental analysis of the pathways in which they participate and the specific roles each may play. Among the endocrine factors GIH from the eyestalk and GSH from ganglia of the central nervous system have received the most attention. Hormone(s) from the mandibular gland, and MF as a specific possibility, are the new arrivals to this scene. Particularly intriguing are some of the relations between the pigmentary effector hormone, RPH/PDH, mentioned above and ovarian weights of *Uca* kept on black or white backgrounds (Quackenbush and Herrnkind, 1983). Ovarian weights of *Uca* kept for 15 days on a black background (16:8 photoperiod) and therefore with a high titer of RPH/PDH in the hemolymph acccompanied by maximum dispersion of melanophore pigment, were very much lower than those from animals similarly maintained on a white background (melanophore pigment slightly dispersed; low level of RPH/PDH in hemolymph) and appreciably lower than ovarian weights of first-day controls. If MF stimulates vitellogenesis, the low ovarian weight of animals on a black background may be explained by the 80% inhibition of MF secretion by mandibular glands cultured with RPH/PDH (Laufer *et al.*, 1987). In eyestalkless *Uca* ovarian weights were as high as those of intact animals kept on a white background, but RPH/PDH concentration in the hemolymph of the former is zero, or below the threshold necessary for dispersing melanophore pigment. What, then, is the relation between RPH/PDH and GIH in vitellogenesis? Do these participate in separate pathways? Have they in common partial sequences of amino acids so that the same physiological response can be effected? Answering the latter question requires knowledge of the amino acid sequence of GIH, a goal that now seems realizable with the reported isolation of pure GIH (Soyez *et al.*, 1987). Parenthetically, the vitellogenesis-inhibiting effect caused by a factor of about 2,000 daltons from *Cancer* (Table 1) might have been due to RPH/PDH under the conditions described above. The eyestalk extracts used in that study had been heated, thereby possibly inactivating a larger GIH molecule.

The present-day view of molt in decapod crustaceans is that it is controlled by the molt gland or Y-organ, located in the anterior part of the cephalothorax

(Echalier, 1955) (Figure 4). This gland secretes ecdysone which is converted to 20-hydroxyecdysone, the physiologically effective hormone. Secretion of ecdysone by the molt gland is believed to be inhibited normally by a hormone (MIH) originating in the medulla terminalis X-organ-sinus gland complex of the eyestalk (Passano, 1953). Removal of this complex by eyestalk ablation accelerates the onset of molt. When internal and external environmental factors dictate this onset of molt, secretion of MIH ceases or is reduced so that the level of ecdysone synthesized by the molt gland and released into the hemolymph rises sharply, bringing about the premolt stage and subsequent shedding of the exoskeleton.

Initial attempts at chemical isolation of MIH were difficult: bioassays *in vivo* attempted replacement of missing MIH in eyestalkless animals with tissue implants or injected extracts, thereby slowing appearance of the anticipated accelerated molt as compared with the intermolt periods of eyestalkless controls. Normally long intermolt periods of many test animals were not conducive to progress with such

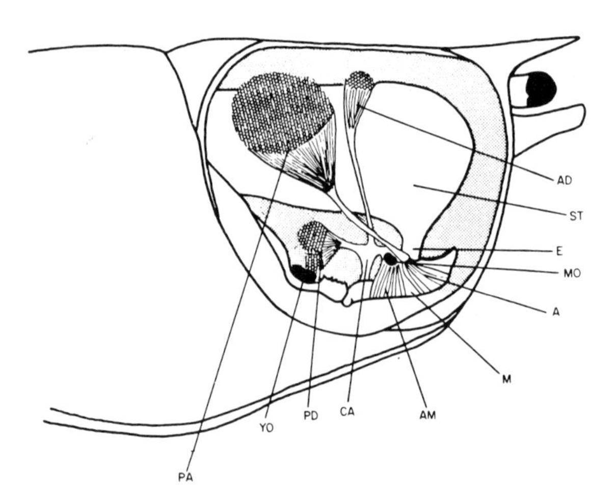

Fig. 4. Lateral view of the opened carapace of *Orconectes*, showing locations of the Y-organ and the mandibular gland. A, anterior adductor muscle of mandible; AD, anterior dorsoventral muscle; AM, adductor major muscle; CA, cephalic apodeme; E, esophagus; M, mandible; MO, mandibular gland; PA, posterior adductor muscle; PD, posterior dorsoventral muscle; ST, stomach; YO, Y-organ. (From Kleinholz and Keller, 1979.)

Table 2

Characterization of molt-inhibiting factors

Material or Source	Test Animal*	Mol wt	Author
Ocypode ES[1]	*Ocypode*	- -	Rangarao (1965)
Palaemonetes ES[1]	*Palaemonetes*	1,000-5,000	Freeman & Bartell (1976)
Pandalus ES[1]	*Orchestia*	<1,000	Soyez & Kleinholz (1977)
Palaemonetes ES[1]	*Balanus*	2,300-3,400	Freeman & Costlow (1979)
Panulirus ES[1]	*Uca*	<5,000	Quackenbush & Herrnkind (1983)
Homarus PP[2]	*Homarus*	- -	Chang (1984)
5-HT[3]	*Cancer*	176	Mattson & Spaziani (1985a)
Vasopressins[4]	*Cancer*	ca 1,000	Mattson & Spaziani (1985b)
Carcinus SG[5]	*Carcinus*	7,200	Webster & Keller (1986)
Carcinus CHH[6]	*Carcinus*	6,726	Webster & Keller (1986)
Homarus SG[5]	*Homarus*	8,700	Chang et al. (1987)
Callinectes ES[7]	*Callinectes*	205	Naya et al. (1988)

1, Eyestalk; 2, Pleopods of female; 3, 5-hydroxytryptamine; 4, Leucine vasopressin most active of the vasopressin-oxytocin group tested in repressing ecdysone synthesis by molt gland in vitro; 5, Sinus gland; 6, Crustacean hyperglycemic hormone from sinus gland; 7, Xanthurenic acid; * tested *in vivo* or *in vitro*.

chemical procedures. Use of short-term cultures *in vitro* of molt glands (Soumoff and O'Connor, 1982) or of sinus glands greatly facilitated bioassay and quickly led to partial chemical characterization of two proteinaceous molt-inhibiting molecules i.e., they repressed secretion of ecdysone by cultured molt glands (Webster and Keller, 1986; Chang *et al.*, 1987). A substantial number of other factors with molt-inhibiting properties when tested *in vivo* or *in vitro* are listed in Table 2. The major problem at the moment is explaining the activity shown by some of these.

The small molecule from eyestalks of *Pandalus* that inhibited molt in *Orchestia* (Soyez and Kleinholz, 1977) was not identified, although it was suggested it might be an idolealkylamine that acted as a releaser of MIH from the sinus gland complex, inasmuch as Fingerman *et al.* (1974) had found 5-HT in eyestalks of *Uca* and believed it might control release of erythrophore pigment dispersing hormone. Mattson and Spaziani (1985a) subsequently reported that 5-HT indeed released MIH in vitro from cultured eyestalk ganglia. Recently, Naya *et al.* (1988) found a small molecule in extracts of crab eyestalks which yielded both elution profiles on gel filtration and electrophoretic patterns on cellulose similar to those reported by Soyez and Kleinholz (1977). This small molecule repressed ecdysone release by cultured cells from homogenized molt glands, and was recognized as 3-hydroxy-L-kynurenine (3-OH-L-K) by a number of physico-chemical characteristics: ultraviolet absorportion, electron impact mass spectrum, circular dichroic spectrum, and nuclear magnetic resonance spectrum. Bioassays with authentic 3-OH-L-K confirmed its inhibitory activity, but this was less than the activity of crude extract of eyestalks. Examination showed that xanthurenic acid, a key metabolite of 3-OH-L-K, was much more active than its precursor, and could be detected in extract of eyestalks by high performance liquid chromatography. The results of further study of this new inhibitor of ecdysone synthesis and its possible relation with the two proteins (Webster and Keller, 1986; Chang *et al.*, 1987) in inhibiting molt will be awaited with interest.

Little more can be added to explain the action of those still insufficiently characterized factors in Table 2 that inhibit molt. The factors of intermediate size (Freeman and Bartell, 1976; Freeman and Costlow, 1979) were found to antagonize the action of 20-hydroxyecdysone in inducing apolysis of mantle tissue *in vitro,* thereby presumably acting directly on the epidermal epithelium.

The large molecular weight peptides from *Carcinus* and *Homarus,* the vasopressins, and the hyperglycemic hormone may owe their common but quantitatively varying ability to suppress secretion of ecdysone by cultured molt glands to similar sequences of amino acids in their molecules. This suggestion remains only speculation until such sequences for the large peptides have been determined. As yet, evidence to support molecular similarity is scant. Vasopressin-like immunoreactivity has been demonstrated in the organ of *Bellonci* in eyestalks of *Palaemon,* but only during ecdysis (Van Herp and Bellon-Humbert, 1982). Immunocytochemical study on the same histological section with antisera raised against pure MIH and CHH of *Carcinus*, on the other hand, showed no observable co-localization of these hormones in sinus glands or in the medulla terminalis X-organs of the eyestalk (Dircksen *et al.*, `1988). Productive pursuit of this line of thinking awaits determination of the amino acid sequences of the two MIHs and of the hyperglycermic hormone.

In summary, it is reasonable to expect that in the next dozen years the amino acid sequences of GIH, MIH, CCH will have been determined and the role of GSH in reproductive physiology better established; GSH might be chemically characterized. With these goals achieved the way is opened to production of such molecules by genetic engineering in sufficient quantities for use in further studies of growth and reproduction.

Literature Cited

Bliss, D.E., J. B. Durand, and J. H. Welsh. 1954. Neurosecretory systems in decapod crustacea. Zeitschrift fur Zellforschung 39:520-536.

Bomirski, A., M. Arendarczyk, E. Kawinska, and L. H. Kleinholz. 1981 Partial characterization of crustacean gonad-inhibiting hormone. International Journal of Invertebrate Reproduction 3: 213-219.

Borst, D.W., L. Kissee, and D. Ramlose. 1987. The synthesis of methyl farnesoate (MF) by mandibular organs of two crabs. American Zoologist 17: 69A (abstract).

Chang, E.S. 1984. Ecdysteroids in Crustacea: role in reproduction, molting and larval development. In. W. Engels *et al.* (eds.) Advances in invertebrate reproduction 3: 223-230. Elsevier Science Publishers B.V.

Chang, E.S., M.J. Bruce, and R.W. Newcomb. 1987. Purification and amino acid composition of a peptide with molt-inhibiting activity from the lobster, *Homarus americanus.* General and Comparative Endocrinology 65: 56-64.

Charniaux-Cotton, H. and G. Payen. 1988. Crustacean reproduction. In. H. Laufer and R.G.H. Downer. (eds.) Endocrinology of selected invertebrate types: 279-303. Alan R. Liss, Inc. New York.

Derelle, E. J. Grosclaude, J. -J. Meusy, H. Junera, and M. Martin. 1986. Elisa titration of vitellogenin and vitellin in the freshwater prawn, *Macrobrachium rosenbergii,* with monoclonal antibody. Comparative Biochemistry and Physiology 85B: 1-4.

Descaraman, M. and T. Subramoniam. 1983. Endocrine regulation of ovarian maturation and cement glands activity in a stomatopod crustacean. Proceedings Indian Academy of Science, Animal Science 92: 399-408.

Dircksen, H., S.G. Webster, and R. Keller. 1988. Immunocytochemical demonstration of the neurosecretory systems containing putative moult-inhibiting hormone and hyperglycemic hormone in the eyestalk of brachyuran crustaceans. Cell Tissue Res. 251: 3-12.

Eastman-Reks, S. and M. Fingerman. 1984. Effects of neuroendocrine tissue and cyclic AMP on ovarian growth *in vivo* and *in vitro* in the fiddler crab, *Uca*

pugilator. Comparative Biochemistry and Physiology 79A: 679-684.
Echalier, G. 1955. Role de l'organe Y dans le determinisme de la mue de *Carcinides* (Carcinus) *maenas* L. (Crustaces Decapodes); experiences d'implantation. Comptes rendus des seances de l'Academie des Sciences 240: 1581-1583.
Fingerman, M., W.E. Julian, M.A. Spirtes, and R.M. Kostrzewa. 1974. The presence of 5-hydroxytryptamine in the eyestalks and brain of the fiddler crab *Uca pugilator,* its quantitative modification by pharmacological agents, and possible role as a neurotransmitter in controlling the release of red pigment-dispersing hormone. Comparative and General Pharmacology 5: 299-303.
Freeman, J.A. and C.K. Bartell. 1976. Some effects of the molt-inhibiting hormone and 20-hydroxyecdysone upon molting in the grass shrimp, *Palaemonetes pugio*. General and Comparative Endocrinology 28: 131-142.
----- and J.D. Costlow. 1979. Hormonal control of apolysis in barnacle mantle tissue epidermis, *in vitro*. Journal of Experimental Zoology 210: 333-346.
Hinsch, G.W. 1979. Vitellogenesis stimulated by thoracic ganglion implants into destalked immature female spider crabs, *Libinia emarginata.* Tissue and Cell 11: 345-353.
Hinsch, G.W. 1980. Effect of mandibular organ implants upon the spider crab ovary. Transactions American Microscopical Society 99: 317-322.
Hinsch, G.W. and D.C. Bennett. 1979. Vitellogenesis stimulated by thoracic ganglion implants into destalked immature spider crabs, *Libinia emarginata.* Tissue and Cell 11: 345-351.
Joshi, P. C. and S. S. Khanna. 1984. Neurosecretory system of the thoracic ganglion and its relation to testicular maturation of the crab, *Potamon Koolooense*. Zeitschrift fur mikroskopische-anatomische Forschung 98: 429-442.
Jugan, P. and D. Soyez. 1985. Demonstration *in vitro* de l'inhibition de l'endocytose ovocytaire par un extrait de glandes du sinus chez la crevette *Macrobrachium rosenbergii*. Comptes rendus des seances de l'Academie des Sciences, Paris 300: 705-709.
Kleinholz, L. H. and R. Keller. 1979. Endocrine regulation in Crustacea. In: E.J.W. Barrington. (ed.) Hormones in Evolution 1: 159-213. Academic Press, New York and London.
Kulkarni, G.K., R. Nagabhushanam, and P.K. Joshi. 1981. Neuroendocrine regulation of reproduction in the marine female prawn, *Parapenaeopsis hardwickii* (Miers). Indian Journal of Marine Science 10: 350-352.
Laufer, H., E. Momola, and M. Landau. 1987. Control of methyl farnesoate synthesis in crustacean mandibular organs. American Zoologist 27: 69A (abstract).
-----, D. Borst, F.C. Baker, C. Carrasco, M. Sinkus, C.C. Reuter, L.W. Tsai, and D.A. Schooley. 1987. Identification of a juvenile hormone-like compound in a crustacean. Science 235: 202-205.
Mattson, M.P. and E. Spaziani. 1985a. 5-hydroxytryptamine mediates release of molt-inhibiting hormone activity from isolated crab eyestalk ganglia. Biological Bulletin 169: 246-255.

----- and E. Spaziani. 1985b. Functional relations of crab molt-inhibiting hormone and neurohypophysial peptides. Peptides 6: 635-640.

Meusey, J. -J., G. Martin, D. Soyez, E. van Deijnen, and J. M. Gallo. 1987. Immunochemical and immunocytochemical studies of the crustacean vitellogenesis-inhibiting hormone (VIH). General and Comparative Endocrinology 65: 333-341.

Naya, Y., K. Kishida, M. Sugiyama, M. Murata, W. Miki, M. Ohnishi, and K. Nakanishi. 1988. Endogenous inhibitor of ecdysone synthesis in crabs. Experientia 44: 50-52.

Otsu, T. 1960. Precocious development of the ovaries in the crab, *Potamon dehaaani,* following implantation of the thoracic ganglion. Annotationes Zoologica Japonenses 33: 90-96.

-----, 1963. Bihormonal control of sexual cycle in the freshwater crab, *Potamon dehaani.* Embryologia 8: 1-20.

Panouse, J.B. 1946. Recherces sur les phenomenes humoraux chez les crustaces. L'adaptation chromatizue et la croissance ovarienne chez la crevette *Leander serratus.* Annales de l'Institut Oceanographique 23: 65-147.

Passano, L.M. 1953. Neurosecretory control of molting in crabs by the X-organ sinus gland complex. Physiologia Comparata et Oecologia 3: 155-189.

Quackenbush, L.S. 1986. Crustacean endocrinology, a review. Canadian Journal of Fisheries and Aquatic Sciences 43: 2271-2282.

-----, and W.F. Herrnkind. 1983. Partial characterization of eyestalk hormones controlling molt and gonadal development in the spiny lobster *Panulirus argus.* Journal of Crustacean Biology 3: 34-44.4

Rangarao, K. (=K. Ranga Rao) 1965. Isolation and partial characterization of the moult-inhibiting hormone of the crustacean eyestalk. Experientia 21: 593-594.

Rao, S., R. Nagabhushanam, R. Sarojini, and K. Jayalakshmi. 1986. Hormonal control of reproduction in the shrimp, *Metapenaeus affinis.* Uttar Pradesh Journal of Zoology 6: 223-240.

Soumoff, C. and J.D. O'Connor. 1982. Repression of Y-organ secretory activity by molt-inhibiting hormone in the crab, *Pachygrapsus crassipes.* General and Comparative Endocrinology 48: 432-439.

Soyez, D. and L.H. Kleinholz. 1977. Molt-inhibitng factor from the crustacean eyestalk. General and Comparative Endocrinology 31: 233-242.

-----, J.E. van Deijnen, and M. Martin. 1987. Isolation and characterization of a vitellogenesis-inhibiting factor from sinus glands of the lobster, *Homarus americanus.* Journal of Experimental Zoology 244: 479-484.

Takayanagi, H., Y. Yamamoto, and N. Takeda. 1986. An ovary-stimulating factor in the shrimp, *Paratya compressa.* Journal of Experimental Zoology 240: 203-209.

Van Herp, F. and C. Bellon-Humbert. 1982. Localisation immuncytochimique de substances apparentees a la neurophysine et a la vasopressine dans le pedoncle oculaire de *Palaemon serratus* Pennant (Crustace Decapode Natantia). Comptes rendus des seances de l'Academie des Sciences, Paris 295: 97-102.

Webster, S.G. and R. Keller. 1986. Purification, characterization and amino acid composition of the putative moult-inhibiting hormone (MIH) of *Carcinus maenas* (Crustacea, Decapoda). Journal of Comparative Physiology B 156: 617-624.